D1189834

THE BUSINESS OF SPEED

Johns Hopkins Studies in the History of Technology
Merritt Roe Smith, *Series Editor*

THE BUSINESS OF SPEED

The Hot Rod Industry in America, 1915–1990

DAVID N. LUCSKO

The Johns Hopkins University Press

Baltimore

© 2008 The Johns Hopkins University Press
All rights reserved. Published 2008
Printed in the United States of America on acid-free paper
2 4 6 8 9 7 5 3 1

The Johns Hopkins University Press
2715 North Charles Street
Baltimore, Maryland 21218-4363
www.press.jhu.edu

Library of Congress Cataloging-in-Publication Data

Lucsko, David N., 1976–
The business of speed : the hot rod industry in America, 1915–1990 /
David N. Lucsko
p. cm.
Includes bibliographical references.
ISBN-13: 978-0-8018-8990-5 (hardcover : alk. paper)
ISBN-10: 0-8018-8990-1 (hardcover : alk. paper)
1. Hot rods—United States—History—20th century. I. Title.
TL236.3.L83 2008
338.4′762922860973—dc22   2008007476

A catalog record for this book is available from the British Library.

*Special discounts are available for bulk purchases of this book. For more information,
please contact Special Sales at 410-516-6936 or specialsales@press.jhu.edu.*

The Johns Hopkins University Press uses environmentally friendly book materials,
including recycled text paper that is composed of at least 30 percent post-consumer waste,
whenever possible. All of our book papers are acid-free, and our jackets and covers are
printed on paper with recycled content.

For my parents,
Susan and Danny Lucsko

CONTENTS

Technically, this project dates back to the early days of my graduate studies, when I first began to contemplate a dissertation project about hot rods. But in a broader sense it dates back much further. Back to a childhood spent playing with Hot Wheels replicas, sketching sports cars on the backs of my school notebooks, and faithfully tuning in to the *Dukes of Hazard*, *Knight Rider*, and *Magnum P.I.* each week. Back to an adolescence spent carefully assembling model cars, handing wrenches to my friend and automotive mentor Walter, and sitting in the stands at the Senoia Speedway with my dad. And back to my teenage years, too, which I spent drooling over automotive periodicals, scouring area salvage yards for treasure, and polishing the bright red paint and chrome accents of my first love. For I am, and have always been, an automobile enthusiast.

But I am not a hot rodder. I have never built a high-performance engine, I have never taken part in a quarter-mile drag race, and only very rarely have I ever deliberately smoked my tires at a stoplight. For mine has always been a passion for restoration and factory authenticity, not modification and individual ingenuity. And the cars I drove and worked on in my youth were vintage, air-cooled, slow-as-molasses Volkswagens, not fire-breathing, raked and blown Chevelles. It's not that I ever disliked high-horsepower modified cars. It's just that my particular automotive interests had always lain—and, right up through the end of my undergraduate years, would continue to lie—elsewhere.

But in the fall of 1998 all of this began to change when I reconnected with an old friend of mine while taking a year off from the academic world before beginning graduate school. Like me, Chris was a VW enthusiast, but his was a passion for the later-model, water-cooled variety. In other words, he knew very little about my older vintage models, and I knew very little about the newer "water-pumper" he drove. However, by the time he and I crossed paths that fall, I had acquired the keys to a water-cooled VW of my own, one of the first New Beetles to hit the streets in our hometown on the southern fringe of metro Atlanta. With my air-cooled cars mothballed (graduate school was almost certainly going to entail a long-term move into the snow belt, and I was not about to subject my aging

Teutonic sleds to the ravages of potholes, parallel parking, and winter road salt), I decided that I ought to try to learn as much as I could about my new daily driver. Chris was more than happy to oblige: before long, I was a counter clerk and graduate school applicant by day and a student of the culture and technology of the late-model VW community by night.

This meant that I often found myself in some peculiar places in the wee hours of the night. In shopping center parking lots, for example, where dozens of modified water-cooled Volkswagens and late-model Hondas—not to mention scores of twenty-somethings—would gather beneath the floodlights to show off their cars and, on occasion, pair off for stoplight contests. In dimly lit alleys in industrial complexes, too, where the VW crowd would sometimes gather on its own to talk shop, to barter parts and services, and to relive its on-road encounters with 5.0 Mustangs and Si Civics.

As you might have gathered, I had a blast that year. And as it came to a close, I found myself wondering what my own cars might look like with custom wheels and stainless steel mufflers, for Chris's world had begun to suck me in. More to the point, I also found myself reflecting on what I witnessed in those alleyways and parking lots: the camaraderie, the bartering, the shop talk, the bench racing, and especially the technological creativity that seemed to me to be the vital core of this pursuit we all referred to as "performance tuning." Well aware that my friends purchased most of the parts and accessories that they used to modify their cars, not from Volkswagen or Honda, but from mail-order catalogs and specialty parts dealers, I also began to wonder about the business and manufacturing end of the equation. Once I finally moved up north and began my graduate work, these random thoughts and musings slowly gelled, and in time they came to be the focus of my research. And the rest, as they say, is history.

♥

Over the past eight years, the generous support of friends, colleagues, instructors, institutions, and family members has enabled me to see this project through from idea to book. This finished product is as much theirs as it is mine, but I alone deserve the blame for any errors of interpretation or fact that might appear in the pages that follow.

First, I thank the Massachusetts Institute of Technology, the Science, Technology, and Society Department, and the Dibner Institute for the History of Science and Technology for their direct financial support of my graduate education. In this regard, I also owe a debt of gratitude to Debbie Douglas of the MIT Museum and David Kaiser of the STS faculty for their summertime assistantships. I also thank the National Science Foundation for the dissertation improvement grant that enabled me to undertake the far-flung research necessary for this project (Grant No. 0322275). Additional travel grants and awards from the STS Department, the Kelly Douglas Fund at MIT, the Society for the History of Technology, and the Southern California Chapter of the Society of Automotive Historians were indispensable as well.

I thank Merritt Roe Smith, Roz Williams, Meg Jacobs, and Bob Post, for their thoughtful advice, their prudent suggestions, and their collective eye for thematic postulation. In particular, I thank Roz for helping me to keep my own enthusiasm for this project and its subject reasonably in check; Meg for always pushing me to think more broadly; and Bob for his resourcefulness, insights, and tireless attention to detail. But my deepest debt is to Roe, my staunchest supporter and my greatest source of intellectual inspiration, for his guidance, for his encouragement, and, not least, for the way his eyes lit up the first time that I mentioned the term *hot rod* back in 1999. Thanks in this regard are also due to the mid-1990s History, Technology, and Society faculty at the Georgia Institute of Technology—especially Steve Usselman, Gus Giebelhaus, Mike Allen, and Mary Frank Fox—for introducing me to the history of technology and inspiring me to make a career of it.

Some of my colleagues at MIT, including Kieran Downes, Deborah Fitzgerald, Brendan Foley, Shane Hamilton, Rob Martello, Jenny Smith, Bill Turkel, Tim Wolters, and the members of the PXY Reading Group, read bits and pieces of my work; for their thoughtful comments and their shrewd editorial advice, I am forever grateful. I also thank the members of the STS departmental staff, especially Deb Fairchild, Kris Kipp, Debbie Meinbresse, Sarah Merrow, and Judy Spitzer for crucial administrative support. Carla Chrisfield, Rita Dempsey, Bonnie Edwards, Trudy Kontoff, and George Smith of the Dibner Institute were also of immeasurable help to me while I was a predoctoral fellow at their facility.

For their support as I revised and expanded my dissertation into this book, I would like to thank my colleagues at *Technology and Culture*, The Henry Ford, and the University of Detroit Mercy, especially John Staudenmaier, Bob Post, Joe Schultz, Suzanne Moon, Reed Benhamou, Marc Greuther, Judy Endelman, and Bob Casey. Extra-special thanks in this regard are due to John Staudenmaier, both for granting me the writing days that I needed over the last three years and, more importantly, for his friendship, his advice, and his willingness to put up with my enthusiasm for snow-free days, carbonated beverages, and the Hotlanta Braves, and also to Bob Post, both for the ongoing support and interest he has shown in this project and for the booklets, magazines, 1940s and 50s photographs, and other research materials that he sent my way. I thank my editor at Johns Hopkins, Bob Brugger, for taking an interest in this project years ago and for his guidance as the book took shape.

I am also deeply grateful to the many librarians, archivists, private publishers, enthusiasts, and speed equipment manufacturers who have helped me in so many ways these past few years. In particular, I thank Dick Dixon and Norma Crowell of Access RPM. Without their assistance, I never would have been able to access the SEMA Research Center in Diamond Bar, California; without their intervention, I never would have been able to attend the 2003 SEMA Show in Las Vegas; and, certainly, without their friendly conversation and shop talk, the time I spent in the desert and on the West Coast in 2003 would have been far less memorable. I thank Don Garlits, Pat Garlits, and Ed Smith of the Don Garlits Museum of Drag Racing in Ocala, Florida, for their gracious hospitality during the productive time I spent in their periodicals archive in 2003. In addition, I am deeply grateful to Dick Wells, Carl Olson, Bob Spar, Vic Edelbrock, Nancy Edelbrock, Camee Edelbrock, Delores "Dee" Berg, Kathy Flack, the late Wally Parks, Matthew Roth, Wayne King, David Boulé, Eugene Ciferno, Brian Brennan, Tom Lieb, Charlie VanCleve, and Doug Thorley, all of whom took time out of their busy schedules either to chat with me informally, to provide me with additional contacts, or, in some cases, to sit down with me for a formal interview.

I am also grateful to Judy Ritchie, Jim Spoonhower, and Shirley Presecan of the Specialty Equipment Market Association (SEMA) for their tireless assistance as I dug through their archival collections; to Ryan Rosales and Margarita Mora of the Cal-EPA Air Resources Board for helping me to locate decades-old board documents at the California EPA headquarters in Sacramento; to Mark Stiegerwald of the International Motor Racing Research Center at Watkins Glen, New York, for tracking down some obscure boxes in the center's holdings; and to Sam Jackson of the National Hot Rod Association Museum in Pomona, California, for allowing me to use his office to conduct a portion of my interviews in April 2003. I thank the staff of the Library of Congress; the Free Library of Philadelphia; the Worcester Public Library; the Davidson Library Special Collections Room at the University of California, Santa Barbara; the Boston Public Library at Copley Square; the Harvard University Library System; the Interlibrary Loan and Retrospective Collections Departments of the MIT Library System; the Benson Ford Research Center at The Henry Ford; and the Library of Michigan. I thank the editor of *European Car,* Les Bidrawn; the editor of the Rod and Restoration Group at Primedia, Kevin Lee; the Free Library of Philadelphia; the Don Garlits Museum of Drag Racing; Jim Orr and Judy Endelman of The Henry Ford; Bob Post; and Amy and Bob Deull for allowing me to reproduce the photographs that appear in this book. Finally, for helping me to ferret out a number of vital periodicals, articles, and documents related to the history of speed equipment manufacturing in the United States, I thank the Deulls, Len Romanick, A. David Wunsch, Vic McElheny, Kem Robertson, Phyllis DeVine (of *The Alternate*), and Charlie Yapp (of *Vintage Ford Speed Secrets Magazine*).

Heartfelt thanks as well to Walter Donila, for teaching me everything I now know about air-cooled Volkswagens, for his willingness to show up on a moment's notice anytime that I encountered an automotive problem that I was unable to address on my own, and for his friendship these last twenty years; to Chris Cox, for introducing me to import tuning, Gran Turismo, and parking-garage rally driving; to the Warrington Drive Thursday-night crew; and to my fellow Atomic Harvesters. I also extend my loving thanks to Charles Robert Hogg Sr., my grandfather and the man whose own enthusiasm for the automobile rubbed off on me at a very early age. In addition, I would like to thank the Topoloskys—my aunt Megan, my uncle Gary, and my cousin Elizabeth—for their frequent hospitality and their boundless generosity; the Deulls—my aunt Amy and my uncle Bob—for introducing me to hydropneumatic suspensions, two-cylinder Panhards, and vintage grand prix racing, not to mention their willingness to loan me thousands of dollars worth of 1950s and 60s periodicals for this project; and my aunt Sandy, for her tireless support and encouragement over the years. I would also like to thank the Taylors—my sister, Carolyn, my brother-in-law, Will, and my niece, Kathryn—for helping me maintain my sanity (for the most part). Mr. Vimbainjer and Korylya deserve a quick wink here too.

Finally, for their love, their support, and their willingness to tolerate my automotive enthusiasm—and the boxes of parts, driveway oil stains, and rusting hulks that came along with it—I thank my parents, Susan and Danny Lucsko. To them, I dedicate this work.

LIST OF ABBREVIATIONS

AAMA	Automotive Aftermarket Manufacturers Association
AHRA	American Hot Rod Association
AI	*Automotive Industries—The Automobile*
AMA	Automobile Manufacturers Association
ARB	Air Resources Board (California)
ARB-A	ARB Archive, California EPA Headquarters, Sacramento, California
c.i.d.	cubic inches of displacement
CARB	California Air Resources Board
CO	carbon monoxide (emissions)
DWF	Dick Wells's private files, Santa Ana, California
E.T.	elapsed time
ECU	electronic control unit
EGR	exhaust gas recirculation
EO	Executive Order (emissions exemption granted by the California ARB)
EPA	Environmental Protection Agency (federal)
FDaO	*Ford Dealer and Owner*
FDaSF	*Ford Dealer and Service Field*
FOaD	*Ford Owner and Dealer*
HC	hydrocarbon (emissions)
HEW	United States Department of Health, Education, and Welfare
HRIN	*Hot Rod Industry News*
HRM	*Hot Rod Magazine*
I/M	Inspection and Maintenance program (California)
MRA	Muroc Racing Association
MVPCB	Motor Vehicle Pollution Control Board (California)
NHRA	National Hot Rod Association
NHTSA	National Highway Traffic Safety Administration
NOx	oxides of nitrogen (emissions)

NSRA   National Street Rod Association
OEM   original equipment manufacturer
PCV   positive crankcase ventilation
Penn-DOT   Pennsylvania Department of Transportation
PHR   *Popular Hot Rodding*
SCTA   Southern California Timing Association
SEMA   Speed Equipment Manufacturers Association (1963–1967)
                Specialty Equipment Manufacturers Association (1967–1979)
                Specialty Equipment Market Association (1979–present)
SEMA-RC   SEMA Research Center, SEMA Headquarters, Diamond Bar,
                California
UCSB-RT   Romaine Trade Catalog Collection, University of California,
                Santa Barbara
UDRA   United Drag Racers Association
VESC   Vehicle Equipment Safety Compact
WD   wholesale distributor
WGR-RC   Root Collection, International Motor Racing Research
                Center, Watkins Glen, New York

THE BUSINESS OF SPEED

INTRODUCTION

By the age of twenty-four, Robert E. Petersen was well on his way.[1] The ambitious Barstow, California, native had left his desert home at the end of World War II, hoping to find a job in the bustling L.A. area. Within a few months, he had landed an entry-level position at MGM Studios in Hollywood, and by the end of his first year there, he had earned a spot on the company's team of publicists. When studio layoffs cut his stint with the company short, Petersen joined with a group of former MGM employees like himself to found an independent consulting firm known as Hollywood Publicity Associates. Brimming with confidence, he eagerly began working on his first commission with the new company, which was to promote a winter exhibition at the Los Angeles Armory. It was the summer of 1947; Petersen was twenty-one.

The show that he set out to publicize that summer was to be the first of its kind anywhere in the world, an automobile expo focusing exclusively on the burgeoning Southern California hot rod craze. The armory was to be filled, that is, not with shiny examples of Detroit's postwar renaissance, but rather with prewar coupes and roadsters that had been modified to extract every last ounce of performance from their often decades-old designs (fig. 1). But as he worked to promote this "Hot Rod Exhibition," he found that there were no dedicated periodicals in which to advertise this unique show—no hot rod newspapers, no hot rod tabloids, no hot rod magazines. Soon Petersen became convinced not only that the sport could use its own periodical but also that *he* could be the one to produce it. Toward the end of the summer, he discussed the notion with a fellow publicist on the exhibition team, Robert Lindsay, and they formed a partnership.

That autumn, Petersen and Lindsay left Hollywood Publicity Associates to begin working on their new venture, *Hot Rod Magazine*.

After securing a $1,000 loan, assembling twenty-four pages of editorial and feature content, and sweet-talking a couple of dozen Los Angeles automotive businesses into purchasing advertisements, the pair contracted with a local printer for a pilot run of 5,000 copies.[2] But with no subscribers and no distribution system, Petersen and Lindsay faced an uphill battle. Literally, they had to sell their new magazine themselves, copy by copy. Nevertheless, their first issue sold out quickly—and exceptionally so. But it wasn't dumb luck: *Hot Rod Magazine's* January 1948 debut coincided with the Los Angeles Hot Rod Exhibition,[3] and for three days the two were able to work the steps of the armory, unloading thousands of copies and spreading the word. Their second issue sold out just as rapidly the following month, and by the end of the first quarter of 1948, it was clear that Petersen and Lindsay had a winner. Within another year, monthly sales topped 50,000, and readers across the continent—indeed, across the globe—could find the latest issue at their local newsstands.[4] In 1950 Lindsay sold out to his partner, and Petersen, at the age of twenty-four, stood alone at the helm of a flourishing publishing empire.[5]

Thirty-two years later, when *Forbes* published its first annual list of the four hundred wealthiest Americans, Petersen comfortably made the cut with an estimated net worth of $100 million.[6] Over the next fourteen years his wealth—and his company, Petersen Publishing—continued to grow at a steady pace, and he became a fixture on the annual Forbes list. Finally, in 1996, a year in which his net worth stood at $450 million, Petersen sold his company for $500 million.[7] By then *Hot Rod* was but one of the thirty-two titles published each month under the Petersen banner, and it had long since ceased to be his biggest seller. Still, *Hot Rod* was nothing short of an unqualified success for its founder and his company, for it put them both on the map. Without it, there never would have been a Beverly Hills address for Mr. and Mrs. Robert E. Petersen, and there never would have been a Petersen Publishing Company.

Hot Rod's remarkable success owed a great deal to the efforts of its top brass. For starters, Petersen himself deserves much of the credit for the magazine's smooth launch. It was he who went out to the many hot rod racing events in the autumn of 1947 to drum up advertising support for its debut.[8] It was he who canvassed the area's hot rod meets, garages, and speed shops to recruit its staff. And it was he who convinced Bob Lindsay to join him in the endeavor in the first place. Together, Petersen and Lindsay tirelessly worked to promote their magazine during its critical formative years. They personally sold every copy of its first issue within weeks of its appearance, managed its expansion as it quickly grew from twenty-four

Figure 1. Publicity photo from the January 1948 Hot Rod Exhibition at the Los Angeles Armory. A similar photo was published in the January 1948 issue of *Hot Rod Magazine.* (Photo by Lee Blaisdell, reproduced courtesy of the Don Garlits Museum of Drag Racing.)

pages of coverage to a robust fifty-plus, and carefully directed its circulation as it swelled some 10,000 percent—from 5,000 to more than 500,000 monthly copies —well before its fifth anniversary.[9] *Hot Rod's* editorial and feature staff also deserves acclaim for its open-ended approach to the problem of orchestrating the "world's most complete hot rod coverage," for without a willingness to adapt to the shifting moods of its core audience, it never could have maintained its status as the leading automotive enthusiast publication for so many years.[10] Let there be no doubt: *Hot Rod's* prosperous run has been anything but lucky.

Yet especially during the magazine's first few years, its *timing* had as much to do with its success as anything else. *Throttle,* a similar publication, had hit the shelves back in January 1941, only to be forced off the market by the beginning of the following year because of war-related shortages of personnel and paper.[11] *Hot Rod* faced no such bad luck with its 1948 debut, of course. But it wasn't Petersen and Lindsay's timing in relation to the war that helped to ensure their endeavor's success so much as their timing regarding their subject. For with respect to hot rodding, their magazine hit the streets at precisely the right moment.

Since the 1910s, enthusiasts across the United States had been building what *Hot Rod* later defined as its central concern, "automobiles whose bodies and engines have been rebuilt in the quest for better performance and appearance."[12]

Early on, modified production cars of this sort were often used on oval racing tracks, but the majority were built for street use. Toward the end of the 1920s, however, a group of gearheads based in Southern California began to take things in a new direction. Their altered rides were built primarily for daily transportation, but they were also raced. And when these enthusiasts competed with their dual-use machines, their track would be either an open boulevard in what was then a relatively undeveloped region or the vast expanse of one of the Mojave Desert's many dry lake beds. In either case, their objective was all-out, straight-line speed, with the clock as much their opponent as the fellows against whom they actually raced. During the 1930s the number of souped-up cars on the streets of Southern California grew tremendously, and numerous clubs and organizations sprang up in support of this peculiar brand of modified motoring. Neither the Great Depression nor the Second World War proved sufficiently jarring to put an end to this activity, although the number of active participants did bottom out during the war. But by the summer of 1945, modified prewar coupes and roadsters had begun to reappear en masse in the greater Los Angeles area, and in the months and years that followed, hot rodding spread like wildfire throughout the United States. *This* is when Petersen and Lindsay introduced their magazine, just as the phenomenon was taking off nationally. *Hot Rod* was the first to catch this postwar wave, and although many others soon would hit the shelves to compete with it, none would ever overcome its first-to-market edge.[13]

Over the past six decades, hot rodding has grown and evolved in ways that few in Petersen and Lindsay's time could have imagined. Today, those who modify their automobiles for improved performance and appearance are a diverse bunch, their interests ranging from 150 horsepower flathead-powered prewar Fords to 500 horsepower 1960s Hemi-powered muscle cars, from 90 horsepower EMPI-equipped VW Beetles to 350 horsepower four-cylinder Honda screamers, and from 400 horsepower Audi 2.0T sedans to 1,000 horsepower Ferrari monsters. Millions of enthusiasts attend thousands of meets, races, and shows across the United States each year, spending in excess of $34 billion annually on custom and performance equipment for their own cars.[14] Scores of periodicals cover every aspect of the sport from virtually every angle, and gearheads looking for a bit more insight into the nuts and bolts of their projects can choose from literally hundreds of performance-oriented automotive manuals. Likewise, those with an interest in the history of particular aspects of the hobby have dozens of popular titles to choose from.[15] Hot rods and hot rodders have appeared in countless movies, television programs, and pop-music classics over the past sixty years as well. The hobby Petersen and Lindsay set out to cover back in the fall of 1947 has proven, in short, to be phenomenally popular and remarkably long-lived.

However, those who build, drive, and write about modified cars have seldom appeared within the vast academic literature devoted to the role of the car in American life.[16] The reason, simply put, is that scholars have for the most part focused their attention on the evolution of mass automobility in the United States. Collectively, theirs is a rich and varied narrative of mass production and mass consumption that includes detailed analyses of the business, manufacturing, labor management, and regulatory strategies of the Big Three[17] automakers; the infrastructural implications of widespread automobile use in rural, urban, and suburban contexts; and, more recently, the experiences of ordinary end users.[18] Hot rodders are not your typical consumers, though. They are *enthusiasts*, end users for whom the car is something more than a way to ferry themselves to work, shuttle their kids to soccer practice, haul groceries, or escape to the mountains on weekends. Enthusiasts do use their cars to accomplish these mundane tasks, of course. But for them, the vehicles in which they do so are cherished, valued more for what they are (and for what they *can* do) than for what they are actually called upon to do. For them, automobiles are not simply a *means*. They are an *end* as well.

That said, untold millions of ordinary consumers are also enthusiastic about their cars for reasons beyond their everyday utility. Some allow the emotional appeal of certain models to override their better judgment when they shop for a new car. Others wash and wax their cars with an obsessive passion, or give them human names, or fill their tanks with premium when low-test would suffice. Many also flock to annual auto shows, devour the monthly automotive magazines, zealously follow local and national motorsports, and savor every glimpse of anything that isn't painted beige or silver. What distinguishes hot rodders is that their passion for automobiles is not entirely passive. For when they flip through the pages of the latest issue of *Motor Trend* or *Car and Driver*, or stroll through retail showrooms, or wander among the manufacturers' displays at big-time auto shows, they do not see finished products. Instead, they see untapped potential: Mustangs that need larger wheels, Jettas that need stiffer springs, and Civics that need more torque. Accordingly, when they buy a new or "pre-owned" car, they are rarely satisfied with the design, engineering, and fit-and-finish work that the original equipment manufacturers (OEMs) put into them.[19] Or more precisely, they are not satisfied with how their OEM-built cars perform. Often enough, all it takes is the addition of a freer-flowing exhaust, a cool-air intake, a set of lowered springs, or a strut-tower brace. Sometimes, though, it takes a great deal more, including everything from bolt-on turbochargers to no-holds-barred engine, transmission, and drivetrain swaps. But whether simple, quickly accomplished, and inexpensive or difficult, time-consuming, and fiscally demanding, all of the modifications that hot rodders undertake involve their active participation in reshaping—indeed, reengineering—their run-of-the-mill cars to meet their personal expectations.

For the most part, these expectations center on the goal of a faster car, but it would be a mistake to say that outright speed is their only aim. For the automotive modifications that hot rodders perform are also acts of creative expression that are intended to demonstrate a certain level of technical mastery. Moreover, rodders recognize, through what amounts to an unwritten code of technical accomplishment, that there is a right way and a wrong way to modify a given car. The difference between those that aren't quite right and those that are spot-on is always in the eye of the beholder, of course, but it tends to boil down to the perceived elegance of the vehicle's modifications. Whether ill-conceived and sloppy or brilliant and meticulously assembled, though, hot rods are above all the end result of their owners' dissatisfaction with the ordinary and their willingness to roll up their sleeves and do something about it.

The overwhelming majority of American automobile owners would never dream of tearing into their engines to install a high-performance camshaft, of swapping out their mufflers for a marginal horsepower boost, or of replacing their brake rotors with cross-drilled and fade-resistant units. Who, exactly, are these über-enthusiasts who do, then? Who are these hot rodders who spend $34 billion each year on their cars? When and where did they first appear, and why? Apart from speed and technical expression, what are their motives? How and where do they learn which sorts of engine and chassis modifications work and which do not? Where do they get their parts? And, perhaps most importantly, are they simply dwellers on the fringe, or have their pursuits influenced the evolution of the broader American car culture in any meaningful ways?

A handful of academic historians have begun to tackle these and other similar questions. H. F. Moorhouse, a sociologist, has published a couple of essays as well as a monograph about what he calls the post–World War II "hot rodding fraternity," works that focus on the collective motivations of 1950s, 60s, and 70s enthusiasts.[20] Robert C. Post's work traces the technological development of drag racing as it evolved from its origins in the hot rod racing of the 1930s and 40s through its elaboration as a high-dollar, competitive sport in the years since.[21] Most recently, John DeWitt has explored the custom-car scene of the 1950s and early 60s (as well as its revival in the 1980s and 90s), convincingly demonstrating that the work of those who created custom cars deserves to be recognized as genuine modern (and postmodern) art.[22] Together, Moorhouse, Post, and DeWitt have done much to open the world of the enthusiast to academic study, but the history of hot rodding as such remains virtually unexplored. We have an excellent historical sociology of the phenomenon, but no rigorous study of its development over time. We have

gained valuable insights into the development of the technological underpinnings of organized drag racing but have no such understanding of the technical
milieu of the ordinary hot rodder. We are able to appreciate well-executed hot
rods and custom cars as artistic expressions, but we still know very little about the
mechanical art of modifying a production automobile.

This book aims to fill in some of these gaps. Doing so properly requires a
suitable perspective, however, one from which to observe the evolution of hot
rodding over its entire history. Fortunately, precisely such a vantage point ex-
ists: the business of speed equipment manufacturing. Long before the postwar
hot rod boom, long before the first issue of *Hot Rod* hit the shelves, and long
before the first dry lakes racing events of the 1920s and 30s, dozens of small shops
across the United States designed and manufactured add-on parts to facilitate the
performance-oriented modification of otherwise run-of-the-mill cars. As early as
the 1910s, therefore, high-performance cylinder heads, camshafts, intake manifolds, and other components were available commercially to those for whom
the standard capabilities of the Model T were insufficient. Toward the end of
the 1930s, with the California hot rod craze in full swing, the manufacture of
these aftermarket parts and accessories began to concentrate in the Los Angeles
area, and by the time the hobby went national in the late 1940s, the overwhelming majority of the high-performance equipment that hot rodders in every state
used on their cars came from Southern California. Consisting of a large number
of very small firms, this industry grew at an impressive rate in the decades that
followed. During the 1960s and 70s speed equipment manufacturing—by then a
$1-billion-a-year business—found itself caught up in the national debate over the
human-safety and environmental impacts of mass automobility. But unlike the
Big Three, this industry managed to deal with the onset of government regulation
in a productive and profitable manner—in fact, the 1960s and 70s rank among
the most successful decades in its history. Today this multibillion-a-year business
serves hot rodders young and old, whether their interests lie in the modification
of American, European, or Asian automobiles.

Dating back some nine decades, the rich and heretofore untold history of
speed equipment manufacturing offers an ideal lens through which to observe
the evolution and elaboration of hot rodding in the United States. On its own
terms, though, the story of this industry is equally appealing. Its firms are small,
and its market is fiercely competitive and almost entirely demand-driven. Furthermore, because the high-performance goods its firms turn out are intended
to enhance extant OEM designs, its product cycles closely follow those of the
mainstream automobile industry. Hot rodders do not necessarily change their automotive interests with each new model-year release from Detroit, Wolfsburg, or

Tokyo, however. The speed equipment industry therefore supplies products not only for newer designs but also for those that have been around ten, twenty, even fifty or more years. At the same time, government regulations define—and, often on a yearly basis, redefine—the legal limits of what consumers and aftermarket companies can and cannot do with their cars. Located at a conceptual nexus where engineering creativity, consumer desires, OEM technology, and government regulations intersect, the business of speed equipment manufacturing warrants a closer look.

What follows, therefore, is a history of the American high-performance automotive aftermarket. Covering the years from 1915 to 1990, it traces the evolution of the industry's constituent firms, characteristic products, industrial organization, and relationship(s) with OEMs, government regulators, and ordinary enthusiasts. Among these points, the last is critical: this book treats the speed equipment industry and its customers as a single topic of inquiry. This is not to say that the one should be equated with the other—far from it, in fact. Instead, this project seriously considers the ways in which the interaction between the two has affected the evolution not only of the industry but of the average enthusiast and his beloved hot rod as well. In other words, this is a user-producer analysis.[23]

This book is also an analysis of the nature of technological enthusiasm. More specifically, it seeks to move beyond the well-established notion that technological enthusiasm is a means-ends affair in which an eager embrace of the means is necessarily predicated on specific ideas regarding the ends it might deliver.[24] Instead, this book highlights the extraordinary power that enthusiasm for a particular technology *in and of itself* is capable of exercising.[25] This is not to deny the importance of the pursuit of fame, fortune, a sense of community, or even the thrill of all-out speed among those who build hot rods and those who manufacture hot rod parts. But it is intended to point toward the importance of a less overtly rational sense of purpose among those who take part in these activities.[26]

The evolution of manufacturing techniques is a key theme here as well. For decades, historians have explored the ideas, processes, and machines that enabled mass production to take off in the United States.[27] More recently, some have also investigated manufacturing systems of a more flexible nature, challenging many of the assumptions of those whose work has stressed the importance of the mass production paradigm.[28] The ensuing debate has been lively and productive; of interest here, however, is the fact that the speed equipment industry fits neatly within the theoretical framework of neither camp. For much of its history, the high-performance aftermarket does appear to mesh perfectly with established models for the way small, batch-oriented manufacturers operate. But by the 1970s and 80s, many of the most successful aftermarket firms had begun to integrate their operations vertically, buy up their smaller competitors, and even mass produce some of their better-selling lines. Over the course of the last seventy-odd years, then, many of these companies have grown from small, custom- and batch-manufacturing concerns into large-scale enterprises. Most remain small and batch-oriented, however, and nearly all have retained their flexible capabilities. In short, this industry defies easy categorization, and in order to make sense of its diversity, this book dispenses with abstractions and focuses instead on the concrete realm of the shop floor. For the story of the speed equipment industry offers a unique opportunity to examine the ways in which an array of real-world manufacturing practices interact within a single industrial sector.

This book also explores the relationship between the automotive aftermarket and the mainstream automobile industry. Simply put, hot rodding is predicated on the existence of the mass-produced car, and so too is the speed equipment industry. Nevertheless, in the relationship between the aftermarket and the OEMs, inspiration and ideas often trickle up as well as down, flowing back and forth in an ongoing exchange that is and has always been *recursive*—in effect, a permanent feedback loop in which decisions taken by the OEMs in response to those of the aftermarket in turn influence future aftermarket decisions. Conduits for this recursive flow include everything from automotive technology itself to the printed text of popular and industry periodicals, but none is as important as personnel. Indeed, the number of engineers and technicians whose careers began in the hot rod industry and ended in Detroit (and vice versa) is staggering, particularly in the decades after World War II.

Finally, and perhaps most fundamentally, this book seeks to reexamine the role of end-user agency in the history of the automobile in America. Recent scholarship suggests that in the early days of mass automobility in the United States, the automobile itself was not a black-boxed technology. Ordinary users thus were free to shape and reshape their mass-produced cars to meet their own often quite specific needs: farmers used them to power washing machines and band saws, while middle-class consumers often modified the bodies of their cars to make them more versatile and comfortable for long-range touring and camping.[29] But those who have studied these phenomena conclude that the automobile industry successfully closed the black box by the 1930s, effectively ending the interactive phase of the history of the automobile and heralding the age of Big Three dominance. However, the evolution of the business, technology, and culture of hot rodding strongly suggests otherwise. For hot rodding's explosive growth *began* in the 1930s and has yet to taper off; moreover, it remains to this day an overwhelmingly participatory phenomenon. Modified automobility has been with us for more than ninety years, that is, and so has end-user agency.

Before we begin, a few brief words on the terminology of the enthusiast community.[30] First, what exactly is or was a "hot rod"? We know that the term first emerged in the 1940s, most likely as a contraction of "hot roadster."[31] But is that really all that a hot rod is, was, or can be—a "hot roadster"? Purists insist that "hot rod" does indeed refer to a very specific type of modified automobile: stripped-down and souped-up prewar Ford roadsters. Sedans of the same era do not count, and neither do coupes or convertibles.[32] Others have been far more liberal. Dean Batchelor, the author of an excellent popular history of the phenomenon, defines as a hot rod "*any* production vehicle which has been modified to provide more performance."[33] All later-model modified vehicles therefore qualify, as long as the principle that guided their construction was that of improved performance. Though I tend to agree with Batchelor, what matters here is not so much the definitive establishment of a universal definition as it is the shifting applicability of the concept over time. In other words, I tend to allow the records to speak for themselves: if a 1950s enthusiast describes his 1957 Chevy as a "hot rod," then so be it.

Far more critical is the question of how to deal with "hot rod" when it is used as a verb. Almost without exception, to "hot rod" a car—that is, the act of "hot rodding" an automobile—is to create precisely the sort of vehicle Dean Batchelor describes, one that has been modified to obtain better performance. It is therefore a *technical* concept: to hot rod a car is quite literally to reengineer the mechanical components of a mass-produced automobile. In more recent years, particularly among enthusiasts of European and Japanese automobiles, the concept of "performance tuning" (or simply "tuning") has joined "hot rodding" in common parlance. Because their meanings are essentially identical, this work treats "hot rodding" and "performance tuning" synonymously.

Synonymous treatment of "hot rodder" and "performance tuner" is impossible, though. "Hot rodder" simply denotes an individual enthusiast who modifies his car for improved performance—that is, one who participates in "hot rodding." "Performance tuner," on the other hand, applies to those firms which design and manufacture performance products for modern, usually imported automobiles. Neuspeed, for example, a company that produces high-performance engine and suspension components for modern imports, is a performance tuner. An enthusiast in his garage carefully constructing a modified automobile with those parts is not, however, even though the activity with which he is engaged is in fact performance tuning and the final product of his efforts is a performance-tuned automobile.[34]

"Performance" itself is problematic too. For as Robert C. Post has pointed out, "performance," in the context of a community of automotive enthusiasts, "has a delicious ambiguity." On the one hand, he explains, "performance" is an engineering concept that refers to the dynamic capabilities of an automobile. Thus, a "high-performance car" is one that is capable of higher speeds, faster (or more stable) cornering, and quicker acceleration than most. On the other hand, "performance," particularly among automobile racers, also implies what Post calls the "imagery of the theater."[35] Dragsters, for example, go through a process known as "staging" prior to a quarter-mile race, a ritual that is similar conceptually to the warm-up laps that take place at an oval-track event prior to the green flag's signal. Similarly, the theatrics that accompany a low-slung, candy apple red custom as it slowly cruises down an urban boulevard—or, for that matter, those that go into competitive automobile show displays—are in every sense real and valid forms of "performance" as well. What's more, fuel efficiency, reliability, ease of maintenance, oil consumption, towing capacity, and even passenger comfort are for many ordinary consumers critical components of a given car's "performance." For the purposes of this book, however, the meaning of the term is relatively simple: *performance refers to the ability to go fast.* Hence, "high performance" implies speed, "performance tuning" is the pursuit of it, and "high-performance parts and accessories" are the means with which to achieve it.

Finally, there remains the issue of the "speed equipment industry." In the context of the era of the traditional hot rod—the 1930s, 40s, and 50s—one is tempted to refer to it as the "hot rod industry," for that is precisely what it was—an industry manufacturing parts for hot rods. But that is not how the manufacturers themselves thought of it at the time. Instead, they tended to identify themselves individually as "speed equipment manufacturers" and collectively as the "speed equipment industry." (Witness the original name of their industrial organization, founded in 1963: the Speed Equipment Manufacturers Association, or SEMA.) However, by the end of the 1960s, "speed" was no longer politically correct, and "specialty" therefore replaced "speed" in official correspondence (and within the SEMA acronym). "High-performance aftermarket," another, by now far more common label for the industry, also emerged in the 1960s. During the 1960s and 70s, "specialty equipment industry" and "high-performance aftermarket" remained synonymous, but sometime after the 1970s, "high-performance aftermarket"—or, more commonly, "the aftermarket"—effectively took over in common usage.

However, "the aftermarket" is a tricky label, the exact meaning of which depends entirely on the context. In the strictest of terms, "the aftermarket" refers to the entire industry that manufactures replacement as well as high-performance automotive parts. Consequently, when an enthusiast refers to "the aftermarket," he might mean the speed equipment industry, but he might also mean the industry that produced the standard-duty replacement water pump he recently installed in his 1997 Caravan. Nevertheless, following the practice of ordinary rodders, hot rod journalists, and high-performance parts manufacturers, I generally treat "the aftermarket" and "the speed equipment industry" synonymously in the chapters that follow. On those rare occasions in which I discuss the replacement-parts aftermarket specifically, the context should preclude undue confusion.

❤

For the moment, though, forget about hot rods, performance tuning, and the speed equipment industry. Forget about Southern California, the dry lakes, and street racing. Forget about manufacturing practices and government regulations. For in the beginning, there were only amateur enthusiasts and millions of identical Model Ts.

CHAPTER ONE

FASTER FLIVVERS, 1915–1927

In the 1910s, automobile production, sales, and use in the United States began to grow at a feverish pace. Total domestic production swelled nearly tenfold between 1911 and 1917, and new automobile registrations rose by more than 400 percent. Per capita automobile ownership doubled every two years, and from New York to Los Angeles, new dealers, service stations, and parking garages cropped up in droves. New highways were in the works as well, as federal, state, and local officials across the country struggled to deal with the dramatic increase in motorized traffic. It was the beginning of a transportation revolution.[1]

At the heart of these developments was the Ford Motor Company. Between 1911 and 1916 more than 1.5 million "universal cars" rolled out of its assembly plants, fully 36 percent of all automobiles manufactured in America during those years. Annual output at Ford grew at an average of 65 percent, and the firm's production costs tumbled dramatically. Sales were brisk, and as early as 1915 better than one in two new registrations each year were for Model Ts.[2] How Ford managed to achieve all of this is now a well-known tale of manufacturing, marketing, and labor relations breakthroughs that needn't be recounted here;[3] what matters, rather, is that Ford's spectacular success did much to usher in the age of mass automobility. And for this feat, history has not forgotten Henry Ford and his revolutionary Model T.

Enthusiasts aside, however, few seem to remember that the Model T also forever changed the face of American motorsports in the 1910s. Before the mass production of the universal car, high-performance motoring was a luxury far beyond the means of all but the very wealthiest of Americans. High-dollar, hand-built racing specials dominated on oval-track and road-racing circuits across the

United States, and on the streets, low and moderately priced horseless carriages simply were no match for high-end Deusenburgs and Benzes. But as the conventional front-engine, rear-drive Model T became available in the early 1910s, it quickly earned an upstart's reputation among performance enthusiasts. As early as September of 1914, the editors of one popular-market periodical, *The Fordowner*, proudly reported that "many a Ford owner has stripped the fenders and body from his car, strapped a pillow onto the gasoline tank, and entered the races at the home town fair or the more pretentious yearly event at the county seat."[4] The following summer, modified Model Ts built for high-speed street use began to surface as well, and the editors of *The Fordowner* quickly found themselves celebrating, on the one hand, the fact that "in every community, no matter how small, some one has changed his Ford as to body and engine, so that he can get more speed than the average touring car," and condemning, on the other, the lawless driving habits these cars encouraged.[5] Abundant and cheap, the Model T enabled thousands of ordinary Americans to begin to enjoy the thrills of dirt-track racing and high-speed highway travel as never before—that is, as *participants*.

Taking part required a knack for things mechanical, however. For the Model T was rugged and versatile but not fast, and only very rarely was it possible to "show the differential to anything on the road" in an ordinary Ford simply by removing the fenders and donning a pair of driving goggles.[6] At the very least, one had to make some basic changes to the car's gangly suspension and humble powerplant. Mechanically inclined enthusiasts could do both, but many more with both the means and the desire could do neither. Here, then, was a market for an entirely new type of product, and in 1915 high-performance parts and accessories for the lowly Ford began to trickle into circulation. By the end of the decade, a handful of firms built and sold hundreds of Model T performance products, and during the 1920s, dozens more would join the fray. Performance tuning and the speed equipment industry began with the Model T; our story begins, therefore, in the era of the universal car.

MODEL T ACCESSORIES AND SPEED EQUIPMENT

More than 15 million Model Ts were manufactured between 1908 and 1927, an international automotive production record that stood for forty-five years.[7] Urban, suburban, and rural residents alike bought them by the tens of thousands each month, putting them to daily use in a variety of contexts ranging from the civilized to the extreme. The Model T was versatile and durable, and it was a bargain. But it was also bare-bones basic, a no-frills economy car designed for utility and value, not comfort. Many of those who purchased new Fords during the 1910s and 20s therefore sought to modify them to better serve their individual needs. For farmers across the United States, this often meant adapting the car's drivetrain for occasional use as a mobile source of power. For urbanites and suburbanites, it sometimes involved extensive changes to the Model T's superstructure for more comfortable long-distance travel and touring.[8] And for owners rural, suburban, and urban alike, it frequently entailed the addition of a floor heater, a set of auxiliary lights, and perhaps a pair of louder horns. The Model T was a perfectly capable car as delivered, but in the eyes of more than a few of its users, it remained a tweak or two away from practical perfection.

This was true not only of the car's interior and exterior appointments but also of its mechanical underpinnings. The Model T lacked a water pump, an oil pump, a fuel pump, a self-advancing ignition system, and, prior to its introduction as an extra-cost option in 1919, an electric starter.[9] Instead, its hand-cranked engine relied on thermodynamic currents to circulate its coolant, a series of crankshaft scoops to distribute its oil, gravity to deliver its fuel, and the user to maintain an appropriate degree of ignition advance while driving. From the very beginning, therefore, a number of accessory firms offered pumps, ignitions, and, for a brief period, electric starters to help the average owner render his Ford more reliable and user-friendly. These mechanical accessory firms, together with the many companies that manufactured add-on comfort and convenience equipment for the universal car, comprised a rich and diverse aftermarket for the Model T that thrived throughout the period in question.[10]

It was within this prosperous milieu of add-on parts and accessories for the Ford that speed equipment manufacturers first emerged in the mid-1910s. However, much of what was produced during their first few years differed very little, conceptually or practically, from the aforementioned general-improvement merchandise. Early advertisements promising "more speed for the Ford" were fairly unambiguous, of course, and often enough red flags like this made it relatively easy to determine which firms actually manufactured speed equipment.[11] Sometimes, though, the line was far less clear. Dozens of companies regularly advertised that their add-on products would improve the "performance" of the Model T Ford, for example. However, closer inspection reveals that by "performance" only a handful of these firms actually meant horsepower, acceleration, and top-speed gains; instead, most simply meant that fuel mileage, oil consumption, or cold-morning starting would improve. Bear in mind, therefore, that the boundary between the high-performance and the general-improvement aftermarkets was often barely there.

That said, companies offering high-performance parts for the Model T first surfaced in 1915; by the time Ford discontinued the car in 1927, several dozen firms had manufactured speed equipment for it.[12] The overwhelming majority of these manufacturers were located in the Midwest, though there was also a sizeable contingent on the West Coast and a smaller group in the East. To those familiar with the early history of the automobile in the United States, this should come as no surprise, for long before the introduction of the Model T, the American automobile industry had come to call the Midwest home. By 1918, six of the top ten states in terms of automobile ownership were located in the Midwest, with California and New York ranking among the top ten as well.[13] Thus, speed equipment manufacturing companies first appeared where the emerging automobile culture was already at its strongest.

Many of these firms began as general machine shops, foundries, and manufacturing subcontractors before they began to dabble in the high-performance market. The Williams Company of Akron, Ohio, for example, opened its doors as a foundry and machine shop in 1888; only after it had done some high-performance subcontracting in the early 1920s did it decide to produce some speed equipment of its own. Others first began as stationary powerplant and engine component subcontractors, as did the Beaver Manufacturing Company of Chicago, whose motor-building operations dated back to 1902. A few had OEM roots too, including the Chevrolet Brothers Manufacturing Company of Indianapolis, which was founded by Arthur and Louis Chevrolet a few years after the latter's automobile company was absorbed by General Motors. Many more first emerged, however, indirectly, as a result of the accumulated racing experience of their founders. Joe Jaegersberger, the man behind the Racine, Wisconsin, firm Rajo, was a dirt track racer long before he brought his aftermarket gear to market in the 1910s. Ed Winfield also began his career as a Los Angeles–area oval-track racer, as did Robert Roof of the Roof Auto Specialty Company of Indianapolis and both Arthur and Louis Chevrolet. Harry Miller and George Riley of Los Angeles, on the other hand, earned their reputations building racing engines. For many of these companies, particularly those with racing roots, speed equipment manufacturing was a full-time business. For others, especially those with machine-shop and engine-manufacturing roots, it was a part-time endeavor, a profitable diversification based on their core competencies.[14]

The range of products that these firms produced is astounding. Dozens at first, and ultimately hundreds, of different high-performance items were available for virtually every part of the standard Ford during the 1910s and 20s. But for the most part, all of these early add-on parts and accessories fell into one of four product categories, each of which corresponded to a basic tuning strategy. The first focused on overall engine efficiency, for one way to extract more power from the Model T engine was to fine-tune its standard parts and components to ensure that they operated at their highest possible potential. Sometimes modifications of this sort were relatively simple and could be carried out by general automotive shops or even by enthusiastic owners, but aftermarket products often played a role as well. Bolting on an oil cooler, for example, enabled one to push the standard motor harder on a daily basis without risking engine damage, as did the addition of a pressure oiling system or a water pump. Also beneficial in this regard was the use of an aftermarket ignition system, for although Ford's "timer" was adequate for low-speed use, it was unable to deliver the precision that higher rpms required. In addition, most aftermarket ignitions were self-advancing, which meant that they required no attention from the driver while the car was in motion.

Tellingly, some of the parts and accessories associated with this first approach weren't really "speed equipment" at all. This was particularly true of add-on ignitions: almost without exception, speed equipment manufacturers and high-performance engine builders alike recommended the use of aftermarket ignitions in performance-oriented Model Ts, but only occasionally did they make them themselves. Instead, they typically suggested systems of the general-improvement type, available in the 1910s and 20s from firms like Atwater-Kent of Philadelphia and American Bosch of Springfield, Massachusetts; so good were their products that they more than adequately served the needs of early performance buffs.[15]

The second general approach to Model T performance tuning focused on reducing friction, vibrations, and unnecessary weight within the engine in order to enable it to achieve higher operating speeds. Ford's three-bearing crankshaft was prone to flex at speeds as low as 1000 rpm—fully 500 revolutions below the engine's peak-power point of 1500 rpm. The Dunn Counterbalance Company of Clarinda, Iowa, for one, sought to remedy this through the use of crankshaft counterweights, small metal wedges that attached to the crank to smooth its rotation. Robert Roof, on the other hand, advocated the more radical procedure of replacing the Ford's standard crank with a five-bearing aftermarket unit. With two additional anchor points among which to distribute the reciprocating forces of the engine, five-bearing cranks operated with substantially less flex, enabling the engine to spin faster. Flywheels, too, could be cut and balanced for smoother motor operation and quicker acceleration, and many speed equipment manufacturers offered completely reworked flywheels or, more commonly, the machining service of balancing and lightening a customer's original flywheel.[16]

Other firms reduced friction and unnecessary weight by modifying the reciprocating assembly (pistons, connecting rods, and related hardware). Ford equipped the Model T with cast-iron pistons, which, though rugged and easy to manufacture, were also heavy and almost always out of balance. One way to attend to this problem was to shave some material from the pistons' bottoms, or skirts, in order to bring them into balance. But even when balanced, cast iron was a fairly heavy material for an engine to reciprocate several thousand times per minute, and many companies therefore produced lightweight aluminum or aluminum alloy pistons as an alternative. The Model T's standard-issue connecting rods were another source of strain within the reciprocating assembly; these could be cut for balance, drilled for lightness, or altogether replaced with aluminum rods.[17]

Equipped with a counterbalanced or five-bearing crankshaft, a lightened and balanced flywheel, and a set of lightweight, precision-balanced aluminum pistons and connecting rods, an otherwise ordinary Model T engine would have been able to achieve much higher rpms than it ever could have in stock trim, at least in theory. But the faster an engine spins, the more air and fuel it requires. Light and balanced though its internal assemblies might have been, therefore, a Model T motor with OEM carburetion, manifolding, and valvetrain components actually would have been unable to rotate much faster at all. Consequently, the third—and by far the most common—approach to Model T performance tuning involved modifications that enabled the engine to consume more air and fuel. Improved carburetion was one way to do this, and although larger carburetors pulled from other makes of cars often worked well, fitting them to a Ford engine could be difficult. Thus, large-bore units designed to fit the Ford were available from a very early date not only from speed equipment companies like Miller but also from general-improvement and replacement-aftermarket firms like Chandler and Stewart-Warner, both of Chicago.[18] Multiple carburetor systems were available from several manufacturers as well.

Larger carburetors allowed more air and fuel to enter the engine's intake passages, or ports, but once there, the Model T's small, low-lift intake valves prevented much of it from actually reaching the cylinders. Fitting valves of a larger diameter was an obvious solution, though opinions varied among enthusiasts and aftermarket companies regarding the best way to accomplish this. Some advocated the use of larger Fordson tractor valves in high-performance Model Ts, but many firms made special valves for this purpose as well.[19] Another way to increase the amount of fuel and air that passed into the engine was to increase the lift or duration of the intake valves, for by forcing the motor's valves to open wider or for a longer interval, more fuel and air could pass through. One way to do this was to modify the camshaft, and skilled machinists and bold enthusiasts sometimes reground Model T cams by hand for this purpose, lobe by lobe. But it was far less risky and much more cost effective to simply replace the original camshaft with a high-performance unit, and many firms sold new or reground "hi-speed" cams during the Model T era.[20] "Multi-Lifts," compound levers that multiplied valve lift, were available from Riley as a further alternative (fig. 2). Reground camshafts

Figure 2. Riley "Multi Lifts." Ford's standard-issue Model T lifters transferred the motion of the camshaft's lobes to the valve stems in a 1:1 ratio. Thus, if a camshaft lobe raised its corresponding lifter by one-half inch, then the corresponding valve would open one-half inch. Riley's aftermarket Multi-Lifts consisted of a series of compound levers that replaced Ford's standard-issue lifters, multiplying the camshaft's action and resulting in a lobe-to-valve-lift ratio of 1.2:1 (or 1.5:1, depending on the model). Consequently, one-half inch of lift at the camshaft resulted, on an engine with 1.2:1 Multi Lifts, in approximately three-fifths inch of lift at the valves. This allowed the engine to consume more air and fuel, resulting in more horsepower. ("The Multi-Lift," *Ford Owner and Dealer*, February 1921, 112. Reproduced courtesy of the Free Library of Philadelphia.)

and multi-lifts could also be used on the exhaust valves, although considerable debate surrounded the issue of whether or not Ford's exhaust system itself required attention.[21]

One final method of improving the flow of fuel and air within these engines was to replace the entire OEM cylinder head with an aftermarket unit. Ford's Model T head was of a very basic design known as an "L-type," or "flathead" (fig. 3). This meant that the Model T's head was a more-or-less flat, rectangular casting

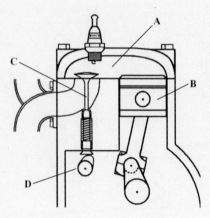

Figure 3. Side-view schematic of one cylinder from a flathead engine. The combustion chamber is at A, its corresponding piston at B, one of its two valves at C, and the engine's camshaft at D.

with four shallow combustion chambers on one side and provisions for the radiator hose on the other. The motor's intake and exhaust ports were mounted in the engine block, as were the valves, which were positioned adjacent to the top of their respective cylinders. The appeal of the flathead design was its simplicity: the cylinder head contained no moving parts, and the valve-actuating mechanism within the block was simple and direct. Its principal drawback, however, was its circuitous intake and exhaust passages, which impeded engine breathing. Early on, a number of speed equipment manufacturers therefore began to produce replacement heads for the Model T that featured overhead valves (fig. 4). Though more complex mechanically, these heads greatly aided the flow of air through the engine by directing the incoming charge (and, in turn, the exhaust gases) through passages that were far more direct than those of an L-head. Many of these high-performance cylinder head conversions simply replaced the standard Ford's eight in-block valves with eight overhead valves, but some actually featured sixteen overhead valves. Whether of eight- or sixteen-valve design, though, most made use of the standard Ford camshaft to operate the relocated valves through a series of pushrods and rockers, though a few used single- or double-overhead camshaft arrangements instead.[22]

Overhead-valve conversions performed well, often boosting the horsepower of the standard engine by 50 percent or more, but they were pricey. Some firms therefore offered high-performance flatheads for the Model T as a cheaper and simpler alternative. Ford's flathead was good enough for everyday use, but its poorly-shaped combustion chambers hindered top-end power, as did its low

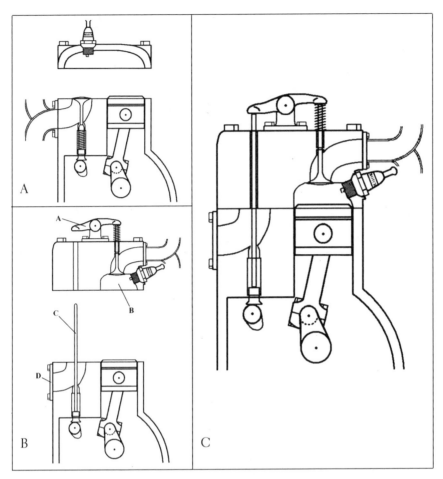

Figure 4. Side-view schematics of an overhead-valve conversion for a flathead engine.
A: The first step in the conversion process, the removal of the original-equipment cylinder
head. B: The aftermarket overhead-valve cylinder head awaits installation: its rocker assembly is at A and its combustion chamber at B, while the original-equipment, in-block
valve has been replaced by a pushrod, at C, and the original block-mounted intake and
exhaust manifolding has been replaced by a cover plate, at D. C: The modified engine.

compression ratio.[23] It was poorly-finished as well, and its rough-cast combustion chamber surfaces dissipated heat unevenly, often leading to power-robbing
"hot spots" in the chambers. Aftermarket flatheads, on the other hand, typically
featured fully-machined combustion chambers that were better shaped (for improved flow) and smaller (for higher compression). Turnbull of Wilmington,
Ohio, was the first to widely market such a head; its "Turko" model debuted
in 1918. Others were slow to introduce similar equipment, however, and by the

end of the Model T era only a handful of manufacturers had done so, reflecting a preference among Model T performance enthusiasts for the far more radical overhead-valve conversion.[24]

The fourth and final approach to Model T performance tuning involved body and suspension modifications. For the former, dozens of streamlined replacement bodies for the Model T were available in the 1910s and 1920s.[25] Aftermarket bodies gave the Model T a racier appearance, and through improved aerodynamics, a few of them aided top-end speed as well. Suspension modifications were far more important from a performance standpoint though. For example, lowering or "underslinging" the Ford chassis brought the body of the car closer to the ground, lowering its center of gravity and improving handling. As early as 1915, articles detailing some of the many ways in which this could be accomplished began to appear in the popular press, and in the late 1910s, numerous manufacturers marketed the special brackets and other hardware that the job required. Stiffer springs could also improve handling, as could modified steering assemblies and other suspension tricks.[26]

Modifications to improve overall engine efficiency, to reduce friction and unnecessary weight in the rotating and reciprocating masses, to enhance engine breathing, and to better vehicle aerodynamics and handling: these were the performance tuning strategies most common in the days of the Model T, and they spawned a diverse array of aftermarket products from a variety of companies during the 1910s and 20s. Depending on the equipment selected (and the care with which it was installed and maintained), modified Fords were often capable of speeds in excess of 70 mph, whereas the standard universal car struggled to reach 45 mph. Perhaps more to the point, modified Model Ts were also fast enough to keep up with a number of more expensive makes of cars both on and off the track.[27]

Dean Batchelor claims in his history of hot rodding that "the vast majority of this equipment was . . . for out-and-out racing cars," and street applications were an afterthought: "The fact that customers soon started buying racing equipment to install on cars that were street-driven and would never see a race-track . . . probably surprised the speed equipment manufacturers. However, anyone capable of creating this equipment had to be resourceful, so it wasn't long before they took advantage of this bonanza and began advertising in both trade and racing publications for milder-tuned race car parts for your sports model."[28] According to Batchelor, then, the industry began by selling racing-oriented products to oval-track drivers, and only after renegade enthusiasts began to use these racing products on the street did the manufacturers begin to market milder products specifically for on-road use.

The majority of the evidence suggests otherwise, however. Almost without exception, early speed equipment manufacturers produced, advertised, and sold equipment intended for street use from the outset. In fact, on-road products were always as important to these companies as their racing lines, if not more so. Detroit Radiator's first ad, for example, gave equal space to its racing and its street-use gears. Likewise, an early advertisement for New York's Walker M. Levett Company boasted of the proven on-track performance of its "Magnalite" piston and connecting-rod assemblies, but the company's actual pitch — "greater flexibility, quicker acceleration, less friction, absence of vibration, easier cranking, more speed and power, less gasoline consumption and virtually the smoothness of a twin six" — makes it clear that the Magnalite line was for street as well as track use. Roof, too, claimed in one of its earliest advertisements that its equipment was intended for both "racing cars and fast road speedsters," and many others made similar claims.[29] Only Craig-Hunt of Indianapolis appears to have actually evolved according to Batchelor's "track first" maxim: in the late 1910s, the company proudly proclaimed that its sixteen-valve cylinder head, which had begun to prove its worth "on the Boulevard or Highway," was initially sold only for racing.[30] But for the most part, speed equipment for the street and for the track emerged together.

Considering the broader context, the importance of the early street-use aftermarket makes perfect sense. In 1915, grassroots Model T–based racing was just beginning to take off in the United States. However, although racing was popular, and although a lot of ordinary people participated in it from the very beginning, it was as spectators that the majority of early enthusiasts enjoyed the visual and aural spectacle of an oval-track race.[31] When it came to their own vehicles, most were content to build a high-performance street car that was merely inspired by the all-out racing scene (fig. 5). This we know from three types of early sources.

The first is advertising. If we consider the sorts of pitches that speed equipment manufacturers made, not only as they first got into the business but also as they grew and further evolved, we find that racing gear almost always shared the page with street-use products. In some, prominent references to oval-track and other racing triumphs at the top of the page led to more detailed discussions of road-going options at the bottom, while in others, street-use products actually received top billing. Either way, their advertisements strongly suggest that early speed equipment companies hoped above all else to use their racing triumphs to generate profitable sales among those who actually planned to use their modified Model Ts on the street.[32]

The second type of published material useful for determining the nature of speed equipment end-use during the Model T era are the technical forums

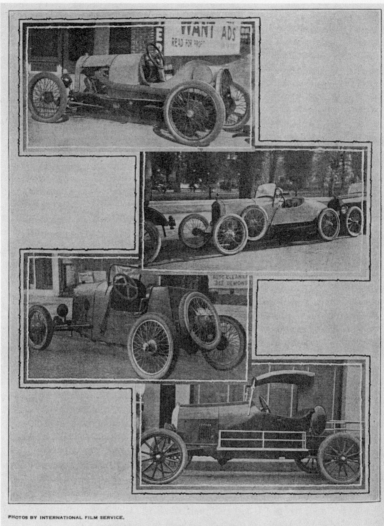

Some California Fords

Examples of Re-built Cars Seen on the Streets of
Los Angeles

Figure 5. Modified Model Ts on the streets of Los Angeles, California, circa 1920.
("Some California Fords," *Ford Owner and Dealer*, May 1920, 82. Reproduced courtesy
of the Free Library of Philadelphia.)

published in popular magazines like *The Fordowner* and *Motor Age*. Though the majority of the readers' letters published and answered in these forums dealt with mundane problems like cold-weather starting, electrical bugs, and general maintenance, many of them dealt with oval-track racing, speedster-style bodies, underslinging, and performance-oriented improvements for the street. Those in *Motor Age* are particularly revealing: nearly every issue of that periodical from the late 1910s contained at least one, if not two or more letters to the technical editor that had something to do with high-performance motoring, and with very few exceptions these letters dealt with street-use modifications.

The same was true of the third type of published source material that is useful in this regard, the "how-to" article.[33] By 1915, *The Fordowner* had begun to run occasional articles designed to bring interested readers up to speed on common performance-tuning methods. At first these articles were of a general nature, providing tuning overviews that dealt with ignition, intake, exhaust, oiling, internal assembly, body, and suspension modifications in one fell swoop.[34] But in time they became more focused and detailed. In October of 1919, for example, a nine-page article on underslinging appeared as, in due course, did lengthy pieces on improving the Ford's oiling system for high-performance use, choosing a suitable high-speed camshaft, drilling and counter-balancing the Ford crankshaft, and selecting a set of aftermarket transmission gears, among others.[35] Most of these were written by Murray Fahnestock, *The Fordowner*'s technical editor, but guests—including speed equipment manufacturers Arthur Chevrolet and Robert M. Roof—authored a few as well. Critically, the overwhelming majority of these pieces were written for those who wanted to build a high-performance Model T for the street. The illustrations accompanying Fahnestock's May 1916 "Ford Speedster" article therefore were of high-performance Fords fully equipped for the highway—headlights, fenders, license plates, and all.[36] Likewise, the text of a piece Fahnestock published the following month focused on the construction of a speedster for street use, as did his October 1919 and May 1923 articles on underslinging.[37] Racing-only pieces did appear from time to time as well. But on balance, road-based applications dominated these how-to features, for the interest of the average enthusiast was firmly rooted in the street.

Of course, one might reasonably object that these pieces actually tell us very little about what end users thought and did, and that these how-to articles better represent a top-down means of knowledge diffusion than a reliable source for understanding Model T enthusiasts. However, buried within the texts of these articles are critical clues that suggest otherwise. "To judge from the number of queries that the *Fordowner* receives from all parts of the country, for advice and instruction as to the converting of Fords into racing type roadsters . . . it would seem that there was a large demand for this type of car," wrote one author in 1915 to justify the instructional piece that followed.[38] Similarly, a 1925 article on adapting Fordson valves for use on Model T engines began as follows: "Judging from the letters which we have received, quite a few of our readers have tried Fordson valves and, in some cases, the results have been rather disappointing."[39] Likewise, Fahnestock claimed in a 1926 article on underslinging that "so many of our readers have tried their ingenuity at lowering Fords that we have been able to compile quite a variety of available methods" for publication.[40] Much of what these early how-to pieces covered therefore trickled *up*, and thus they do provide at least a glimpse of what ordinary enthusiasts thought and did. In this period, they thought about oval-track racers, but they actually built hopped-up road cars. Speed equipment manufacturers knew this, and in the era of the Model T, many of them therefore proudly boasted of their on-track exploits in their advertisements but actually earned their living on the highways and byways of 1910s and 20s America.

Nestled in the cradle of automobility and staffed by engineers, machinists, and racers of extraordinary capability and creativity, this early high-performance aftermarket prospered for more than a dozen years. In order to better understand its origin and evolution, however, we need to shift our focus from the macro to the micro—that is, we need to examine more closely a few of its firms. Let us turn, therefore, to the endeavors of Robert M. Roof, Joe Jaegersberger, and Arthur and Louis Chevrolet.

Robert M. Roof and Laurel Motors

In the summer of 1916, inspired by the growing popularity of Model T–based racers, an Anderson, Indiana, man by the name of Robert Maurice Roof began to design an engine accessory to boost the performance of the universal car. Convinced that the Ford's L-type cylinder head was unsuitable for high-speed use, Roof hoped to produce an improved high-performance unit to replace it.[41] Chief engineer at a foundry and machine shop that specialized in stationary diesels and the recipient of several U.S. patents, Roof was by no means new to engine design. Thirteen years earlier he had opened an industrial-application gasoline engine firm in Muncie, the Robert Roof Machine Company, and in 1909 he developed a successful air-cooled aircraft engine as well. In 1911, however, Civil War veteran and former Indiana Governor Winfield Taylor Durbin convinced the accomplished twenty-nine-year-old to sell his company and move to Anderson. There he went to work for Durbin's nephew, William, at the Anderson Foundry and Machine Works, where he developed a new line of heavy diesel engines in the early 1910s. But Roof was more than just an engineer who happened to work on internal combustion powerplants; he was also an amateur racer. In 1908 he built and raced a six-cylinder oval-track machine, and throughout the late 1900s and the early 1910s, he closely followed the Midwestern racing scene in his free time. In 1916, therefore, he brought considerable engineering experience as well as an intimate understanding of the ins and outs of oval-track racing to bear on his new cylinder-head project.[42]

What he came up with was clever but not entirely novel. Featuring sixteen overhead valves and hemispherical combustion chambers, Roof's design borrowed heavily from the successful Peugeot racing engines of the period.[43] Designed to bolt directly to any Model T engine block without modification, his new head used the standard Ford camshaft to operate its overhead valves: long pushrods passed through the factory valve openings, linking the original camshaft lifters with the new rocker arms. Sophisticated but easy to install, the conversion was a sure-fire hit in Roof's opinion. In the winter of 1916–17, he therefore founded a side venture called the Roof Auto Specialty Company to produce the equipment, with himself as president and William Durbin as secretary and treasurer. Shortly thereafter, his high-performance cylinder head, the "Model A," made its debut.[44]

Years later, Roof recalled that "in 1916, the fastest speed obtained from racing Fords was 60 miles an hour"; however, cars fitted with his new "Model A" overhead valve equipment in 1917 were capable of speeds as high as 78.[45] More importantly, oval-track racecars equipped with his new head won often and big. According to company advertising, in fact, "In every official race for Ford cars in 1917 where entered, one of the drivers with ROOF-PEUGEOT CYLINDER HEADS on his Ford won. And it was common custom for others with Roof 16 Overhead Valves to follow second, third and fourth."[46] Even allowing for self-promotional exaggeration, Roof's new head was definitely a winner.

Indeed, even before the 1917 racing season began, demand for the conversion among oval-track racers and road-going speedster owners quickly outpaced the manufacturing capacity of the Roof Auto Specialty Company. In the summer of 1917, Roof and Durbin therefore began to search for another partner—an investor, perhaps, or possibly an independent company interested in their product and willing to merge to produce it in greater volume. Fortunately, Durbin's uncle had a lead. Laurel Motors, a Richmond, Indiana, company founded in the mid-1910s, was a would-be automobile manufacturer still struggling to bring its first vehicle to market when Durbin and his nephew contacted its founder, Charles Hayes, in 1917. The Durbins had close ties to the local banking community and were a persuasive duo. Soon they convinced Hayes not only to make the fifty-seven mile move to Anderson but also to give up the new car business and focus instead on speed equipment for the Model T. For his part, Roof agreed to dissolve his company and assign his patents to Laurel in exchange for a share of its ownership. Hayes assumed the post of president, Roof those of vice president and chief mechanical engineer, and William Durbin that of second vice president. Roof also resigned his engineering position at the Anderson Foundry and Machine Works, but Durbin, whose duties at Laurel were to be light, continued as the general manager of the foundry. By January 1918, the transition was complete, and Laurel Motors was producing, marketing, and selling Roof's creation.[47]

Creations, actually. For in the fall of 1917, Roof had developed a second type of high-performance cylinder head for the Model T known as the "Model B." This new design was also of the hemispherical, sixteen-overhead-valve variety, but it was designed for ordinary touring cars and trucks and therefore was of milder tune than the original head. From the outset, then, Laurel offered high-performance cylinder heads both "for racing cars and fast road speedsters"—the "Model A"—and "for regular Ford touring cars and converted trucks"—the "Model B."[48] In 1918 the company further expanded its line of speed equipment for the Ford, adding speedster-style bodies, aluminum alloy pistons and rings, iron pistons and rings, underslinging brackets, racing gears, carburetors, ignition systems, crankshaft counterbalances, camshafts, and wire wheels to its catalog (fig. 6).[49] Of these, however, Laurel manufactured only the bodies, underslinging brackets, and cylinder heads. The rest of the parts were distributed by Laurel but produced by other companies. A brochure from 1923, for example, indicates that Laurel sourced its counterbalances from Dunn, its pistons from Green and McCadden, its carburetors from Miller and Zenith, and its ignitions from a variety of other manufacturers.[50] Laurel Motors therefore operated not only as a manufacturer but also as a wholesale and resale distributor of speed equipment.

Wartime shortages slowed the company's growth, but this was more than offset in November 1918, when the War Department expressed an interest in Roof's sixteen-valve designs for light truck and tank use. Over the next several years, Laurel maintained a contractual relationship with the military that resulted in a sixteen-valve, four-cylinder tank engine (the Laurel "Model J") in 1921. Between its military contract and the strong demand for Model T equipment, Laurel grew steadily in 1919, and in 1920, the company announced a major expansion of its facilities. By then, its distribution network spanned from coast to coast, and Laurel products were available in nineteen foreign markets as well.[51]

In the 1920s, Roof continued to expand the Laurel line, adding a large-capacity oil reservoir, a pressure oiling system, and a cross-drilled five-bearing crankshaft by 1923; high-speed camshafts and a line of superchargers followed in 1925 and 1926,

Figure 6. Laurel Motors advertisement for its overhead-valve cylinder head conversions, designed by Robert M. Roof. The cylinder head at the bottom of the page is a 16-valve unit, and the cars at the top are Model T Fords rebuilt with Laurel "racing" bodies. (Laurel Motors Corporation advertisement, *The Fordowner*, April 1918, 79. Reproduced courtesy of the Free Library of Philadelphia.)

respectively. Roof also developed the company's first high-performance products for another make of car, introducing two sixteen-valve heads for the four-cylinder Dodge in 1921.[52] But he devoted the majority of his time and effort to the further refinement of his cylinder-head conversions for the Model T. In 1921 he discontinued the racing-only "Model A" in favor of a new design based on the general-purpose "Model B." Dubbed the "Model BB," Roof's new head featured two spark plugs per cylinder and was intended for speedster and racing use. The following year, Roof introduced a line of high-performance cylinder heads equipped with eight overhead valves. Simpler in design than his B and BB sixteen-valve heads, the new "Roof 8" was intended to be a low-cost alternative to the company's pricier options; at $65 for the touring car and truck model and $75 for the racing and speedster units, the new equipment was precisely that.[53] In the spring of 1923, he also added a new flagship sixteen-valve design, the "Type C," which was for racing use only.[54] By 1923, therefore, Laurel offered five different types of high-performance cylinder heads for the Model T as well as two for the Dodge. But Roof was not yet satisfied. In the fall of 1923 he introduced (and patented) a double-overhead-camshaft adaptation for the Type C head, followed in 1924 by a single-overhead-camshaft version of the Roof 8 equipment known as the "Victory Eight" (also patented).[55] Two years later he redesigned his Roof 8 equipment to fit the new "improved Ford" engine of 1925; this resulted in the "Model 40."[56]

Period literature often featured detailed descriptions of Roof's cylinder heads as well as those of his competitors, and these descriptions reveal quite a lot about their manufacture. Take the 1922–24 Roof 8 line, for example, which began as one-piece steel-alloy castings. Skilled machinists then tapped and drilled them for installation on a standard Ford engine, and they also machined smooth the otherwise rough combustion chambers. Eight removable valve guides then went into the machined casting along with eight large-diameter steel valves. The entire rocker mechanism was of drop-forged steel, and each of the eight rocker arms was carefully machined and then hardened before its installation on the rocker shaft. Four large brackets held the rocker assembly to the head. Hand-machined and hand-assembled from cast and forged components, the entry-level Roof 8 was a complex unit that required substantial time and manpower to complete. The same was also true of Laurel's other cylinder-head conversions.[57]

Machining, finishing, and hand-assembly work was done in-house at Laurel, but all of the rough castings and most of the heavy forgings were obtained from subcontractors.[58] Through mid-1924, the Anderson Foundry and Machine Works handled most of Laurel's casting and forging, with William Durbin acting as the principal mediator between the companies. In July of that year, however, the cozy relationship between Anderson and Laurel unraveled when Durbin left the foundry to pursue another business interest. At the same time, Laurel itself began to falter, even as the prolific Roof continued to unveil new products. The reason was simple: Charles Hayes founded an airplane-manufacturing company in 1923, and he steadily lost interest in Laurel. In 1926, he sold his stake in the company to Arthur Sinclair of the St. Louis–based Zenith Company, and Roof soon did the same. In 1927, Laurel's operations moved to St. Louis, and shortly thereafter, the Laurel name was dropped entirely.[59]

Together with Myron Reynolds, one of Charles Hayes's partners in his 1923 airplane-manufacturing venture, Robert Roof went on to found the R&R Manufacturing Company of Anderson, Indiana, in 1927.[60] But by then the era of the Model T was coming to a close, and not until the universal car's replacement began to hit the streets in sufficient quantities to support a high-performance aftermarket of its own would R&R thrive as Laurel once had. The following chapter revisits Robert Roof in the context of the Model A and V8 industries of the 1930s and 40s, but for the time being, we will turn our attention to one of his chief competitors during the Model T period, Joe Jaegersberger.

Joe Jaegersberger and Rajo

Three years after Roof's first high-performance cylinder head hit the market, a formidable competitor emerged in Racine, Wisconsin. This was nothing new for Roof, for Craig-Hunt had introduced a sophisticated single-overhead camshaft, sixteen-valve conversion in 1917, just as Roof's own unit was beginning to make a name for itself.[61] But Craig-Hunt never contemplated a milder version of its cylinder head to seriously compete with Laurel in the street-use market. Instead, J. R. Craig and W. L. Hunt had their sights set on the new car business, and as they tried to bring a low-priced car to market in the late 1910s and early 1920s, their interest and presence in the high-performance aftermarket waned appreciably.[62] For Roof, therefore, Craig-Hunt was at worst a relatively minor nuisance.

Joe Jaegersberger's Racine Auto Equipment Company was altogether different. Harboring delusions neither of competing with Highland Park nor of paying its bills with racing speed equipment sales, its focus from the outset was the high-performance, street-use aftermarket. The company's first product was the 1919 Model 30 Valve-in-Head, an overhead-valve conversion distributed through the Trindl Sales Corporation of Chicago.[63] In the winter of 1919–20, the Racine Auto Equipment Company reorganized as the Rajo Motor Company,[64] and in the fall of 1920, the firm assumed control of its own sales through a network of regional distributors. The Model 30 cylinder head bolted directly in place of the Model T flathead, and it bore a striking resemblance to the Roof 8 line brought out by

Laurel in 1922: long pushrods linked the OEM camshaft with a rocker assembly that operated its eight overhead valves. Cheaper and far less complex than any other kit on the market, the Model 30 was intended for use on trucks, touring cars, and road-going speedsters.[65]

All-out racing was never far from mind at Rajo, though. Jaegersberger, the company's president, was a moderately successful oval-track driver and racing-engine mechanic who was active in several minor circuits in the 1910s. In 1919, however, his driving career came to an abrupt end when he was seriously injured in a dirt-track accident; shortly thereafter, he began to work on his first overhead-valve conversion for the Model T. For Jaegersberger, it was a natural move: twenty years earlier, he had studied engine design under Gottlieb Daimler's son at the Daimler factory in Germany.[66]

Shortly after the Model 30's debut, a racing version therefore joined the Rajo line, and the company was never shy about its on-track exploits. In 1920 and 1921, the firm boasted of the many victories posted by famed track racer Frank Cobb in a Rajo racing special, and in 1922, the company proudly celebrated Noel Bullock's victory at Pikes Peak in a Rajo-equipped speedster.[67] In its advertising, though, the company primarily used these triumphs not to sell its racing gear but rather to sell its road-going speed equipment. Thus, its November 1920 ad began with a brief description of Cobb's victories in several South Dakota races but continued with the claim that "Ford pleasure cars and trucks have equal possibilities"—if you equip your Ford with the Rajo Model 30, that is, you too will be able to beat all comers on the byways of your hometown (fig. 7).[68] Similarly, Rajo's 1922 campaign featuring Bullock's Pikes Peak victory urged the reader to "put your Ford in the champion class."[69] Other advertising for the company promised more power, higher top speeds, greater engine flexibility, and lower fuel consumption, emphasizing that Rajo overhead-valve equipment would "make your FORD a Real Car," comparable with "the high-priced car class as regards mechanical perfor-mance."[70] The Rajo Model 30 was racing-inspired, but it was designed above all else to enable the owners of the affordable Model T to run neck and neck with higher-dollar makes on the street.

In 1924, four years after the Model 30's debut, Jaegersberger updated and ex-panded the entire Rajo line. The centerpiece of this expansion was the "Model A," which was designed "for those who wish[ed] added power and speed, with the least possible complication and noise."[71] This new product was a hybrid de-sign known as an "F-head": four overhead intake valves operated in conjunction with the standard in-block exhaust valves. In this design, long pushrods and a set of rocker arms linked the intake lobes of the Model T camshaft with the over-head valves, and standard lifters operated the in-block valves. By far the simplest

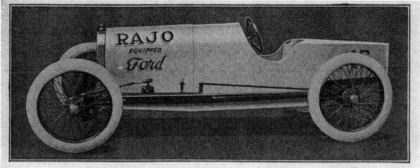
Figure 7. Rajo advertisement featuring the racing triumphs of a Ford equipped with the firm's overhead-valve equipment. The text explains that the "Rajo-Valve-in-Head" offers "equal possibilities" for street-driven Ford cars and trucks. (Rajo advertisement, *Ford Owner and Dealer*, November 1920, 113. Reproduced courtesy of the Free Library of Philadelphia.)

conversion then available, the "Model A" was intended as an entry-level alternative to the company's new "Model B," which replaced the Model 30. "Designed for speedsters and racing cars," the Rajo "Model B" featured two spark plugs per cylinder as well as a full set of eight overhead valves operated, as before, by the standard Ford camshaft.[72] Rounding out the new line of Rajo conversions was the "Model C," an eight-overhead-valve conversion with one spark plug per cylinder and simpler intake and exhaust manifolding.[73] In 1926, Rajo attempted to capture a larger share of the market, proudly proclaiming that "quantity production has made it possible to quote a new 'Low Price'" for its Model C-35.[74]

Rajo's approach to the manufacture of its high-performance cylinder heads was broadly similar to Laurel's.[75] However, what exactly Rajo meant by "quantity production" is unclear, for its new C-35 was cast, machined, and assembled exactly like its predecessors.[76] Thus, while it is conceivable that the subcontracted production of the basic cylinder head casting—as well as that of some of the individual moving parts—might have been increased for the C-35, a lot of hand finishing and fitting was still required. Perhaps Rajo simply hired more machinists and assemblers to handle the volume. In any event, its "quantity production" maneuvering for the C-35 would be its last: by 1928, the company had vanished, one of many speed equipment manufacturers that disappeared when Ford discontinued the Model T.[77] Used Rajo conversions remained popular well into the 1930s, however, especially among California enthusiasts who raced their Model T–based roadsters on the Mojave Desert's dry lakes.[78]

Jaegersberger himself continued to be involved in the speed equipment industry for many years to come. In the 1930s he designed and built a four-overhead-valve conversion for the Model A Ford. Known as the "Ramar" Valve-in-Head, it was similar to Jaegersberger's Rajo "Model A" equipment for the Model T Ford.[79] Years later, Jaegersberger also surfaced as a manufacturer of high-performance cylinder heads for the inline six-cylinder GMC and Chevrolet engines of the 1940s and 50s, for which he revived the Rajo name.[80] In its original incarnation, though, Rajo came and went with the Model T.

The Chevrolet Brothers

Born in Switzerland, Louis Chevrolet was a talented mechanic and racecar driver who first made headlines as a member of Buick's factory racing team in 1909. In 1911 he teamed up with William C. Durant, the former head of General Motors, to design and produce an automobile for the American market. But at $2,150, the 1912 Chevrolet Classic Six was far too expensive for its capabilities; after less than a year, Durant withdrew the Six, bought out Louis's interest, and began to reorganize the company in preparation for the introduction of a cheaper model. Durant's retooled Chevrolet Motor Company went on to become the cornerstone of a renewed General Motors empire in the 1920s.[81]

After the Classic Six debacle, Louis returned to his Indianapolis home, where he joined with his brother Arthur to design, build, and race oval-track cars. In 1919 four of their specially prepared "Frontenac" machines qualified for the Indianapolis 500. Three of these cars failed to finish the race due to suspension-related difficulties, however, and the fourth, driven by a third brother, Gaston, finished tenth.[82] Disappointed with the results but proud that they had qualified for the prestigious race, Arthur and Louis returned to the drawing board in the winter of 1919–20 to refine their design. The following year, one of their Frontenacs captured the checkered flag at Indianapolis, a feat they repeated in 1921.[83]

Now world famous, the Chevrolets turned their attention to the budding Model T speed equipment business and decided to produce an overhead-valve conversion inspired by their Frontenac racing mills. Together they organized the Chevrolet Brothers Manufacturing Company, and in the fall of 1921, they introduced the Frontenac high-performance cylinder head for the Model T. An eight-overhead-valve conversion not unlike the Rajo Model 30 and the soon-to-be-released Roof 8, the Frontenac was designed to bolt directly in place of the Ford's original flathead.[84]

From the outset, the Chevrolet Brothers offered three distinct versions of their equipment, the "Model T," "Model S," and "Model R."[85] The first two were for touring cars and speedsters, and the last for Model T–based racers. The size of their combustion chambers was the chief difference among them: the T featured large chambers for a moderate compression ratio; the S, slightly smaller chambers for a mild compression boost; and the R, tiny chambers for maximum performance.[86] The T and the S were equipped with steel alloy valves with carbon-steel stems for longevity, while the R used tungsten steel valves better suited for racing. All three had one intake and three exhaust ports and came complete with a manifold designed to mate with the original Model T exhaust system. Two different intake manifolds were available, one for use with the Ford carburetor and the other for use with larger-bore aftermarket units. The standard Ford camshaft operated the overhead valves via pushrods and a rocker assembly, and an aluminum cover kept the dust out and the oil in. Though designed to bolt in place with a minimum of fuss, the Frontenac's external dimensions exceeded those of most of its competitors, and a bit of finesse was sometimes needed to install one in a Ford with an unmodified body.[87]

Then, in the winter of 1921–22, the new-car bug bit Louis once again. With the backing of A. A. Ryan of Stutz, he organized a side venture in Indianapolis, the

Frontenac Motor Company. Their plan was to introduce a "popular price" sports car inspired by the Chevrolet brothers' winning racecars, but the project died quietly the following spring.[88] Meanwhile, the Chevrolet Brothers Manufacturing Company was prospering. In the early 1920s a far more exotic racing cylinderhead conversion designed by a Japanese American associate of the Chevrolets, James Sakayama, joined the firm's product line. Like its lesser brethren, this new unit featured eight overhead valves, but it used a pair of chain-driven overhead camshafts to operate them. A twin-port version of the "Model R" debuted in 1926, and by the end of the 1920s the company also produced water pumps, pressure oilers, crankshafts, pistons, gears, and other speed equipment for the Model T as well as products for four-cylinder Chevrolets and Whippets.[89]

Arthur and Louis continued to participate in organized motorsports throughout the 1920s, using their on-track victories to generate interest in their street-use products. In 1922, two Frontenac racers fitted with "Model R" cylinder heads qualified for the Indianapolis 500, and through the following spring, the company's advertisements boasted of this accomplishment and promised similarly blistering performance to its street and track customers.[90] In 1923 another Frontenac racer using the standard "Model R" conversion finished fifth at the Indianapolis 500 "at an average speed of 82.25 miles an hour." Impressed with the Frontenac's triumphs, *Ford Owner and Dealer* invited Arthur Chevrolet to guest-author a "Secrets of Speed" article that fall, an offer he accepted.[91] Toward the middle of the decade, Arthur and Louis also began to build and race dirt-track cars, and *their* success led the popular Ford magazine to offer Arthur yet another chance to share his wisdom.[92]

10,000 Frontenac conversions for the Model T were built before the company folded in the early 1930s, more than double Rajo's output and triple that of Laurel. Yet even the pushrod-action T, S, and R kits were just as sophisticated as were those of the company's rivals. These were not cut-rate heads, that is, for as with the Laurel and Rajo conversions, the manufacture of even the lower-end of the Frontenac line of cylinder heads was a time-consuming and labor-intensive process.[93] Fortunately for the Chevrolets, sales of the kits more than made up for the cost of their manufacture, at least for a while. Upon the introduction of the Model A Ford in 1928, Arthur and Louis quickly came out with a double-overhead-camshaft racing conversion for the new car, but sales of their Model T staples went into a downward spiral. Thus, when the American economy collapsed in 1930, it brought their weakened company down along with it.[94] Used Frontenac conversions remained popular among amateur dirt-track racers and early California hot rodders for a few more years, but in time, as the Model T forever passed from the cutting edge of the performance scene, so too did the Frontenac.[95]

THE EARLY SPEED EQUIPMENT BUSINESS

Laurel, Rajo, and the Chevrolet Brothers all produced high-performance overhead-valve conversions for the Model T. All three also manufactured their kits in roughly the same manner, from subcontracted castings to hand-machined and -assembled finished products. All three operated during the same period, advertised in the same venues, and maintained close ties to the racing scene while focusing the majority of their efforts on street-use products. What's more, their most popular designs—the Roof 8, the Rajo "Model B," and the Frontenac "Model T"—were nearly identical eight-overhead-valve conversions. Finally, all three were based in the Midwest, the cradle of American automobility and the region where the majority of the nation's automobiles were owned and operated at the time.

Beneath the surface, however, these firms were actually quite different. First, their backgrounds varied considerably, even though Robert M. Roof, Joe Jaegersberger, and Arthur and Louis Chevrolet were all involved in organized motorsports before entering the speed equipment business. Roof had considerable training in the foundry and machinist trades, and the company he teamed up with in 1917, Laurel, had been trying to break into the new-car market for several years before Roof and the Durbins persuaded its management to focus on the high-performance aftermarket instead. Louis Chevrolet also dabbled in the new-car market both before and during his time as a speed equipment manufacturer. Nevertheless, neither his short-lived experience with William Durant nor the stillborn Stutz-Frontenac project had much at all to do with the success of his Frontenac cylinder heads. Instead, his conversion's prosperous run was largely due to the fame that he and his brothers had earned racing—especially big-time Indianapolis 500 racing. Jaegersberger dove right into the aftermarket business following his days as a circle track star, but his racing experience was on the less-prestigious local circuits. Rajo therefore had to build its national reputation from scratch.

Second, the business strategies of Laurel, the Chevrolet Brothers, and Rajo differed considerably. Laurel manufactured not only cylinder heads but also underslinging brackets, oil and water pumps, high-speed camshafts, and a host of additional performance components for the Model T. At the same time, Laurel also acted as a distributor for other speed equipment manufacturers, something Rajo and the Chevrolet Brothers never tried. Arthur and Louis Chevrolet did, however, manufacture a number of other high-performance components for the Model T in addition to their line of Frontenac conversions. Rajo, on the other hand, focused almost exclusively on overhead-valve conversions for the Model T.

In fact, Jaegersberger's company is the only one of the three that never branched out, even on a limited basis, to produce speed equipment for other makes of car.[96]

Finally, all three vanished by the early 1930s, but for very different reasons. Laurel crumbled from within: sales were strong when Charles Hayes lost interest in the firm, Roof sold out to Sinclair, and the entire enterprise became a part of the Zenith Company. Rajo, on the other hand, vanished shortly after the Model T was discontinued, whereas the Chevrolet Brothers' business survived the end of the Model T era but was later crushed by the Depression.

In spite of their differences—actually, precisely because of them—Laurel, Rajo, and the Chevrolet Brothers were typical speed equipment manufacturers of their day. Small, Midwestern, closely associated with oval-track motorsports, and heavily focused on the production of high-performance components for the Model T, their stories are broadly similar to those of most of the other aftermarket companies that operated during the 1910s and 20s. Let's pause for a moment to consider the wider implications of these characteristic traits.

Take, for example, the fact that the early speed equipment industry was largely concentrated in the Midwest. The key word here is "largely," for as it happens, the geographic concentration of the early industry was far too loose for it to have mattered. "The Midwest" is a big place, and the companies of the 1910s and 20s were scattered over thousands of square miles. Therefore, only inasmuch as Midwestern racing circuits *might* have put some of these companies in occasional contact *might* this overall "Midwestern concentration" have mattered. In other words, this was certainly not a regional cluster in Philip Scranton's sense—it was nothing like the Grand Rapids furniture industry, the Philadelphia textile district, or even the early 1950s Los Angeles hot rod industry.[97]

A second characteristic of the early speed equipment industry that warrants emphasis at this point is that of the larger milieu of aftermarket accessory manufacturers to which it belonged. Unlike the California companies that would supplant them in the 1930s, these early high-performance firms openly identified with the replacement-parts and general-improvement companies with which they shared their advertising space. These early speed equipment companies never formed their own industrial organization, never established their own trade press, and only rarely sought to distinguish the nature of their businesses from those of their general-improvement counterparts.

Third, the design, manufacture, and sale of high-performance components in the 1910s and 20s was possible only because of the revolution in mass automobility that Ford spearheaded. It was no coincidence that the speed equipment industry first emerged in the years immediately following the introduction of the mass-produced Model T. Nor for that matter was it merely by chance that the majority of the high-performance components available during the period were made to fit the universal car. And it certainly was no accident that when these early manufacturers decided to branch out, they mostly chose to produce add-on parts for Chevrolets and Dodges—numbers two and three, respectively, in the late 1910s and early 20s.[98] For speed equipment manufacturing was a byproduct of the mass automobility revolution, and aftermarket firms focused on those makes of cars for which the largest possible market existed. This was true in the era of the Model T, and it would remain so for many decades to come.

Fourth, and finally, the manufacture of over-the-counter speed equipment for the Model T was an enterprise that was *enabling*. Without this equipment, there can be little doubt that far fewer enthusiasts would have been able to modify their cars at all. Many could (and did) hop up their powerplants themselves, tweaking the carburetor or the ignition system and in some cases even venturing to mill the cylinder head for a compression boost. But most relied instead on the high-performance components that they bought from their local dealers and bolted in place.

With the end of Model T production and the introduction of the Model A, much of this would change. The thriving accessory market that served so many Model T owners for so many years lost its footing, and the Model A proved to be less receptive to add-on bits and pieces than its predecessor. Among speed equipment manufacturers, considerable turmoil accompanied this shift: although many of the original high-performance aftermarket manufacturers soldiered on into the 1930s, many more did not. However, performance tuning and the speed equipment industry as a whole began to thrive as never before in the late 1920s and the 1930s, even as the Great Depression decimated the American economy.

CHAPTER TWO

WESTWARD HO, 1928–1942

By the mid-1920s, the Model T was out of date. Sales were slipping, and Chevrolet, whose cars were more expensive but also far more modern and better equipped, began to outsell Ford. In the spring of 1927, Ford therefore announced that it planned to replace the universal car with a new model. For six months engineers wrangled over the design of the new car and struggled to retool the firm's facilities for its production, a costly transition that revealed fundamental flaws in the company's manufacturing strategy. Still, hundreds of thousands of Americans placed deposits at their local dealerships in anticipation of the updated Ford, and others put their purchase plans on hold entirely. In November, the new car finally began to trickle out of the Highland Park plant, and in December, it made its official debut in New York.[1]

With a stiffer frame, a more powerful engine, a conventional transmission, and many creature comforts never before available from the company, the Model A was a runaway success. By the end of 1930 more than three million of them were on the road, and Ford appeared to have regained its competitive edge.[2] So good was the new car, in fact, that the once-bustling Model T accessory industry rapidly declined in the late 1920s and early 1930s. Some have claimed that this was due at least in part to the fact that the generation of automobiles to which the Model A belonged was less amenable to end-user tinkering than the cars of the 1910s and early 1920s; thus, with the passing of the Model T, the age of open-ended automobility began to come to a close.[3]

To be sure, the mechanical underpinnings of the new Ford were a marked improvement over those of the Model T, and many of the tricks that aftermarket manufacturers and amateur enthusiasts used to enhance the performance of the universal car were incorporated into the design of its replacement. With more cubic inches, a stronger engine block, a stiffer crankshaft, aluminum pistons, improved carburetion and exhaust manifolding, a more refined L-head, pressure oiling to most of its vital components, and a standard water pump, the Model A developed forty horsepower in stock trim—fully double that of an unmodified Model T. None of these changes discouraged enthusiasts, however. Instead, they welcomed the new and improved Ford because they realized that if a twenty-horse-power Model T could end up as a forty- or fifty-horsepower screamer through the careful application of add-on products, then surely a forty-horsepower Model A could end up far stronger—and faster—with a tweak here and a new part there.[4]

THE MODEL A ERA

Not surprisingly, over-the-counter speed equipment for the new Ford quickly appeared, and by 1930 hundreds of high-performance components were available for it. The transition from the era of the universal car was anything but smooth, however, and many Model T aftermarket companies never made the switch to the Model A. Some went out of business, others continued to produce Model T accessories for a few more years, but many simply turned to other pursuits. In their place, a number of new firms joined the industry during the era of the Model A.

It warrants mention that although the period in question here, the era of the Model A, spanned from 1928 through approximately 1937, the Model A itself was only produced until 1932. That year, Ford debuted its flathead V8, an engine that would ultimately come to dominate enthusiast activities for the better part of two decades. However, Ford also introduced a revamped four-cylinder car, the Model B, in 1932, and for a number of years most performance enthusiasts actually preferred it to the all-new V8. Critical in this regard was that most of the high-performance parts designed for the Model A fit the Model B as well as the four-cylinder Model C, introduced in 1933.[5] Thus, the "era of the Model A" was actually a period within which the efforts of the high-performance aftermarket and those of the enthusiast focused on *three* different four-cylinder Fords, the A, the B, and the C.

Approximately three dozen companies manufactured high-performance aftermarket parts during this period, which means that the speed equipment industries of the Model T and Model A eras were roughly the same size.[6] Compared with the geographic distribution of the high-performance industry of the 1910s and 1920s, however, that of the Model A era clearly indicates that a regional shift was underway: the Midwest's share was dwindling, while that of California, especially Southern California, was expanding.

Part of the reason for the growing importance of Southern California firms during the late 1920s and the early 1930s was that an entirely new type of automotive racing was emerging in that region at the time. Scattered throughout the Mojave Desert northeast of Los Angeles are a number of dry lake beds. Perfectly smooth and perfectly flat, some of these "lakes" are but a few hundred feet across, but many measure several—even a dozen or more—miles from end to end. During the late 1920s, young enthusiasts began to drive to the dry lakes on the weekends, where they would strip the fenders, windshields, and other heavy parts from their modified roadsters and charge out across the arid landscape full-tilt, often in chaotic clusters of five or more cars at a time. These enthusiasts raced not in an oval but in a straight line: their only object was to find out just how fast their daily rides could go. This, many would vociferously argue in the decades to come, was the birth of hot rodding proper.[7]

Whatever it was, it steadily grew in popularity in the late 1920s and 30s. Most of those involved in the new activity performed their own modifications, and some of those who did so sourced their high-performance parts from junkyards—and in some cases, from cars left unattended on dimly lit streets.[8] Many others relied on over-the-counter speed equipment, however, and what they bought tended to come from shops in metropolitan Los Angeles rather than from the East Coast, the Midwest, or even the Bay Area. Data from 1930s dry lakes meets, for example, suggests that among those who raced four-cylinder Model A Fords on the lakes, speed equipment sourced from Southern California was far more popular—often by a factor of more than ten to one—than that from any other part of the country.[9] And because those who raced at Muroc, Rosamond, Harper, and the other dry lakes also raced on public roads in Los Angeles, Riverside, Orange, San Bernardino, Ventura, and San Diego Counties, we can safely infer that on the streets of Southern California, locally sourced gear was also prevalent.

Proximity had never mattered all that much within the industry before: Laurel, the Chevrolet Brothers, Rajo, and many others managed to sell their speed equipment to oval-track racers and road-going enthusiasts all across the country during the era of the Model T. However, this new period was different, and part of the reason lies in the technologically iterative nature of the dry-lakes style of racing. Whether charging across the desert alone or in a pack of cars, lakes racing was essentially a solo act. One raced not against his opponents per se, but against the clock. That is, what mattered was not so much whether you could consistently cross the finish line in first place, but whether your times improved. Thus, with every pass, lakes racers aimed to beat their personal best top speeds, and at the end of each run, they sought to further tweak their vehicles so as to be able to go a bit faster the next time around. Oval-track racers also sought to improve their cars between each race, of course, and lakes racers did indeed care about how their times stacked up against those of their competitors. But on the lakes, iterative improvement assumed an importance far greater than it had ever held in any other form of motorsports, and this translated into a sizeable advantage for area companies. For in light of the overarching goal of continuous improvement, it made perfect sense to buy locally. According to Don Montgomery, for example, "when the decision was made to buy a high performance head, the rodder could, for example, just go into 'town' (L.A.) and talk to George Riley. And after installation of the hop up parts the car could be driven by so George could see how it ran. The rodder gained technical help and advice while the manufacturer quickly learned the good and bad points about his product."[10] Thus, as a Southern California enthusiast improved his car, he had relatively easy access to those who designed and made his aftermarket parts—if he bought them locally. In return, the high-performance companies of the area enjoyed the benefits of frequent contact with those who actually used their components on a daily basis. Proximity to the action therefore worked to the mutual advantage of enthusiasts and entrepreneurs, giving the California industry of the early 1930s a bit of a boost.

Meanwhile, the popularity of oval-track racing continued to grow throughout the United States. Production cars equipped with high-performance parts were fast becoming the norm on local racing circuits too, and during the late 1920s and early 1930s, dozens of new oval tracks sprang up all across the country, including many on the West Coast. In short, however popular street and dry-lakes racing were in Southern California in this period, more traditional motorsports more than held their own, there and elsewhere. Thus, while many of the California-based equipment manufacturers found it profitable to focus on the dry-lakes scene, many of those in the Midwest and on the East Coast continued to sell high-performance gear developed through—or inspired by—oval-track racing.[11]

But then, two short years after the Model A's introduction, just as manufacturers in California, the Midwest, and the Eastern states alike were beginning to unveil their lines of products for the new car, the American economy collapsed. Within three years, the Depression knocked the Chevrolet Brothers, Morton & Brett, and Ramar out of the picture. Miller's firm also briefly went into receivership, as did Zenith, and many other companies curtailed their advertising and their research and development expenditures.[12] But the majority survived the Great Depression intact, and many actually prospered and expanded during the 1930s. Moreover, dozens of altogether new firms sprang up toward the end of the decade, particularly in Southern California. For in the end, high-performance enthusiasm did not wane at all during the 1930s, and although the period was wrenching for some performance-parts companies, for many others it was more or less business as usual. Consider the mixed experiences of Winfield, Miller-Schofield, and R&R Manufacturing.

Winfield

Ed Winfield's days as an enthusiast and speed equipment manufacturer spanned more than five decades, from the early 1910s into the 1960s. Born near Los Angeles in 1901, he began working in a blacksmith's shop at the age of eight, where he learned the basics of forging and metalworking. Four years later, he enrolled in an automobile shop class at the local YMCA, and in 1914 he went to work for Harry Miller. Miller, by then a top-notch manufacturer of high-performance carburetors and custom-built racing engines, first assigned the boy to his carburetor department. At the age of eighteen, however, the talented young mechanic was promoted into Miller's prestigious racing-engine department, where he hand-assembled custom mills for those among the oval-track crowd who could afford them.[13]

In his free time, Winfield raced. His career on the boards began in 1916, when the fifteen-year-old convinced the operator of a local track that he was actually twenty-one and therefore eligible to run. Behind the wheel of a Model T–based racer that he built, Winfield steadily climbed through the ranks, becoming the star of the Legion Ascot track in Los Angeles by the age of twenty-seven. At twenty-eight, Winfield married, promising his new bride that he would never race again. But by the early 1930s, not only was he was back on the boards, but he had also begun to race on the dry lakes, where he set a class record of 119.60 mph at Muroc in 1933 that would stand for more than a dozen years.[14]

Back in the mid-1910s, however, when he was just beginning to race and when he was still an entry-level technician in the Miller carburetor department, he had an idea. Miller's carburetor was an excellent design in theory, but in practice it was difficult to set up and even trickier to maintain. While working on them in Miller's shop, Winfield therefore developed a way to simplify them. But he kept his mouth shut. He later claimed—tongue in cheek, perhaps—that his silence on the matter was simply because no one in the carburetor department at Miller ever asked him if he had any ideas for improving the design. Subsequent developments suggest that it was anything but innocent humility that kept the young technician from sharing his thoughts, however. For by 1919 Winfield had prepared a prototype high-performance carburetor of his own, and in 1921 he left the Miller engine department to found the Winfield Carburetor Company with his brother, Bud, in Glendale.[15]

For the first few years, Winfield's company was a small-time operation. Focusing on the racing market, Ed and Bud produced a limited number of carburetors for use on Miller and Deusenburg engines. But after a couple of years, they decided to adapt the design for use on modified Fords, and their business took off. So successful was their new carburetor within the racing aftermarket, in fact, that it led the Winfields to contemplate quantity production of the unit for street use. In 1924 they took the plunge and contracted with an area firm, the Hammel-Gerke Company of Los Angeles, to market and distribute their product. Featuring a one- or a one-and-one-quarter–inch bore, the Winfield carburetor offered better performance than that of the Model T; and with but two moving parts (the float mechanism and a rotary throttle), it was simpler and easier to maintain too. Through Hammel-Gerke, Winfield also sold street- and track-use carburetors for Dodges and Chevrolets, but it was the model for the Ford that paid the bills.[16]

In 1926 Winfield broke with Hammel-Gerke and assumed control of his own marketing and distribution.[17] The following year he released a new carburetor after extensive on- and off-track testing. A progressive, double-barrel design, the new model allowed for more economical operation at lower engine speeds, with a healthy reserve of power available on demand at higher rpms. It was an interesting way to avoid compromising driveability for peak performance and vice versa, and it was highly effective. Because the new design sacrificed virtually nothing at the top end of the powerband, Winfield was able to "test" it prior to its release to the general public on the most prestigious and demanding of American stages, the 1926 Indianapolis 500, where all of the first ten finishers used the new double-barrel design.[18] During the next two racing seasons, Winfield's carburetors continued to prove their on-track mettle, and Ed prepared to take his company to the next level.

In 1930, two years after the introduction of the Model A, Winfield brought out an entirely new downdraft carburetor, the Model-S. As had come to be his custom, Winfield tested his new design at Indianapolis, where it was used by nine of the ten drivers who finished the 1930 event. But it was with the new Ford in mind that Winfield actually conceived the Model-S, and he marketed the new unit aggressively to Model A owners.[19] Popular among enthusiasts and racers, especially those who ran on the dry lakes and surface streets of Southern California, the Model-S—together with its later derivative, the Model-SR—solidified Winfield's reputation as one of the premier performance-carburetor manufacturers in the country.[20]

But Ed Winfield was ambitious, and his success in the racing and street-use speed equipment industry convinced him that the time had come to try his hand in the profitable OEM-supply end of the market. Consequently, when he introduced the Model-S, Winfield also opened a new sales and distribution office in Detroit. Through this office he worked hard to win the approval of the mainstream industry, submitting samples to a number of companies for review, but the response was disappointing. Buick, for example, reported that although the Model-S was a superior carburetor that would surely make their cars snappier performers, the additional per-unit cost was far too prohibitive for them to make a commitment with the Los Angeles-based company.[21] In time, Winfield abandoned his effort to become an OEM supplier, but he did maintain his secondary Detroit address for a number of years to come.

Meanwhile, back in California, Winfield branched out into other areas of the high-performance industry. In 1931 he introduced a line of high-compression cylinder heads for the Model A. A cast-iron flathead, Winfield's design was available either as a 6:1 compression "yellow head" (nearly two points higher than the standard Ford's 4.22:1 head) or as a 7:1 "red head." Both produced substantial horsepower gains, although the latter, designed for use with higher-octane leaded fuels, outperformed the other by a margin of roughly 30 percent.[22] A single casting with no moving parts, these heads were relatively easy to manufacture, especially when compared with overhead-valve conversions. At his facility in Glendale, Winfield's employees simply tapped the raw castings (obtained from a local foundry), machined the combustion chambers, and boxed them up. At a cost of $40 for the head itself, or $75 for the head along with a Model-S carburetor, the Winfield flathead was an instant and enduring hit.[23] Dry lakes racers in particular preferred the yellow and red heads not only over all other high-compression aftermarket flatheads but also over most of the overhead-valve conversions available for the Model A.[24] Cheap and troublefree, the Winfield flathead sold by the thousands during the 1930s.

Carburetors and cylinder heads for the Model A were his staples, but Winfield also produced a steady trickle of reground camshafts. High-performance cams had been an on-again, off-again hobby for Winfield since he was a teenager, and in the early 1920s, he began to regrind Model T camshafts by hand on a custom basis.[25] By the early 1930s, he had begun to devote more time to the endeavor, experimenting with countless intake and exhaust lobe profiles for street, track, and dry-lake applications. His radical, long-duration camshafts proved to be solid performers, especially on the lakes, and demand for his bumpsticks quickly outstripped the supply. To boost production, Winfield designed and built a special machine to copy his cams. Similar conceptually to the key duplicators found in hardware stores, Winfield's machine traced the profile of a single hand-ground lobe and ground that profile onto a target lobe. The working cam was then shifted and the process repeated until all of its lobes had been reground to match the hand-ground pattern. Unfortunately for enthusiasts eager to purchase one of his cams, Winfield trusted no one but himself to handle their production, even on the duplicator. Accordingly, delivery was exasperatingly slow.[26] Nevertheless, the solid reputation of the Winfield camshaft made it worth the wait for a number of 1930s enthusiasts.

Winfield's reluctance to trust anyone else with his camshaft operations stemmed less from a concern about product quality than it did from a desire to keep his lobe profiles and grinding methods secret. Indeed, Southern California hot rodders of the early 1930s who went by his facility were by all accounts lucky even to be admitted into the lobby, let alone the workshop.[27] But for those that did get in, Winfield proved to be an excellent mentor and collaborator, and a number of young and mechanically inclined enthusiasts, including Kong Jackson, Eddie Meyer, and Ed Iskenderian, all received their early hands-on training in the Winfield shop.[28] Winfield also came to trust a fellow equipment manufacturer, George Riley, with whom he worked to develop an F-head conversion for four-cylinder Fords and to whom he eventually licensed his camshaft profiles for sale under the Riley name.[29]

In the mid-1930s Winfield also began to grind camshafts for Ford's new V8, and he began to experiment with induction systems for the new mill, too. He even designed, tested, and patented a complete fuel-injection system for the flathead V8 in 1939, one so advanced that it is rumored not only to have inspired hot rodder Stuart Hilborn to design his own fuel-injection equipment in 1940 but also to have been the basis for the Chevrolet system introduced just after Winfield's patent expired in 1957.[30] After the war, Winfield continued to produce aftermarket components for four-cylinder and V8 Fords for a number of years, all the while maintaining his custom racing-engine shop, turning out small batches of camshafts, and working on several Indianapolis 500 projects. In the 1930s, though, carburetors and flatheads were his staples, and they made his name a household word among hot rodders.

Ed Winfield was an enigma, a jack-of-all-trades whose activities over the course of several decades defy thematic categorization. He was a track, street, and lakes racer, but he was also a manufacturer. He ran a high-volume production facility that turned out thousands of flatheads and tens of thousands of Model-S and Model-SR carburetors, yet he also maintained a custom-oriented department that handled not only small-batch camshaft production but also individual racing-engine projects. His company was based in Los Angeles, but he maintained a foothold in the Midwest. He was jealously secretive with some, yet remarkably open with others, and his shop ultimately served as a place where a select few who would later go into business for themselves first learned the ins and outs of high-performance manufacturing. Finally, the most prosperous days of his entire career were not in the boom years of the 1910s, the 1920s, or even the 1950s.

Instead, they were in the early to mid-1930s, in the midst of the Great Depression, a time during which a number of his competitors were less fortunate.

Miller-Schofield

Toward the end of 1928, Harry Miller teamed up with George Schofield, Fred Keeler, G. E. Moreland, and Gilbert Beesmeyer to form a new manufacturing company in Los Angeles. Keeler, Moreland, and Beesmeyer were from the Lockheed, Moreland, and Beach Aircraft Companies, respectively, and Schofield was an independent financier. For his part, Miller ran a famous shop that made expensive hand-assembled racing engines, hand-ground camshafts, custom-built race cars, and other high-end oval-track equipment in the 1910s and 20s. Miller also manufactured carburetors, and his workshop served as an informal school for high-performance technicians: Ed Winfield started off at Miller, as did Fred Offenhauser, whose own engines dominated American racing from the mid-1930s through the late 1960s.[31]

The new company, popularly known as Miller-Schofield but chartered as Schofield Incorporated of America, was to manufacture lightweight aircraft engines as well as select aftermarket components for the Model A Ford.[32] By the end of 1929, Miller's team had prepared a line of high-performance carburetors, camshafts, alloy pistons, and overhead-valve cylinder heads, and in early 1930, the firm began to advertise for dealers and distributors. Offering special introductory pricing to prospective retailers, these early advertisements proudly proclaimed that Miller-Schofield's substantial fiscal and physical resources had enabled "Miller products, which were formerly made in limited quantities only for the racing profession and for wealthy sportsmen," to be "produced in quantities at prices which appeal to the general public."[33] Impressive dynamometer results were also provided. The Miller carburetor alone reportedly bumped the Model A's horsepower from forty-one to fifty at 2400 rpm, with a peak of fifty-seven horsepower at 3000 rpm. Coupled with the company's overhead-valve equipment, the gains were even more pronounced: eighty-six peak horsepower at 3200 rpm, more than double that of an unmodified Model A.[34]

Though its carburetors, camshafts, and pistons performed well, the Miller-Schofield Valve-in-Head quickly came to be the firm's signature product. Designed in 1929 by Leo Goosen, the star of Harry Miller's engineering staff, this conversion was relatively conventional, featuring eight overhead valves operated by the OEM camshaft via pushrods and rockers. Unique, however, was Goosen's attention to detail. The head's intake and exhaust ports mimicked the shape and layout of the standard Ford's in-block ports, which meant that factory Model A intake and exhaust manifolds could be used with the Miller-Schofield conversion.[35] In addition, Goosen's design called for the use of standard-issue Buick rocker arms, shafts, and springs, which made the assembly less of a chore to manufacture (and to service and maintain).[36] Large-diameter Miller racing valves and a healthy 6.75:1 compression ratio ensured that the bolt-on conversion outperformed its rivals. Complete with an aluminum rocker arm cover, the Miller-Schofield Valve-in-Head retailed for $137.50 in early 1930, and the company produced an average of fifty units daily.[37] Goosen also designed a gear-driven, double-overhead camshaft head for the new company. Equipped with eight oversized valves, this complex conversion for the Model A was intended "exclusively for dirt track racing car and sports use" and sold, complete with an American Bosch ignition and two Miller carburetors, for a jaw-dropping $500. Understandably, the company's standard Valve-in-Head was far more popular among cash-strapped enthusiasts.[38]

But it wasn't popular enough. By the end of April 1930, the company was on the brink of insolvency, and in June, it went into receivership. Period evidence strongly suggests, however, that Harry Miller fled the sinking ship before it actually went under. Toward the end of April, the firm submitted an advertisement to the editors of *Ford Dealer and Service Field* that highlighted, as had each of its predecessors, Harry A. Miller's personal involvement with the company.[39] But by the end of the following month, Miller-Schofield had revised its pitch for what would prove to be its final advertisement, deliberately avoiding the use of the name "Harry A. Miller" while still claiming that the company's "Miller Hi-Speed Head" was "[m]anufactured by the famous builder of World Champion racing cars and engines."[40] To readers, of course, this probably seemed to be little more than a matter of stylistic whim. But further contextual evidence suggests otherwise, for Harry Miller ran an advertisement of his own in June of 1930, a full-page spread introducing a new overhead camshaft conversion for the Model A Ford in which the following announcement appeared in bold italics: "I [Harry Miller] . . . wish to take this opportunity to announce to the motoring public that I have absolutely no connections with any other head for Ford engines."[41] Miller, that is, had severed his ties with Schofield of America and gone his own way. Significantly, his advertisement also claimed that the production of his *new* cylinder head was already underway. Even if we assume, therefore, that Miller designed his new conversion earlier in the spring when he was still affiliated with Miller-Schofield, he still would have had to have left the ailing company no later than the beginning of May in order to have his new design — and his advertisements — ready for a June launch.

In any event, Miller-Schofield was no more. George Schofield went on to produce a limited run of high-compression flatheads for the Model A Ford in a partnership with an area high school shop teacher in the early 1930s before slipping into anonymity.[42] As for Leo Goosen, he continued to work with Harry Miller on a number of racing engine projects during the mid- to late 1930s. Then, in 1941, he teamed up with Ed Winfield and a handful of others to design an advanced engine and chassis combination for the Indianapolis 500; and in the 1960s he worked with Bob DeBisschop, Dale Drake, and champion racer Louis Meyer to successfully apply turbochargers to the aging Offenhauser racing engine.[43] Miller himself went ahead with the production of his overhead-camshaft conversion for the Model A in 1930, although he had to sell off some of his independently controlled plant and machinery to do so.[44]

A sophisticated design, the new Miller conversion utilized a single overhead camshaft to operate four large overhead intake valves, while a second specially ground camshaft mounted in the standard position used all eight of the Ford's original in-block valves to expel the exhaust gases.[45] Complete with two camshafts, a chain-drive system, intake and exhaust manifolds, metric spark plugs, and a special Miller-Adamson carburetor, the kit sold for $165 and was intended primarily for street-use.[46] Pricey and perhaps excessively complex, the conversion sold poorly, and very few of them were ever made.[47] Miller soon turned back to his all-out racing roots, teaming up with Preston Tucker and the Ford Motor Company to produce a run of ten V8s for the 1935 Indianapolis 500. By the end of the decade, he had abandoned speed equipment altogether and was back in the custom racing business for good.[48]

Meanwhile, the defunct Miller-Schofield concern quickly attracted a new group of investors. In the summer of 1930, Harlan Fengler, a veteran Los Angeles board-track racer, and Crane Gartz, the heir to a prosperous plumbing supply company, formed a partnership and chartered a new firm, Cragar.[49] In the fall, the Cragar Corporation purchased the remaining Miller-Schofield patterns, tools, and inventory. Under Fengler's direction, the company then established a manufacturing facility in Hollywood, equipped with "precision machinery worth hundreds of thousands of dollars"—or so they claimed—and staffed by machinists and technicians formerly employed by Miller-Schofield. Ultimately, Fengler hoped to produce a full line of speed equipment for the Ford, but at first he focused on the much more modest goal of re-introducing the popular Miller-Schofield Valve-in-Head. Production of the kit, now known as the Cragar Valve-in-Head, began in earnest early in the spring of 1931.[50] At $112.50, the conversion was a bargain, especially compared with its virtually identical predecessor, and it sold well.[51] That summer Cragar also began to produce custom racing engines and even complete race cars as a low-volume complement to its overhead valve sales. But in the end, in spite of their products' popularity, Gartz and Fengler were unable to sustain their fledgling start-up in the difficult economic climate of the time, and in 1932 Cragar folded.

Leo Goosen's design remained popular among enthusiasts, however, especially those in Southern California. Before long yet another investor therefore decided to try his luck with it. Only this time, the interested party didn't have the financial backing of a wealthy manufacturing heir, a large Los Angeles bank, or a group of aircraft companies. Instead, all he had was a quarter-acre automotive salvage yard, the confidence of a handful of young hot rodders, and the guts to take a risk. His name was George Wight.

Back in 1923, Wight had opened a small automotive wrecking yard in the tiny Southern California town of Bell. Then 46, Wight had been an amateur oval-track racer for a number of years, and Bell Auto, his new business, was supposed to be a way for him to leave his racing days behind and settle into a more stable and far less dangerous way of life. As an ex-racer, Wight knew speed equipment when he saw it, and he soon amassed an impressive collection of high-compression heads, overhead-valve equipment, magnetos, carburetors, high-speed cams, and other performance parts that he had rescued from the wrecks that passed through his lot. Soon he set up a couple of shelves in his office to display this equipment, which he then began to sell to shoestring-budget enthusiasts. Word spread quickly, and by the middle of the decade, used aftermarket parts accounted for the majority of Bell Auto's business. In 1928 Wight built a small brick building on his site to house his growing inventory, and he also set up a machine shop so that he could recondition some of the more heavily worn components that he salvaged. In 1931 he also began to sell new speed equipment, acting as a distributor for Riley, Winfield, and, as fate would have it, Cragar.[52]

Since his shop was more or less a gathering spot for local rodders, Wight knew that his success depended on their favor. With the aid of Gilmore Oil, he therefore organized a series of dry-lakes events at Muroc in the early 1930s, raising his profile considerably.[53] With his finger on the pulse of the budding Southern California hot rodding community, Wight knew exactly what most enthusiasts wanted, and they in turn knew that Bell Auto was an excellent place to get it. Accordingly, when Cragar failed in 1932, Wight knew that it couldn't possibly have been for lack of demand. Curious, he contacted Crane Gartz, and when he learned that the defunct company's patterns, fixtures, and leftover inventory all were up for sale, he made an offer. Negotiations dragged on into the winter, but by the beginning of 1933 the sale was complete. Cragar, once a large and seemingly well-heeled Hollywood-based company, was now a wholly-owned subsidiary of a tiny salvage yard outside the city.[54]

Determined to succeed at what so many others had failed, Wight reconfigured his machine shop with the Cragar jigs and tools. He then signed a contract with a local foundry for raw cylinder-head castings, and by the spring of 1933, the Cragar Valve-in-Head was back on the market. With the help of several machinists formerly employed by the Cragar Corporation, Wight soon managed to achieve an average daily output that enabled him to drop the price of the conversion, complete with a revised intake manifold, to $90.[55] In the mid-1930s, Wight introduced a second model, the "Improved Cragar." Designed primarily for all-out racing, the new head featured a 7.5:1 compression ratio and larger ports, and it sold for $100.[56] Sales boomed, and by the end of the decade, Cragar overhead-valve conversions were second only to the simpler Winfield flatheads among Southern California rodders.[57] Wight had done it: Leo Goosen's design was now a commercial as well as a racing success.

George Wight died during World War II, but in 1945 his business passed into the hands of one of the many young enthusiasts who frequented the Bell shop during the 1930s, Roy Richter.[58] Under his direction, Bell Auto became one of the first mail-order speed shops in the late 1940s, and its Cragar subsidiary continued to prosper as well, ultimately developing into one of the most prosperous of postwar brands. But its success originated not with the hundreds of thousands of dollars worth of assets at the disposal of Schofield Incorporated or its successor, the Cragar Corporation, but rather with the small-time grassroots contacts of George Wight and his successor, Roy Richter.

R&R Manufacturing

Close contact with the street and lakes racers of early-1930s Southern California was critical to Cragar's success under George Wight, and it also worked to the advantage of Ed Winfield, George Riley, and other Los Angeles–area speed equipment manufacturers during the era of the Model A. For those located well to the east of the Rockies, however, no such arrangement was possible. For them, the traditional oval-track and street-use markets had to suffice, just as they had in the 1910s and 20s. But in the context of a depressed economy, many Midwestern and East-Coast firms found it difficult to stay afloat on the basis of the ever-dwindling profits that these particular segments of the automotive aftermarket generated. This was true for the Chevrolet Brothers, Morton & Brett, the Forster Brothers, Ramar, and several others, all of whom went under well before the beginning of Roosevelt's first term. But for other, ostensibly similar companies, Eastern and Midwestern oval-track racers continued to provide a steady stream of income throughout the 1930s. This was true for Green Engineering and

Dreyer, for example, and also for the R&R Manufacturing Company of Anderson, Indiana.

Together with Myron Reynolds, a local investor, Robert M. Roof founded R&R Manufacturing in the summer of 1927, about a year after he sold his interest in Laurel to the Zenith Corporation. Roof, a talented machinist and self-taught engineer, was an expert in the art of Model T performance tuning, and he quickly brought out several low-cost items for the universal car that fall.[59] Shortly thereafter, Ford unveiled its all-new Model A, and Roof spent the better part of the following year developing a line of speed equipment for it. By the end of the summer of 1928, he had introduced a dual carburetor system for the Model A engine, and within another year, a high-performance exhaust manifold as well.[60] Intake and exhaust equipment for the Model A sustained the company through its first two years of operation, but by the end of 1929, Roof had introduced a complete range of internal and external components for the new mill.

In April of 1930, *Ford Dealer and Service Field* ran a feature article by Roof and Murray Fahnestock on Model A performance modifications that featured a number of add-on parts for the new Ford that were available from R&R Manufacturing. These included special brackets for installing a magneto, high-pressure water and oil pumps, cross-drilled crankshafts, down- and updraft carburetors, as well as an all-new overhead-valve conversion. Roof was a seasoned veteran in the design and manufacture of high-performance cylinder heads, having done so for more than a decade at Laurel. His new head, an eight-overhead-valve pushrod model, was similar conceptually to many of those that he had built before. Where it differed from its predecessors, however, was in the unique position of its intake ports: the incoming charge passed from the intake manifold straight down through the top of the head, past the valves, and into the cylinders. This resulted in higher intake velocities, which made the new design a strong performer. Fordson valves, compound valve springs, removable valve guides, high-lift rocker arms, metric spark plugs, and a choice of up- or downdraft carburetion rounded out the package, which doubled the horsepower of the Model A mill. When equipped with a gear-driven overhead camshaft attachment, also available from R&R, the gains were even more impressive.[61]

The following year, R&R introduced yet another cylinder head. Designed as a low-cost alternative to the company's eight-overhead-valve equipment, this new "Cyclone" head featured four overhead valves centered directly over their respective hemispherical combustion chambers. Operated by the standard Ford camshaft via long pushrods and rockers, these valves measured a full two inches in diameter and were forged from a steel alloy. Handling the spent gases were four steel exhaust valves mounted in the stock location in the block. Complete with down- or updraft Winfield carburetors, the Cyclone F-head retailed for $89.50.[62]

Neither the Cyclone nor its costlier siblings were designed or marketed for street use, though. In fact, none of R&R's equipment for the Model A motor was suitable for everyday duty, and Robert Roof was not afraid to admit it. His eight-overhead-valve equipment, for example, was "especially designed for dirt track racing," as were his Cyclone, his dual carburetor conversions, his water and oil pumps, and his magneto brackets.[63] Never did the company's advertisements mention truck, touring car, or speedster applications, and never did Roof brag of their on-road passing or hill-climbing capabilities. Instead, R&R Manufacturing catered solely to oval-track racers, and as the 1930s progressed, Roof spent an increasingly significant portion of his time at R&R working on custom engine, chassis, and complete racing car jobs for individual customers.[64] Narrowly focused on the oval-track scene, R&R Manufacturing sought to weather the Depression by relying on its founder's core area of expertise.

It was a gamble that paid off. By the early 1940s, R&R Manufacturing was one of only a handful of high-performance companies east of the Rockies that had survived the 1930s intact. Continuing to specialize in speed equipment and custom-built racing engines for oval-track use, R&R thrived well into the postwar period. And to the very end, it did so with virtually no involvement in the burgeoning hot rodding phenomenon.[65]

⏷

The era of the Model A was a time of transition for the American high-performance industry. Ford's retreat from its single-model philosophy made for a far less stable aftermarket environment: speed equipment firms would no longer be able to rely on the long-term OEM product stability that had characterized the Model T years. At the same time, Midwestern manufacturers were losing their near-monopoly within the industry as the growth of lakes and street racing on the West Coast fostered the emergence of a powerful and close-knit core of Southern California companies. With one foot in the flourishing 1920s accessory market and the other in the turbulent, enthusiast-driven 1930s rodding scene, the industry stood, in 1930, on the cusp of a critically transformative period.

At the time, however, none of this was apparent to anyone involved in the high-performance community. In fact, especially during 1928–31, when production of the Model A Ford was in full swing, the new period did not seem very new at all. For although the Model A was a more sophisticated, powerful, and capable automobile than its predecessor, it was still a low-cost, high-volume, four-cylinder Ford. The same performance-tuning tricks that worked so well on the universal car could also be applied to the new model, and the speed equipment that was available for the Model A was very similar to that which had come before: overhead valve conversions, improved carburetors, high-speed camshafts, and so forth. Manufacturing, marketing, and distribution channels remained virtually the same, as did the industry's overwhelming and customer-driven preference for Fords. And with nearly two million Model As rolling out of Ford's assembly plants each year, few had any reason to suspect that the car was destined for anything but a long and prosperous run.

But in the end, its tenure proved to be neither. Less than two years after Henry and Edsel Ford unveiled the Model A at the Waldorf Hotel in New York, the stock market collapsed, bringing the booming economy of the late 1920s to its knees. By the fall of 1931, the ensuing economic depression had begun to affect the speed-equipment industry, forcing a number of firms into receivership and casting a pall over many of those fortunate enough to stay afloat. Shortly thereafter, when Ford announced its plans to discontinue the Model A just a few short years after its introduction, the illusion was over for most: the era of the Model A had turned out to be nothing at all like that of the Model T. Moreover, whatever relief some aftermarket firms found in the essential mechanical similarities between the Model A and its four-cylinder replacements, the 1932 Model B and the 1933 Model C, also quickly proved to be fleeting when Ford discontinued its production of four-cylinder engines altogether in 1934.

But the market for Model A/B/C speed equipment did not vanish overnight, and neither did most of the companies that produced it. In fact, the Great Depression and the end of four-cylinder Ford production notwithstanding, the Model A equipment industry soldiered on through the 1930s and early 1940s, and in some cases well into the postwar era. This in turn had a measurable impact on the 1930s emergence of its eventual successor, the V8 industry.

SOUTHERN CALIFORNIA AND THE V8 INDUSTRY

In 1932 customers willing to part with an additional $50 could drive away from their local dealers in a Ford with a V8 engine. A first among inexpensive cars, this V8 displaced 221 cubic inches and produced sixty-five horsepower, fully 25 percent more than the company's "new" four-cylinder engine, the Model B.[66] Even more impressive was the manner in which the V8 delivered its punch. With four additional cylinders, the motor benefited from twice the number of power strokes per revolution than its predecessors, making it smoother, quieter, and considerably more refined than its four- and six-cylinder contemporaries.[67] Moreover, the new mill was fundamentally similar to Ford's previous designs, for it featured a pair of L-type cylinder heads and an in-block valvetrain.[68] Smooth, powerful, easy to maintain, and cheaper than most of the competition's six-cylinder models, the new V8 sold well—so well, in fact, that its success persuaded Ford to drop the four-cylinder Model C in 1934. Thus, between 1932 and 1934, the Ford Motor Company, its dealers, and the overwhelming majority of its customers enthusiastically hopped aboard the V8 bandwagon.

Far less enthusiastic, at least at first, were ordinary high-performance buffs. They did recognize that with sixty-five horsepower as delivered, the V8 had tremendous potential. And when Ford bumped this figure to seventy-five and eighty-five horsepower in 1933 and 1934, respectively, a handful of enthusiasts did begin to tinker with the new mill.[69] But by and large, V8s remained a curiosity among lakes, street, and oval-track enthusiasts until the second half of the decade. And even then, its adoption among shoestring-budget racers was tentative.

Part of the reason was the V8's cost. In 1932, the cheapest V8 model that Ford offered, the roadster, set its buyers back $460. Compared with estimates that place the total cost—including extensive powertrain modifications—of hopping up a used Model A in the early 1930s at around $400, the new V8s were simply too expensive, especially in the context of a depressed economy.[70] Besides, a well-modified $400 Model A would easily outrun a stock $460 V8 roadster, and enthusiasts knew it. Consequently, not until V8 models began to appear on used-car lots and in wrecking yards in the mid-1930s did they begin to become an attractive option. But even then, most enthusiasts were slow to abandon their Model As, primarily because there was no speed equipment available for the new V8. Indeed, apart from a few simple tweaks, those with V8s were largely limited to the factory horsepower rating of sixty-five to eighty-five, while their buddies' hopped-up Model As often sported well over one hundred. Cheap, abundant, and supported by an experienced speed equipment industry, the Model A remained the rodder's car of choice for much of the decade.

Ultimately, oval-track racers were the first to adopt the V8 in considerable numbers. Sponsored stock-car teams went over to the new engine as early as 1932, of course, but among average racers, it did not become a feasible choice until the middle of the decade, when V8 speed equipment for dirt- and board-track applications began to emerge. High-compression cylinder heads, high-speed camshafts, dual carburetor manifolds, and even complete racing engines were available from R&R Manufacturing and Green Engineering by 1935, for example; and in 1936, the popular Ford press began to run occasional stories on the ins and outs of V8 tuning for oval-track use. V8s even began to appear in limited numbers at Indianapolis in the mid-1930s, and with the advent of the tiny, 136 cubic-inch, sixty-horsepower V8-60 in 1937, eight-cylinder Fords soon dominated the smaller tracks of the period as well.[71]

But on the dry lake beds and open boulevards of Southern California, the circumstances were different. There enthusiasts who used their modified automobiles both for racing and for everyday transportation found little appeal in the oval-track equipment that was available for V8 applications, and most stuck with their tried-and-true Riley-, Winfield-, or Cragar-equipped four cylinders. What began to change *their* minds, in the mid- to late 1930s, was the overall growth of competitive dry-lakes racing, together with the experimental bent of its participants.

The American Automobile Association and several other independent groups sponsored a number of dry-lakes events in the mid- to late 1920s, but permanent organization of this peculiar form of racing did not come until the early 1930s. The first to bring some order was the Muroc Racing Association (MRA), formed by a group of enthusiasts with the backing of the Gilmore Oil Company in 1932. Events held under MRA sanctioning in the early 1930s were not much safer than the impromptu runs of the mid- to late 1920s, though, for several cars still raced at once, with only the leading driver enjoying a view of the course unobstructed by the dust of his competitors. Though it certainly gave participants a powerful incentive to improve their cars from meet to meet, this style of racing was exceedingly dangerous, and many lost their cars—and some, their lives—in a cloud of dust at MRA events. Then again, safety was not among the top priorities of most enthusiasts. Instead, what mattered was going fast, and with average speeds in the neighborhood of 90 to 110 miles per hour, MRA events were enormously popular.[72]

The MRA was a loose-knit umbrella organization that *sanctioned*, but typically did not *sponsor*, the events held in its name. Instead, the task of organizing and staffing these meets fell to local hot rod clubs. The first of these clubs appeared in the early 1930s, and they usually consisted of a couple of dozen teenagers and twenty-somethings who drove (and raced) modified production cars. Criteria for membership varied: some clubs had few or no prerequisites, while others required certain dry-lakes benchmarks—the ninety-mph barrier, for example—to be reached by prospectives before they would be admitted. Club members often pooled their resources to purchase or rent a garage for use as their headquarters, and these clubhouses quickly became important centers for the generation and dissemination of performance-tuning knowledge among enthusiasts. Membership in one of these groups was a source of pride, and most of their members' cars sported license-plate sized plaques bearing the clever, if sometimes vulgar, names of the clubs to which their owners belonged: *Throttlers*, *Idlers*, *Bungholers*, and so forth.[73]

The number of clubs and would-be participants grew quickly, particularly in 1935 and 1936. By the end of 1937, in fact, there were simply too many hot rods and too many hot rodders at the already-chaotic MRA events, and casualties began to mount. At the same time, the local authorities were losing their patience with these enthusiasts and their propensity to supplement their lakes runs with illegal street races. Here a lack of consistency was at least in part to blame: the MRA sanctioned but did not sponsor, so the precision with which speeds were measured at its events often varied. Since stoplight contests were a direct and reliable way to settle competing claims based on time slips earned at different events, they were relatively common. With all of this in mind, "representatives from five Southern California car clubs met at the *Throttlers'* Hollywood clubhouse" on November 29, 1937, and "agreed that an organization of several clubs would be of benefit to all club members and dry lakes racing." Together, those gathered chartered a better-organized and more-centralized body, the Southern California Timing Association (SCTA),[74] and on May 15, 1938, the new group held its first event, at Muroc.

Several hundred cars showed up to race that day, and more than 10,000 spectators gathered to watch. Mechanically, the SCTA's only requirements were that the cars' engines had to be American-made and that the parts used to modify them had to have been legitimately obtained (though $100 worth of borrowed equipment was allowed). The course was seventy-five feet wide and three-and-a-half miles long: drivers had a mile and a half to get their cars up to speed before entering a quarter-mile time trap, and then they had a mile and three-quarters to slow down. Through careful scheduling, rigorous time management, and a crack-of-dawn start, everyone was able to run the course twice before the punishing afternoon heat brought the event to a close.[75]

The meet was a smashing success, but before the SCTA and its members could assemble at Muroc for a second event, the United States Army intervened. Muroc was to become an air base, and the last thing the military wanted to do was to share its new proving grounds with these ragtag rodders. Although the army did occasionally allow an SCTA meet to take place at Muroc during 1941 and early 1942, for the most part, the meets of the late 1930s and the 1940s took place on smaller dry lakes that were somewhat farther afield—typically Rosamond, Harpers, or El Mirage.[76]

Much to the chagrin of the local authorities, the SCTA's efforts on these dry lakes small and large did little to curb the popularity of street racing.[77] Still, most enthusiasts recognized that the lakes were now a reasonably reliable venue at which to settle scores and to prove one's mettle as a shoe. Moreover, most of those who raced knew full well that the dry lakes were an excellent place to learn new tricks. Also, for the younger brothers and the uninitiated friends of those who raced, a stroll through the pits at a dry-lakes event could be transformative. There one could observe first-hand what hot rodders did and begin to learn how they did it by asking questions, listening in, and helping out.[78]

The last of these points is significant. Writing about postwar stock-car racing, Ben Shackleford argues that "the pits" at stock-car events help to communicate masculine expectations and skills among working-class American men, even though the overwhelming majority of them never have an opportunity to walk through the pits themselves. Stock-car racing, after all, is a *spectator*, not a *participatory* sport.[79] If this is indeed the case — if, that is, stock-car pits do in fact serve to reinforce dominant notions of masculinity even when experienced from afar (in the stands through binoculars or in one's living room through television coverage) — the implications for the goings-on at these prewar lakes meets are enormous. For at the lakes, the pits (such as they were) were open, and spectators were free to mingle with the racers and to poke and prod at the queued cars (fig. 8). And with the exception of Veda Orr, the only female member of the SCTA and the only woman to race at the dry lakes prior to World War II, those who circulated in the dry-lakes pits were men. Wives and girlfriends often came along to watch but seldom to turn wrenches or to race. Thus, when H. F. Moorhouse writes about the existence of a "hot rodding fraternity," he is right: it *was* a fraternity, and, particularly in the prewar years, its organizational nexus was the SCTA.[80]

At the SCTA's late-1930s and early-1940s events out in the Mojave Desert — and for that matter at the mid-1930s events of its predecessor, the MRA — modified motoring was an increasingly competitive endeavor. And this led to quite a bit of mechanical experimentation among enthusiasts. Most ran modified Model As because they were abundant, cheap, and easy to modify with over-the-counter speed equipment. However, others raced everything from hopped-up Model Ts to four- and six-cylinder Chevrolets, Dodges, Buicks, Oldsmobiles, Plymouths, and Austins. A handful of others owned V8 Fords, and a few even ventured to run twelve-cylinder Lincolns, straight-eight Packards, and sixteen-cylinder Cadillacs. Though there was some aftermarket equipment available for Chevrolets and Dodges, for the most part, enthusiasts who did not own four-cylinder Fords had to come up with their own tuning tricks and powertrain modifications. Scouring area junkyards for stronger crankshafts and larger carburetors, fabricating intake and exhaust manifolds in their backyard workshops, and enlisting the aid of local automotive repair shops for things like cylinder head milling and porting, these hot rodders came up with some creative ways to boost their cars' performance. Nevertheless, the majority of those who raced cars other than Model A Fords in the mid- to late 1930s did so, not by choice, but because some other brand of car happened to be what was available to them. Competition *forced* them to be creative with their hand-me-down Plymouths and Oldsmobiles, in other words, but their preference would have been the simpler and more familiar four-cylinder Ford.

Figure 8. The pits at a dry-lakes meet, where drivers and spectators mingled freely among the cars. This photo was taken at El Mirage in 1947; prewar pits looked much the same (although the "lakester" in the foreground would not have been present at a prewar meet). (Photo by Lee Blaisdell, reproduced courtesy of the Don Garlits Museum of Drag Racing.)

The exceptions were those who *chose* to trade their Model As for V8 Fords. By the middle of the decade, a critical mass of L.A.–area enthusiasts had managed to get their hands on used V8s, and they began to experiment with the engine — not because they had to, but because they believed in its potential. Undaunted by the dearth of over-the-counter V8 speed equipment, they tried anything and everything they could think of to make their flatheads fly. They built their own intake manifolds, to which they fitted multiple carburetors. They fabricated their own free-flowing exhausts. They hired local machine shops to weld and mill their cylinder heads for more compression.[81] They bored their engines' cylinders and offset-ground their crankshafts for more displacement. They tried all sorts of different ignition systems. They reground their own camshafts, lobe by lobe. They fitted larger valves and stronger valve springs sourced from other makes of cars. Some even tried overhead-valve conversions designed for commercial vans and trucks.[82] For the most part, though, their modifications were entirely home-made, hand-crafted, and unique.

By the time the SCTA began to bring order to the lakes in 1937–1938, some of the better-built of these V8 Fords were able to post times competitive with the four-cylinder cars. Nevertheless, four-cylinder hot rods continued to dominate the dry-lakes racing scene, and not until 1941 would the flathead V8 power a majority of the cars at SCTA meets.[83] The reason for the delay was simple: no matter how fast a handful of home-built, V8-powered hot rods were, they did not inspire widespread emulation until the average enthusiast was able to *purchase* ready-made, bolt-on speed equipment for the new motor. As it turned out, this would not be possible until the end of the decade, for the V8 gear that a handful of Midwestern and East-Coast firms began to produce mid-decade was specifically designed for oval-track racing applications, and it rarely appealed to West-Coast enthusiasts who drove their cars both on the lakes and on the streets. By the time the V8 began to become popular in the Los Angeles area, most of the remaining Midwestern and East-Coast manufacturers had long since ceded the Southern California lakes and street scenes to the local firms that enjoyed the competitive advantages of proximity to—and regular contact with—the area's enthusiasts.

As for the local companies, their delay in introducing V8 speed equipment in the 1930s appears to have involved what can only be described as the V8's "newness." When the Model A first appeared, many speed equipment manufacturers quickly discovered that the same sorts of tricks that worked on the Model T also worked on the new car. All they had to do was adjust their designs so that they would bolt directly to the newer mill. The V8, on the other hand, was an entirely different type of motor, and when it was introduced in 1932, very few aftermarket firms knew what to do with it. It was of course a flathead engine, and basic tweaks like raising the compression ratio or adding additional carburetors still worked, but more fundamental modifications would require a bit of research and development. In the context of a depressed economy, there wasn't much incentive for these companies to do so, however, for most West-Coast enthusiasts weren't able to obtain the expensive V8 models anyway. And with demand for their Model A equipment holding strong, California companies simply plodded along for much of the 1930s without giving the V8 much thought.

Thus, when hot rodders began to tinker with the flathead V8 in substantial numbers toward the end of the decade, these existing four-cylinder companies were no more prepared to develop parts and accessories for the new mill than were the enthusiasts themselves. For example, when 1930s hot rodder (and later speed shop owner) Karl Orr went with fellow enthusiast Vic Edelbrock to visit an area camshaft grinder, Pete Bertrand, to obtain a pair of bumpsticks for their roadsters, Bertrand told them that he was not yet ready to offer a line of cams for the new motor. "Fellows, I don't know, Ed Winfield don't know, Kenny Harmon [*sic*] don't know, no cam grinder knows yet what exactly is going to make a V-8 run," he explained. "We've got to experiment a little."[84] Experiment they did, and in due course Bertrand, Winfield, and Harman all had V8 camshafts ready for sale. But it took some doing, and as they worked on the problem in the late 1930s, they were effectively in the same boat as the ordinary enthusiasts who were tinkering with their manifolds and porting and relieving their blocks by hand (fig. 9).

With the extant California manufacturers struggling to come up with effective V8 tuning programs in the late 1930s, a window of opportunity opened for some of the more talented of the amateur enthusiasts. Hot rodder Tom Spalding began to sell converted ignitions for the flathead V8 engine in 1936, for example, when he was still in high school. Later, he explained that "it was the A-V8 that was responsible for getting me into the ignition business. The V8 would cut out at about 4500 rpm, so I built the first dual-point / dual-coil ignition in the machine shop, while still a sophomore in high school. I purchased the dual-coil from Coberly Lincoln / Mercury in L.A. and fabricated the rest myself . . . My first ignition system ran great, and the engine would turn 5500 to 6000 rpm. The system caught on, and soon I was building them for other racers at the lakes."[85]

Racer Tommy Thickstun, on the other hand, found his niche in the manifold business. Early in 1939, he sketched a design for a dual-carburetor intake manifold for his V8 roadster and hired a local pattern-maker and foundry to produce a prototype, which he then polished and assembled in his own workshop. When it proved to be a strong performer, he called the foundry back and ordered a batch of castings, and his finished manifolds soon began to sell well among fellow racers.[86] Vic Edelbrock, who owned a local automotive repair shop and drove and raced a 1932 Ford V8 roadster, was one of Thickstun's dealers. Toward the end of 1939, after failing to convince Thickstun to work to further improve his product, Edelbrock designed his own aluminum intake manifold (the "Slingshot") and began to market it to other enthusiasts.[87] Two years later, in 1941, speed shop owner and former racer Phil Weiand produced a manifold that combined what he believed to be the best features of the Thickstun and Edelbrock designs, and once it had proven its mettle on the lakes, the Weiand "hi-riser" manifold fast became a hot seller among enthusiasts.[88] Likewise, manifolds by Dave Burns, Joe Davies, Jack Henry, Wayne Morrison, Eddie Miller, and Mal Ord emerged from the Southern California lakes and street racing scene in the late 1930s, as did camshafts from Ted Cannon and Harry Weber, ignitions from Joe and Tommy Hunt, and exhaust systems by Sandy Belond.[89] A handful of other enthusiasts made similar products and other odds and ends for the flathead V8 by the beginning of the 1940s, and one, Lee Chapel, even produced a cylinder head for Chevrolets.[90]

Figure 9. Porting and relieving a flathead engine block with a hand-held grinder. (Photo by Lee Blaisdell, reproduced courtesy of the Don Garlits Museum of Drag Racing.)

Figuring conservatively, something on the order of seventeen or eighteen new enterprises grew out of the efforts of these V8 tinkerers in the late 1930s and the very early 1940s. Together with the established California manufacturers that eventually produced V8 equipment of their own, these new and enthusiast-based companies also helped to make the flathead engine the top choice of area rodders by the beginning of World War II. But they were not the only Depression-era newcomers. Eddie Edmunds, an enthusiast based in Portland, Oregon, began to make manifolds for the flathead V8 in the late 1930s.[91] Aluminum Industries, an Ohio foundry, likewise introduced a line of high-compression heads for the V8 Ford at the time.[92] McCulloch Engineering of Milwaukee developed a popular centrifugal supercharger for the flathead mill in 1936 too.[93] For the most part, though, the fledgling V8 industry belonged to Southern California, and by 1941 Golden State manufacturers dominated the American speed equipment business by an overwhelming margin of more than three to one.

Moreover, the newly dominant Southern California V8 industry of the period was geographically concentrated in the strictest sense: by the very early 1940s, there was a critical mass of speed equipment manufacturers operating within the Los Angeles area that shared employees, customers, and even basic producer-industry contacts.[94] Competitors both on and off the lakes, these enthusiast-entrepreneurs were the seeds of the postwar American hot rod industry.

♥

By the early 1940s, Southern California hot rodding had developed into a full-fledged, self-contained phenomenon.[95] It had its own particular style of racing. It had its own close-knit core of enthusiasts. It had its own clubs and organizations. It had its own lingo. It had its own industry. And as of January 1941, it also had its own limited-circulation periodical, *Throttle*.[96] But then, in the blink of an eye, it all came to an end. Shortly after the events of December 7, 1941, *Throttle* closed its doors, the military permanently sealed the fences at Muroc, and most young rodders were called to serve either in the armed forces or in local, critical-category war-supply jobs. Fledgling manufacturers mothballed their patterns, tools, and jigs, and enthusiasts parked their roadsters for the duration.

CHAPTER THREE

FROM HOT RODS TO HOT RODDING, 1945–1955

During World War II not a single roadster stormed across the dry-lake beds of Southern California. Area clubhouses stood vacant, as did prewar hot spots like George Wight's speed shop in Bell. From time to time the rumble of a cut-out exhaust echoed here and there as servicemen on leave fired up their A-V8s for old time's sake. But for the most part, Los Angeles was abuzz with war materials production, not hot rodding.

Following the end of the war, however, the sport resumed with a remarkable vigor, as if the interest and momentum it had generated by the end of 1941 had somehow strengthened during the inactive wartime years. By the end of 1945, the SCTA had regrouped, rods had reappeared en masse, weekend time trials at the dry lakes had begun anew, and a handful of high-performance companies had resumed production. By 1947, the ranks of Southern California's dry-lakes racers reached the thousands, and by 1948, the number of active SCTA clubs was at an all-time high.[1] Enthusiasts across the United States also joined in, and by the end of the 1940s, rodders back East actually outnumbered California's pioneers by a wide margin.

Nevertheless, hot rodding as it emerged in the remaining forty-seven states was but an extension of the original California phenomenon. To this day, in fact, Southern California remains the epicenter of American performance tuning—at least in several crucial respects. But in the late 1940s and early 1950s, no qualification would have been necessary: Southern California was *the* center of American hot rodding, for it was the source of trends, know-how, literature, and aftermarket parts. Never before and never since has modified motoring in the United States had so clear and indisputable a regional bent as in the decade after World War II.

Petty reservations aside, in other words, this—for California enthusiasts and manufacturers alike—was *their* time.

But between 1945 and 1955, the hobby grew substantially. And as it did, new
participants, new goals, new forms of organized competition, and new automotive and high-performance technologies combined to transform what was once
a relatively homogenous and reasonably well-defined leisure-time activity into a
multifaceted pursuit capable of satisfying the interests of a diverse array of performance-oriented enthusiasts. Some of these new twists resulted from the hobby's
importation into climates (social, political, and literal) less favorably disposed
to modified motoring than the sunny, open boulevards of Southern California.
Others—indeed, the majority—originated in the L.A. basin. And although a vo-
cal minority of purists lamented these shifts, particularly as the 1950s wore on,
realistically, there was no going back. For hot rodding had changed, for better or
for worse.

POSTWAR RODDING

On April 28, 1946, the SCTA held its first official dry-lakes meet since 1942.
Although SCTA-sanctioned time trials had resumed at El Mirage the previous
fall, this April meet was special, for it signaled a return to normalcy for the region's
enthusiasts.[2] That is, this meet counted: it was the first of five events scheduled
for 1946 at which official SCTA standings points were up for grabs. Hundreds of
freshly discharged war veterans and eager teenagers representing dozens of newly
reconstituted hot rod clubs therefore made the trip up to the Mojave Desert to
run their cars, catch up with old friends, and get reacquainted with their rivals.
There were a few new faces in the crowd, but few new tricks. Plenty of day-
dreaming and bench racing had of course gone on among the many enthusiasts
stationed abroad during the war, thanks in no small part to Veda Orr, who wrote,
published, and distributed a mimeographed newsletter to her fellow SCTA members in Europe and the Pacific.[3] Still, the cars were largely as they had been four
years earlier, as was the sport itself.

Somewhere along the way, though, modified roadsters had acquired a new
nickname: *hot rods*. No one knows exactly when or why it first was used, but what
is clear is that in the weeks and months immediately following the end of the
war, "hot rod" became the dominant moniker not only among the enthusiasts
but also among their critics. Some have claimed that "hot rod" was originally a
derisive term, an abbreviation of "hot roadster" meant to belittle these homemade
machines—and to imply that their owners were up to no good. And indeed, many
a journalist, police officer, and small-town official made liberal use of the phrase

in precisely that manner. But in the first printed use of the term, which appeared in the fall of 1945, "hot rods" were not a menace, but rather the artifactual end results of the technological creativity of their young owners.[4]

In any event, by the time the SCTA reconvened at El Mirage in April of 1946, all who took part knew exactly what a "hot rod" was. It was a domestic roadster— almost certainly a Ford, but possibly a Chevrolet or a Dodge—manufactured in the late 1920s or early 1930s. It was devoid of fenders, running boards, and anything else its owner deemed unnecessary. It was fitted with slightly larger tires in the rear than in the front. It was equipped with a modified engine; Ford and Mercury V8s were by far the most popular, though four-cylinder Model A and Model B mills still had sizeable followings.[5] It was homemade, with the overwhelming majority of its builder's attention and budget having gone into its engine and transmission rather than its suspension, steering, brakes, paint, or up-holstery. It was a work in progress, a never-ending project always in the throes of a redesign or reconceptualization of some sort. It was used both for dry-lakes racing and for daily transportation, and with the possible exception of the presence (on the street) or absence (at an SCTA event) of its headlights and windshield frame, it would have looked much the same in either context (fig. 10). Finally, it had California tags.[6]

In addition, everyone present at that April meet knew exactly what a "hot rod-der" was. A bona fide hot rodder was a mechanically inclined young man who built, drove, and raced a "hot roadster." He was almost certainly a resident of the greater Los Angeles area, and he was also very likely a dues-paying member of an SCTA-affiliated club. He sourced his parts from wrecking yards, speed shops, and, increasingly, from some of the handful of his acquaintances who manufac-tured high-performance equipment of their own design. He gathered with his friends at clubhouses, speed shops, automotive repair shops, and local diners to share ideas, to plan events, and to bench race. If he had begun to participate back in the 1930s, he was likely on his second or third roadster, on his fourth or fifth engine, and on thin ice with his girlfriend or wife for spending so much time and money on his car. If he was a newcomer, he was probably just beginning to try out, on his first hot rod, some of the many ideas he had learned from his friends and associates. Performance was his top concern, and money spent on chrome-plated grilles, elaborate paint jobs, or anything else that would not have helped propel him to a faster time at the lakes would have been, in his opinion, money wasted.[7]

In short, hot rodding—active participation in the construction, driving, and/ or racing of a hot rod—was a close-knit, narrowly construed, and relatively ho-mogenous pursuit in the months and years immediately following the end of the Second World War, and peer pressure ensured that no one strayed too far from

Figure 10. A typical 1940s dual-use hot rod, captured with its owner at El Mirage in 1947. This rod has a flathead V8 with aftermarket carburetion and exhaust manifolding, and its headlights and windshield frame have been removed for a lakes run. (Photo by Lee Blaisdell, reproduced courtesy of the Don Garlits Museum of Drag Racing.)

its informal technological standards. Ak Miller, a prominent hot rodder and parts manufacturer, faced considerable harassment from fellow enthusiasts when he began to drive a car equipped with an inline Chevrolet six-cylinder engine in the late 1930s. Similarly, "California Bill" Fisher, a 1950s six-cylinder Chevrolet guru, confessed in a tuning manual in the early 1950s that he too once scorned those who drove anything other than a Ford.[8] Meanwhile, the SCTA's rulebook saw to the official exclusion of those whose vehicles failed to conform to the activity's codified aesthetic norms. Until the end of the 1940s, that is, the SCTA's rules declared that roadsters—and *only* roadsters—could race under its sanctioning at the dry lakes. Thus, coupes, convertibles, phaetons, and sedans, no matter how fast, were officially excluded.[9] Consequently, the roughly finished L-head V8-powered Ford roadsters that had only just begun to dominate the sport in the early 1940s still ruled the roost in 1945 and 1946.

But as the ranks of the hobby swelled, and as its influence spread beyond the borders of the Golden State, it wasn't long before these prewar biases began to lose their grip. Within ten years, that which was strict, homogenous, and local-ized in 1945 came to be loosely defined, diverse, and nationally organized. And thus, although the late 1940s and the early 1950s did in fact constitute the era of the 1930s-style California hot rod, period enthusiasts both in Southern California and in the rest of the country were by no means stuck in a late-Depression-era mindset.

For one thing, the participants themselves changed (fig. 11). Hundreds of new faces were in evidence at the lakes as early as the spring of 1946, and in the years to come, popular perceptions of hot rodding as a "teenage craze" began to assume a considerable amount of truth.[10] Of course, the returning veterans who had begun to race on the streets and on the lakes in the 1930s were not to be outdone by their new and younger rivals. Dean Moon, a returning veteran who went on to found a successful high-performance business, actually served a brief jail term in the late 1940s for racing his hot rod on public streets.[11] But Moon's experience was exceptional: postwar rodders often raced on the streets, but far more often it was the younger ones that did so. Older participants, somewhat more mature, tended instead to channel their enthusiasm toward the lakes or, in a growing number of cases, toward the further refinement of the roadsters in which they rumbled to work each day.[12] Indeed, for many of these older enthusiasts (a few were push-ing thirty-five), hot rods were no longer their chief purpose in life. Instead, their cushy postwar jobs, their new suburban homes, and their growing families were their primary concerns, and roadster-building often slipped from all-consuming passion to free-time hobby among them.[13]

Of course, hot rodding grew and changed in the years immediately following the end of the Second World War not only because it attracted new blood in Southern California but also because it quickly spread throughout the United States. As it did, new participants from Massachusetts to Denver and from Billings to New Orleans added their own unique twists. Although some East Coast parti-sans have claimed that these variations are illustrative, not of the ways in which hot rodding changed as it took root outside of the Golden State, but rather of the fact that the activity "occurred simultaneously throughout America right after World War II,"[14] most of the evidence suggests the opposite: Southern California was indeed the wellspring of the hobby, and localized differences within the sport sprang up only as it grew beyond the confines of the L.A. area. For as Arnie and Bernie Shuman explain in their marvelous book celebrating New England's postwar rodding scene, of which they were a part, "Each region tried to be like [Southern California], but—like chili—there were regional renditions, not all by preference."[15] Harsh winters made coupes, sedans, and proper convertibles acceptable alternatives to the traditional open roadster across the Midwest, the

Figure 11. Participants and spectators at a dry-lakes meet at El Mirage in 1947. (Photo by Lee Blaisdell, reproduced courtesy of the Don Garlits Museum of Drag Racing.)

Mid-Atlantic states, and New England.[16] Tighter regulations in many states also made it much more difficult to register and drive a fenderless hot rod with the requisite booming exhaust. In addition, over-the-counter speed equipment was often tough to come by outside of Southern California, especially in the late 1940s, leaving many would-be rodders with little choice but to hack away at their projects unassisted. Finally, there were no dry lakebeds in downtown Chicago, no open rural boulevards in Boston's Back Bay, and no isolated, finely surfaced roads in Southern Georgia on which area enthusiasts could safely test the capabilities of their freshly built rods. And as a result, even the older, ostensibly more mature enthusiasts based outside of Southern California often raced their souped-up rides on public thoroughfares.

Official efforts to combat the "hot rod problem" by cracking down on anyone behind the wheel of anything even remotely resembling a hot rod thus sprang up across the United States in the late 1940s. To be fair, many an outlaw enthusiast had it coming—rings of high-performance-parts thieves, organized groups of street racers that blocked off public roads for stoplight contests, and dishonest rodders who used their connections to circumvent their states' inspection programs in order to register their sub-par machines, to name but a few.[17] And in fact, many of those in positions of prominence within hot rodding circles often called for harsher penalties for those among them irresponsible enough to engage in im-promptu downtown races with their poorly constructed "shot rods."[18] But by the end of the 1940s, many rodders had grown suspicious of their local sheriffs' poli-cies. They began to complain that hot rod owners—even those who fastidiously obeyed the law—were targeted unjustly, and many feared for the future of their pastime. One can readily appreciate, with the benefit of hindsight, the rodding fraternity's growing sense of unease in an era when sensationalistic journalists often penned misleading missives detailing the alleged misdeeds of misfit rodders who raced about in their ungainly "jalopies."[19]

This was particularly true for hot rodders based in states like Minnesota, which launched an official "war on hot rods" in 1950; New Jersey, where a legal loophole allowed new cars to be fitted with dual exhausts but prohibited reconstructed automobiles of an earlier vintage from being so equipped; and Massachusetts, where the "registry police" had the authority to declare in-use automobiles non-compliant for even the most minor of equipment infractions.[20] New York was re-portedly the worst: well into the 1950s, Long Island clubs in particular often held their weekly meetings in unmarked buildings under the cover of darkness, lest the local police detect the presence of more than one hot rodder in one place at one time and declare the gathering evidence of gang activity.[21] On balance, East Coast states were far less welcoming for rodders than were Washington, Oregon, Colorado, and California. But even in California, official attempts to solve the "hot rod problem" through the legislative process caused their share of headaches among enthusiasts, as when officials in Sacramento passed a law in 1951 requir-ing the use of fenders on all cars registered for street use. Although creative hot rodders quickly found that "cycle fenders" allowed them to comply with the law without seriously compromising their cars' clean lines (fig. 12), the message was clear: rodders needed to watch their backs, no matter where they lived.[22]

By 1950, rodders coast to coast had a serious image problem on their hands, but those who lived in Southern California spearheaded much of their collective response. For starters, many L.A. rodders began to try to generate some positive publicity through their clubs by sponsoring special events—voluntary automo-tive inspections, neighborhood clean-ups, and the like—with the cooperation of local authorities. Others inaugurated the practice of requiring their members to carry a set of official club business cards so that if they were in a position to assist a stranded motorist, that motorist could be given a reminder of the name of the group to which the helpful stranger in the hopped-up car belonged.[23] With the assistance of the police, some also organized "reliability runs," day-long cruises

Figure 12. A hot rod equipped with "cycle fenders" over its front wheels, allowing it to comply with the law without losing the fenderless look. (Photographed at Akron, New York, in 2004. Reproduced courtesy of the photographer, Robert Deull.)

somewhat akin to low-speed road rallies in which minor violations of the traffic code were heavily penalized and orderly, by-the-books motoring generously rewarded.[24] Many L.A. clubs revised their charters to include strict penalties for members caught in the act of an illegal street race. Moreover, on a more regional level, the SCTA joined with area speed shops, equipment manufacturers, enthusiast publications, and hot rod clubs to sponsor a yearly exposition at the Los Angeles Armory to showcase to an often skeptical public the technological creativity and the mechanical prowess of the average rodder.[25]

In the long run, though, organized drag racing was by far the most important public-relations innovation of the period. Held at first on remote Santa Barbara boulevards cordoned off with the cooperation of the police, this novel form of racing had graduated to airstrips, where, by the summer of 1950, hundreds of spectators and competitors would gather for the tire-smoking action. In these contests, racers paired off, lined up, and, when signaled, accelerated in a straight line for a quarter-mile before frantically bringing their roadsters to a halt. Drag racing therefore closely resembled the impromptu stoplight contests that landed so many rodders behind bars, especially in the mid- to late 1940s. Thus, although it first emerged in Southern California, the notion of an off-road, safe, and legal alternative to street racing was especially important for enthusiasts in the rest of the country, where the absence of dry lakebeds had long left rodders with little choice but to risk it all on the streets.[26]

It wasn't long, therefore, before popular magazines began to wax enthusiastic about this new form of racing. *Hop Up* frequently encouraged its out-of-state readers to follow the lead of their California peers by working with their local authorities to establish similar contests in *their* hometowns. In addition, *Hop Up*'s editors explicitly endorsed the new activity as the *only* reasonable way to put an end to street racing, and they even went as far as to advocate the dragstrip as the *proper* place for rival clubs to settle grudges.[27] But it was the editor of *Hot Rod*, Wally Parks, who did the most to promote the new activity outside of California. In 1951, with the cooperation of his employers, Robert Petersen and Bob Lindsay, Parks transformed his monthly editorial space into a soapbox from which he began to promote the idea of a "National Hot Rod Association" (NHRA) to organize and sanction dragstrip contests across the United States. In March of that year, *Hot Rod* printed what appeared to be a letter to the editor from an enthusiast by the name of Bob Cameron Jr. In this letter, Cameron argued forcefully for the creation of a national association for hot rodders interested in drag racing; Parks's editorial reply was to announce the creation of just such an organization, the NHRA, with himself at the helm. Later, Parks admitted that the letter from Cameron was a hoax: it was actually the work of Lee O. Ryan, a *Hot Rod* staffer who wanted to help launch the NHRA.[28] Organized drag racing ultimately developed into a colossal, semi-professional activity with its own raison d'être, thanks in no small part to the subsequent growth of the NHRA.[29] But in the early 1950s, the primary purpose of the new association was to provide hot rodders with an alternative to illegal street racing. And in that aim, Parks and his collaborators were spectacularly successful.

By 1953, in fact, the NHRA boasted more than 12,500 dues-paying members nationwide, and affiliated clubs across the country were renting airstrip time or building their own dragstrips with the cooperation of civic authorities. As they did so, street racing declined appreciably.[30] Apparently impressed, California police in particular praised their state's hot rodders and urged enthusiasts in other parts of the country to follow their example by behaving responsibly on the streets and letting it loose at the drags.[31] Enthusiasts in Ohio and suburban Boston also received the unsolicited applause of their local authorities, and even on Long Island some hot rodders began to make inroads with their police by organizing off-road drags and promoting civic programs through their clubs.[32] Never one to rest on his laurels, Wally Parks embarked on an 18,000-mile "Drag Safari" in 1954, traveling from coast to coast in a contemporary Dodge wagon (complete with an aluminum camping trailer emblazoned with the NHRA logo) both to enlist the support of other hot rod clubs and to coordinate the inauguration of more drag strips.[33] The result of all of this was that hot rodding became much more visible, much less offensive, and at least marginally respectable in the early 1950s.

However, none of this would have been possible had it not been for the emergence of the postwar publications.[34] By far the most famous and influential among them was *Hot Rod*, which appeared in January of 1948, but the honor of having been the first postwar enthusiast magazine actually went to a by now less-well-known monthly, *Speed Age*, which debuted in the summer of 1947.[35] By the early 1950s, *Hop Up*, *Car Craft*, *Rod and Custom*, and a number of others were available to enthusiasts on newsstands coast to coast. In addition, technical magazines like *Popular Mechanics* and *Mechanix Illustrated* ran occasional stories on hot rods, as did general-interest automotive periodicals like *Road & Track*, *Motor Trend*, and *Car Life*.

Collectively, these publications helped to foster a sense of community among the nation's far-flung rodders, enabling enthusiasts in North Dakota, Massachusetts, West Virginia, and Louisiana to follow the West Coast trends.[36] On the other hand, they made it possible for the Southern California boys to monitor the progress of their sport beyond the L.A. basin. They also provided folks like Wally Parks with a national forum in which to promote their ideas and their visions for the future of the hobby.[37] Furthermore, their advertisements connected distant enthusiasts with the growing Southern California speed equipment business, contributing to the growth of a national market for high-performance parts and accessories.[38] Finally, and perhaps most importantly, they were unparalleled as a source of detailed technical information.

In 1948 alone, *Hot Rod* ran eighteen full-length technical features covering everything from proper cam selection to the hows and whys of crankshaft stroking.[39] Over the winter and spring of 1948–49, it also ran an extensive series of articles which detailed every imaginable procedure involved in the creation of a traditional flathead V8–powered Ford hot rod; and in 1949, it ran the first of literally dozens of practical "engine conversion" articles (detailed discussions of the specific modifications that worked the best for specific engine types), which appeared each month well into the 1950s.[40] For its part, *Hop Up* tended instead to dwell on the theoretical aspects of performance tuning, running a number of articles in the early 1950s that focused on combustion chamber efficiencies, disparate engine tuning philosophies, and the fundamental chemistry of the internal combustion process.[41]

Combining the practical and the theoretical was *Car Craft*, which succeeded *Honk* in 1953. Explicitly marketed as a hands-on guide for the do-it-yourselfer who was not afraid to pick up a welding torch or a hand-grinder, *Car Craft* tackled complex performance-tuning procedures that did *not* involve the use of bolt-on speed equipment. These included hand-polishing intake and exhaust ports; fabricating a free-flowing exhaust manifold, a high-performance ignition system, or a set of pushrods; and modifying an OEM intake manifold to accept multiple carburetors. For indeed, however much the availability of over-the-counter speed equipment had rendered the task of modifying a production automobile simpler over the years, there remained—and remain to this day—many sophisticated tuning options that require individual ingenuity and skill. For those with the courage to attempt them, the hands-on *Car Craft* was an ideal place to look for advice and assistance.[42] General-interest magazines like *Motor Trend*, on the other hand, offered more distilled features that examined the process of performance tuning from a beginner's point of view.[43] Moreover, most of these publications carried a forum of one sort or another, a monthly column in which the editors would address specific technical questions submitted by their readers. Together with the yearly annuals that many of the more prosperous periodicals brought to market, these monthly technical features represented a veritable gold mine of hot rodding know-how.

For those for whom the monthly periodicals did not suffice, a number of how-to books were also available. Ed Almquist, an East Coast enthusiast who went on to found one of the largest speed equipment manufacturing companies outside of California, published the first such book in 1946. Four years later, Roger Huntington brought out a hugely successful volume titled *Souping the Stock Engine*, an encyclopedic tome that balanced its detailed technical essays with a healthy dose of humor to educate its readers in the ways of intelligent hot rodding. Daniel Roger Post of California published more than a few how-to books of his own during the period, most of which focused on streamlining and custom bodywork.[44] There were, of course, many others, and between the magazines and these how-to manuals, postwar enthusiasts across the United States had ready access to printed sources of information. Particularly for those who lived in isolated areas, the printed word assumed a prominent role in the distribution of technical knowledge within the sport. In larger towns and cities, clubs and speed shops remained a vital link, but the publications helped to bring a measure of the wisdom, if not the skill, of well-known rodders to enthusiasts everywhere.[45]

Another critical source of postwar change within the hot rodding community were the new engines and the new high-performance technologies of the period. By the early 1950s, Cadillac, Chrysler, Oldsmobile, Buick, Dodge, Lincoln, DeSoto, and Studebaker all offered newer overhead-valve V8s. In addition, Ford and Willys offered all-new overhead-valve six cylinders.[46] Hot rodders across the

United States immediately began to experiment with these new OEM mills, and soon Chrysler's overhead-valve, hemi-chamber V8 emerged as a new favorite, particularly among those building all-out competition engines.[47] For street and mixed-use applications, Studebaker, Oldsmobile, Buick, and Cadillac V8s— along with Chevrolet's older, overhead-valve inline six—were among the most popular alternatives to Ford's flathead mill.[48] However, the countless engine-conversion articles published during these years clearly attest to the fact that hot rodders were willing to try anything at least once: Pontiac and Hudson sixes, Hudson straight eights, Auburns, industrial-application four-cylinder Fords, and even the lowly Rambler thus received a share of the how-to coverage along with the much more common Cadillac, Buick, Chrysler, and Studebaker V8s.[49] Over-the-counter speed equipment was available for most of these mills as well, making it easier than ever before for rodders with hand-me-down Chevys, Oldsmobiles, and Dodges to join their Ford-owning peers in the engine-hopping craze.

Throughout the period in question, however, none of these new mills even began to approach the level of popularity that the aging flathead Ford and Mercury V8s enjoyed. By 1954, Chrysler's hemi-chamber V8 had largely nudged aside the flathead Ford among serious racers, but among average rodders, the flathead mill remained the most popular choice throughout the 1950s.[50] Reader-submitted technical questions dealing with flathead-tuning issues in *Hot Rod, Hop Up, Rod and Custom*, and the other enthusiast periodicals continued to easily outnumber those about the newer engines, and the Ford-oriented how-to pieces greatly out-numbered those for other makes. Because it was cheap, abundant, simple, and easy to repair, and also because it had been the focus of nearly two decades' worth of performance tuning and speed equipment manufacturing, the flathead V8 was the engine of choice for period enthusiasts determined to get greasy.

Postwar rodders also enjoyed the benefits of at least two other critical technical developments of the 1940s and early 1950s. The first and perhaps the most important was the advent of high-octane leaded fuels. Whereas prewar enthusiasts would have been lucky to come across a service station selling gasoline with octane ratings in the low 80s, early 1950s enthusiasts had easy access to mid-90s and even 100-plus-octane fuel.[51] This enabled postwar rodders to bump their compression ratios to 9:1, 10:1, or even higher without the risk of serious engine damage through low-octane detonation.[52] And since a higher compression ratio typically makes for a more powerful engine, the 100-octane super fuels of the 1950s were a welcome advance for ordinary rodders.

So too was the second development: cheap, reliable, and effective superchargers (fig. 13).[53] In the 1930s Milwaukee-based McCulloch produced a centrifugal supercharger for Ford and Mercury V8s that sold quite well, but it was unable to produce levels of boost—and, by extension, power increases—commensurate with its cost.[54] But by the late 1940s and early 1950s, a number of aftermarket companies produced superchargers and supercharger accessories that were much more effective. Los Angeles hot rodder and camshaft grinder Jack McAfee began to manufacture a line of adapters that enabled enthusiasts to fit war-surplus GMC superchargers to their motors in 1950, for example.[55] Italmeccanica, an Italian company, also began to market a bolt-on blower; and in the early 1950s, McAfee assumed control of Italmeccanica distribution under the brand name SCOT.[56] In 1953 McCulloch, now of Los Angeles, brought out an improved version of its centrifugal kit, and by 1955 the Judson brothers of Conshohocken, Pennsylvania, had introduced an effective unit of their own design.[57] Otherwise unmodified engines fitted with any of these add-on blowers could make upward of 40 percent more horsepower; modified motors equipped with one could make far more. Consequently, supercharged engines fast became the norm among serious drag racers and dry-lakes competitors, and even on the streets, one often heard the distinctive whine of an aftermarket blower in the 1950s.

Largely absent from the postwar rodding scene was the once-ubiquitous overhead-valve conversion, for after World War II very few manufacturers bothered to produce them. Among the exceptions were George Riley, who made a few sets for the flathead V8 in the late 1940s, and Lou Madis, who turned out a trickle of overhead-valve conversions for Ford's flathead six-cylinder engine in the early 1950s.[58] By far the most famous of the handful of available postwar kits was that of Zora Arkus-Duntov, a German-educated, Russian-born engineer based in Manhattan who manufactured a hemi-chamber overhead-valve conversion for the Ford V8 in the late 1940s. Few hot rodders were able or willing to plunk down the requisite $500 for one of his "Ardun" kits, however, forcing Arkus-Duntov to move laterally into the heavy truck aftermarket in an attempt to save his business. This too failed, and at the end of the 1940s, Arkus-Duntov sold his brainchild to a Southern California speed equipment company, C&T Automotive. There additional development substantially bolstered its marketability as genuine hot-rod equipment, but the Ardun still found precious few customers.[59]

Part of the reason for the Ardun's failure was its timing. Shortly after its reintroduction by C&T Automotive, Chrysler's similar (but cheaper) mass-produced hemi-chamber V8 hit the market, effectively pulling the rug out from under C&T's complicated kit. Most other postwar overhead-valve conversions were similarly unsuccessful, largely because of the widespread availability of OEM overhead-valve V8s in the late 1940s and 1950s. Thus, whereas rodders in the 1910s and 20s had little choice but to base their projects on the abundant flathead four-cylinder Fords and Dodges of the period, their postwar counterparts had a

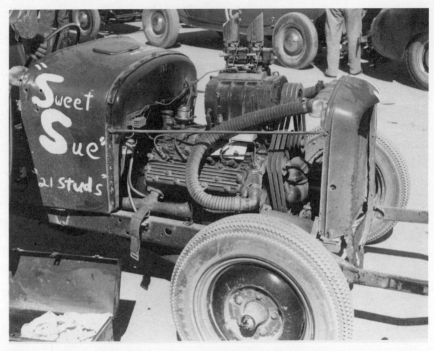

Figure 13. A flathead-powered hot rod with a belt-driven supercharger, captured at El Mirage in 1947. (Photo by Lee Blaisdell, reproduced courtesy of the Don Garlits Museum of Drag Racing.)

plethora of modern, mass-produced designs to choose from. After all, why spend $500 to convert a flathead engine when for but a tiny fraction of the cost, one could easily shoehorn in a late-model, overhead-valve mill instead?

Why, indeed? Hundreds of hot rodders asked themselves that very question in the late 1940s and early 1950s, and as a result, engine swapping became one of the most popular trends among postwar enthusiasts. Forums in the periodicals teemed with reader-submitted letters on the subject, letters suggesting that many would-be engine swappers were willing to go to great lengths to bring a dash of modernity to their modified roadsters.[60] Feature-car articles from the early 1950s confirm that the possibilities were nearly endless: rodders put Cadillac engines into Studebakers, Studebaker engines into postwar Fords, Buick V8s into Mercurys, Chrysler Hemis into '32 Fords, Oldsmobile V8s into Model A Fords, Dodge mills into Model Ts, and so on.[61] *Hot Rod, Hop Up, Rod and Custom*, and the rest of the monthlies also ran detailed features on what to look for when shopping for a late-model, overhead-valve V8 in a wrecking yard, or on the snags and pitfalls one should expect when mating certain engines with certain chassis.[62]

In addition, a number of aftermarket manufacturers introduced special brackets, flywheels, bellhousings, and other parts to facilitate the engine-swapping process. However, very few enthusiasts went to the trouble of dropping, say, a Dodge V8 into their Model A roadster without also taking the time to modify the new engine. In other words, most of those who performed engine swaps in the early 1950s did so, not in lieu of, but rather in addition to, other well-established tuning tricks. Enthusiasts thus began by swapping out their engines, and then they tapped their local speed shops for high-compression pistons, dual-coil ignitions, high-lift rockers, high-speed cams, multiple-carburetor manifolds, and free-flowing dual exhausts.[63]

Between the new engines, the new techniques, and the new cross-brand combinations, considerable technological variety had thus come to characterize much of the postwar rodding scene by the early 1950s. One must be careful not to overstate the case, however. "Studillacs," "Fordillacs," "Fordebakers," and "Bu-Mercs" generated a lot of buzz, but they were nowhere near as popular as the tried-and-true flathead roadster. Variety there was, that is, but not until the end of the decade would the era of the L-head-powered dual-use hot rod truly come to an end.

In fact, readily obvious though the increasing mechanical diversity of the period was, it had less to do with the gradual decline of the traditional dual-use Ford hot rod than did the mounting technical and competitive specialization that took place within modified motorsports in the late 1940s and early 1950s. In 1950 or 1951, for example, it would have been possible for a Los Angeles enthusiast to use his roadster not only on the street and at the lakes but also at the local drag strip. Rather quickly, though, the mounting popularity of drag racing rendered it fiercely competitive. As a result, many drag racers modified their dual-use, street-lakes roadsters in ways that rendered them more competitive in the quarter-mile but much less tractable on Sepulveda Boulevard or the surface of El Mirage. These modifications included, among others, the deletion of their cooling systems, the removal of their bodies (and, frequently, the addition of dozens of "speed holes" in their frames) to save weight, the use of miniscule front tires and massive rear slicks, and the assembly of radical and temperamental racing engines that were impossible to use on the street. For those who did not want to sacrifice their automobiles' everyday utility at the altar of the quarter-mile, special racing classes were established, but for serious racers, specialized dragsters became a virtual necessity.[64]

At the same time, dry-lakes racing began to assume a distinctive air of its own. It too became increasingly popular, but unlike the drags, which emphasized acceleration, the lakes were all about top speed. In time, racers enamored of dry-lakes

competition therefore modified their cars with aerodynamics and sustained engine rpms in mind, and before long, many of them trailered special-purpose,
highly-modified, streamlined "lakesters" to the desert time trials rather than continuing to attempt to convert their daily-driven roadsters into competitive lakesters
on arrival (fig. 14).[65]

By the early 1950s, however, the number of dry-lakes meets began to decline,
and by the end of the decade the activity had ceased to be a major aspect of the
hot rod racing scene. Common wisdom—both among popular historians of the
sport and among those who lived through the period—holds that as drag racing's
popularity swelled, the popularity of dry-lakes time trials waned.[66] Though there
is some truth to this, it was the very popularity of the lakes events themselves that
ultimately led to their relative decline. By the end of the 1940s, a number of organizations sponsored regular time trials on the dry lake beds, including the Bell
Timing Association, the Russetta Timing Association, and, of course, the SCTA.[67]
Multiple dry-lakes sanctioning associations enabled enthusiasts in Southern California to race their cars far more often than ever before, but the lakes themselves
began to suffer from overuse. Cracked and brittle, the surface of El Mirage in
particular fast became an inappropriate venue for high-speed trials.[68] At this point,
some did indeed abandon the dry lakes altogether in favor of the local drags, but
many died-in-the-wool enthusiasts were unwilling to do so. They were willing to
race less often, however, and at a place much farther from Los Angeles: Bonneville, Utah. By the end of the period in question, the SCTA's yearly time trials on
the Bonneville Salt Flats were, along with a limited number of scheduled events
at California's dry lakes, essentially the last vestige of that peculiar form of racing
that had once been all the rage.[69]

But in the late 1940s and the very early 50s, the lakester joined the dragster for a
brief while as a highly specialized type of vehicle that was beginning to contribute
to the decline of the traditional dual-use roadster. So too did several others—full-
size oval-track roadsters and tiny, purpose-built "midget racers," in particular.[70]
Meanwhile, street-driven hot rods changed as well, as many enthusiasts began
to sink more time, effort, and money into their roadsters' paint jobs and interior
appointments as their double-duty days began to wane. Tellingly, these street-only
hot rods soon were known instead as "street rods."[71]

Other rodders, wary of their local police, constructed "sleepers"—vehicles
whose run-of-the-mill looks concealed red-hot power plants.[72] And in the early
1950s, some abandoned their prewar roadsters altogether in favor of newer cars
that they would customize. Their bodies redesigned through roof-lowering "chop
jobs," generous amounts of lead body filler, and lavish paint jobs, these custom
cars were heavy and slow, but that was precisely the point. Built to cruise and not

Figure 14. Specialized, streamlined, racing-only "lakesters" at an El Mirage meet in 1947. *Top*: lakester 97c is push-started prior to a run (its engine, fully enclosed and situated just behind the driver, had no self-starter). *Bottom*: lakester 16 charges through the traps. (Photos by Lee Blaisdell, reproduced courtesy of the Don Garlits Museum of Drag Racing.)

to race, they were a very different type of car. Although some prewar roadsters ended up as radically altered customs, most of the customized cars of the 1950s were based on modern, postwar, closed-body automobiles. Customs were popular, and a number of do-it-yourself guides and "restyling" magazines quickly hit the shelves to cater to custom-car buffs in the early to mid-1950s. Some of the better-off custom-car enthusiasts paid top dollar to have their Mercurys, Cadillacs, and Buicks restyled by well-known, professional customizers like George Barris and Gene Winfield, but the majority of the customs built in the 1950s (and, for that matter, in the years since) were the work of amateur enthusiasts in their home garages. For interpretive flexibility was alive and well not only functionally and mechanically but also aesthetically.[73]

Many hot rodders built customs, but customs were not hot rods. For that matter, many of the performance-tuned vehicles used on the streets in the early to mid-1950s weren't hot rods either—at least not in the strictest sense. Many a rough-hewn, prewar roadster with an exposed and highly modified engine continued to see daily use throughout the period in question and beyond, but increasingly these traditional hot rods were joined (and soon outnumbered) not only by the occasional street rod, sleeper, and custom but also by postwar coupes, sedans, and convertibles sporting powerplants as highly modified as any of those nestled behind the grilles of traditional hot rods. Many enthusiasts referred to these performance-modified, postwar-bodied automobiles as "street machines," while others of a more inclusive disposition insisted that they too were "real" hot rods.[74] Either way, by the mid- to late 1950s, modified postwar cars were a critical component of the rodding scene, and modified prewar cars were less and less important.

FRAGMENTATION

During the late 1940s and the early 1950s, the world of the hot rodder was fragmenting. Whereas hot rodders once drove and raced dual-use, prewar roadsters almost without exception, they now drove customs, street rods, sleepers, and/or street machines on the streets and raced specialized dragsters, lakesters, midgets, and/or track-roadsters on the side. In other words, the hobby was breaking up into a number of smaller niches, or, as Don Montgomery has so often put it, the sport was "splintering."[75]

In his frequent ruminations on the subject, though, Montgomery fails to fully consider the broader implications of this process. For he claims that with the decline of the traditional, dual-use hot rod in the 1950s, the activity known as "hot rodding" began to disappear as well. And intuitively, he seems to have a point. After all, how can there be "hot rodding"—or, for that matter, "hot rodders"—if in fact there are no longer "hot rods"? What Montgomery misses, however, is that as this fragmentation took place, "hot rodding," both as an organized activity and as the act of modifying a production automobile, was beginning to assume a considerable degree of conceptual independence from its artifactual origin, the "hot rod."

By the mid- to late 1950s, therefore, one no longer needed a traditional, by-the-books, prewar-roadster-based "hot rod" in order to participate in "hot rodding."

Instead, all one needed was an interest in performance tuning and some sort of production vehicle with which to experiment. Thus, although the broader applicability of the noun "hot rod" would continue to be a subject of considerable debate among enthusiasts and popular chroniclers for decades to come, the same was by no means the case with "hot rodding." Few in 1955, for example, would have given a second thought to a *Speed Mechanics* feature titled "Rodding a Rambler." Likewise, no one in 1962 would have questioned the meaning of the title of Petersen Publishing's latest how-to manual, *Hot Rodding the Compacts*. Similarly, few would have furrowed their brows at the appearance in 1984 of a book called *How to Hotrod and Race Your Datsun*. Finally, no one who happened to come across *European Car's* 1997 reference to the "extreme 911 hot-rod procedure[s]" that went into the construction of a particular 710-horsepower Porsche 911 would have considered it an inappropriate use of the term.[76] For as a direct result of its 1950s splintering, "hot rodding" assumed a much broader applicability, one that it retains to this day.

H. F. Moorhouse claims that this conceptual broadening actually resulted from a concerted effort by the editors and publishers of enthusiast periodicals to expand their readership. In other words, he argues that it was the likes of Robert Petersen, Bob Lindsay, and Wally Parks who caused this 1950s fragmentation through a deliberate attempt to bring more enthusiasts into the hot rodding community in order to strengthen its civic and political clout and, of course, its capacity to absorb ever-larger monthly runs of *Hot Rod*, *Hop Up*, and the like.[77] Publishers did have an interest in growing their readership, of course, and if you were to thumb through a representative sampling of editorial columns from the various magazines, you would indeed find that their use of the concept of "hot rodding" broadened during the 1950s. However, these magazines were not simply fabricating new niches in order to boost their readership. Neither, for that matter, were enthusiasts blindly following their editors' collective lead, changing the way they conceived of their pastime as a result of what they read in the opening pages of *Hot Rod* each month. Instead, *enthusiasts themselves* drove this process of fragmentation. For they were the ones who decided to build dedicated dragsters and single-purpose lakesters. They were the ones who installed supercharged Buick V8s into late-model Mercurys and Cadillacs. They were the ones who towed their midget racers to the racetrack on the weekends with their beautifully finished street rods and customs. They were the ones, in other words, whose conception of "hot rodding" became more inclusive; and for their part, the publications simply followed suit.

In fact, the magazines' editors often had to defend their relatively traditional definitions of what a "hot rod" was—or of precisely what "hot rodding" encompassed

—to readerships that were often far more progressive than they were.[78] From time to time, of course, angry purists fired off missives to the periodicals, accusing them of "having sold the hot rod sport down the river" with their frequent coverage of vehicles that did not fit the traditional definition of what a "hot rod" was.[79] But Wally Parks, writing in defense of his magazine's position on the matter in 1954, was absolutely right when he maintained that it was the ordinary rodders themselves who were moving on and that the magazines, by contrast, were simply trying to "keep apace."[80]

John DeWitt interprets the evolution of the sport during the 1950s rather differently than Moorhouse. According to DeWitt, the story of the 1950s is about the decline of on-road high-performance tuning and the emergence in its place of the custom car and the "Kustom Kulture." With the disappearance of the traditional dual-use hot rod, DeWitt maintains, those with an interest in serious performance moved into the ranks of the drags and lakes competitors, while those left on the street focused their attention on aesthetics to the virtual exclusion of tire-smoking performance. The end result, in his view, was that the mechanical creativity that had long characterized the activity of hot rodding all but vanished from the realm of the everyday car, replaced in toto by a surge in automotive artistry.[81] Unfortunately, DeWitt's interpretation relies on the assumption that with the decline of the traditional dual-use "hot rod" in the 1950s, performance-oriented, on-road "hot rodding" must therefore have disappeared as well. However, by the mid-1950s, "hot rodding" no longer depended on the existence of traditional "hot rods." Thus, although artistic customization and the Kustom Kulture did indeed thrive in the 1950s, so too did performance-oriented hot rodding.

What was happening within the hobby during the late 1940s and the early 1950s was not the end of hot rodding, as Montgomery suggests, and neither was it the absorption of its creative energies into the Kustom Kulture, as DeWitt maintains. Moreover, it was by no means a top-down process of expansion driven by profit-seeking periodicals, as Moorhouse claims. Instead, it was a wholesale grassroots broadening of what "hot rodding" was. And though it certainly made the likes of a Robert Petersen very happy, no one reaped the benefits of this transformation more than did the entrepreneurs behind the California speed equipment industry.

CHAPTER FOUR

THE CALIFORNIA HOT ROD INDUSTRY, 1945-1955

During the late 1940s and early 1950s, as hot rodding spread throughout the United States and grew increasingly diverse, so too did the high-performance automotive aftermarket. Going into the war, a couple of dozen tiny backyard Southern California businesses were the core of the industry; heading into 1946, the same was true. But by the end of the 1940s, there were nearly one hundred speed equipment manufacturers in greater Los Angeles alone, and by the early 1950s, the largest of these had grown to become multimillion-dollar-a-year operations with large production facilities and national distribution networks. Retail outlets known as speed shops and mail-order houses cropped up all across the country too, bringing manifolds by Edelbrock, heads by Weiand, and cams by Cannon to enthusiasts building hot rods in Miami, Denver, St. Louis, Chicago, and Boston. At the same time, dozens of entrepreneurs established new equipment manufacturing companies all across the United States.

The result, by 1955, was that what had been a tiny, insular, and almost exclusively Southern Californian business ten years earlier had grown to become a booming national industry in nearly every respect. In fact, if we include in our conception of "the industry" the distribution centers, mail-order houses, speed shops, custom engine builders, and engine machine shops that catered to hot rodders, the share of the speed equipment business that belonged to Los Angeles companies dwindled from nearly 90 percent just before World War II to less than half—possibly as little as 10 to 15 percent—by 1955.[1]

Nevertheless, this was undeniably the era of the California hot rod industry. In distribution, sales, and end-use, the industry did indeed go national between 1945 and 1955, but in the actual design and manufacture of high-performance components, Southern California companies continued to dominate throughout the period in question and well beyond. Consider the numbers: by mid-1948, there were 94 speed equipment manufacturers in the state of California, 86 of which were located in the Los Angeles area; by comparison, only 11 were based in other states.[2] Three years later, in mid-1951, there were 132 active manufacturers in the United States: 107 in California as a whole, 98 in the Los Angeles area, and approximately 25 in the rest of the country combined. Finally, by mid-1954, the ranks of the manufacturing end of the industry had swelled to 158, 122 of which were headquartered in the Golden State (111 in Los Angeles) and 36 elsewhere.[3] Thus, percentage-wise, Southern California companies did lose a bit of ground during the period in question: their share of the industry stood at 82 percent in 1948 and 70 percent in 1954. Nevertheless, it is remarkable that after a decade's worth of coast-to-coast growth and diversification within the hobby as a whole, seven out of ten aftermarket manufacturers still hailed from a single metropolitan area. More remarkable still is that, as late as 1954, Chicago — Los Angeles's closest rival throughout the 1950s — was home to but 8 percent of American speed equipment companies. The story of production volume is similar. Almquist Engineering in Milford, Pennsylvania; Gotha in Harvey, Illinois; and Mallory in Detroit were high-volume firms by the mid-1950s, but they were by no means representative: Southern California companies like Weiand, Offenhauser, Iskenderian, Harman & Collins, Edelbrock, Fenton, Jahns, and McCulloch were actually responsible for the vast majority of the bolt-on, high-performance components sold in the late 1940s and early 1950s.[4] Any way you look at it, therefore, postwar aftermarket *manufacturing* was a Southern California enterprise.

MANUFACTURERS AND MANUFACTURING

With very few exceptions, these postwar Southern California companies were either carryovers from the enthusiast-based backyard industry of the late 1930s, existing automotive parts manufacturers that diversified into the high-performance market as hot rodding began to boom, or altogether new companies founded by prewar enthusiasts upon their return from active military duty. Among the first group, the carryovers, Thickstun, Edelbrock, Weiand, Burns, Davies, Cannon, Weber, Belond, Hunt, and Spalding were the most prominent. These firms were the heart of the fledgling California high-performance industry when the United States entered World War II, and in the late 1940s, their founders returned to their prewar positions of prominence within the bustling aftermarket. For some of them, this meant starting over from scratch. Tom Spalding, for example, had abandoned his tiny ignition-conversion business to serve in the armed forces, and on returning to his home in Monrovia, California, he faced the unwelcome task of reestablishing his name among enthusiasts.[5] Tommy Thickstun's case was similar, albeit tragic. After leaving his manifold business to serve as an aviation engineer in the war, the thirty-four-year-old returned home, only to die of a heart attack while on vacation in 1946. A lifelong friend of Thickstun's, Bob Tattersfield, subsequently built the business back up.[6]

For others, like Harry Weber, the return of peace meant reconversion. During the war, Weber ceased to produce his signature reground cams and dedicated his small Los Angeles machine shop to the production of tank and aircraft camshafts for the Allied effort. When the war ended, the Weber Tool Company resumed production of high-performance flathead V8 cams.[7] Vic Edelbrock also spent the war in Southern California, but he closed his prewar shop in 1942 and went to work for a local machinist in an essential-industry production job. After the war he opened a new shop in Los Angeles, but he was not content to simply resume his prewar status as one among many manufacturers of high-performance components. He therefore "sank his savings into machinery" immediately after V-J Day, equipping his new business lavishly at what would prove to be precisely the right moment: within a few short years, Edelbrock would be the largest equipment manufacturer in the country.[8]

Firms like Witteman, Jahns, and Grant, on the other hand, were among a handful of well-established Southern California automotive parts suppliers that joined the speed equipment industry after World War II in order to capitalize on the hot rod craze. Witteman was a valvetrain specialist that introduced a line of adjustable camshaft tappets for the popular Ford and Mercury flatheads in 1949.[9] Similarly, Jahns, an OEM-specification replacement-piston manufacturer that had been in business since 1912, tested the waters by unveiling a line of aluminum pistons in 1950. Bill Jahns liked what he saw, and for many years his firm remained a key player in the performance-piston market.[10] Grant had also started off in the OEM-replacement business, turning out piston rings for Fords, Chevys, and other domestic makes for a number of years before beginning to aggressively market a line of specialized rings for high-performance applications shortly after the war.[11] There were others too, like Mitchell Mufflers, which opened during World War II as a munitions contractor and moved into the high-performance exhaust business in the late 1940s; and Huth, which began during the Depression as a muffler-repair shop before its founder decided to join the booming postwar performance-muffler trade.[12]

However, the firms in these first two groups—the carryovers and those that diversified—were vastly outnumbered by those in the third category, new start-ups. In fact, reading through the scores of published interviews, memoirs, and period "meet the manufacturer" shop-tour pieces that ran in *Hot Rod, Hop Up*, and *Drag News*, it's hard to believe that the stories of how these entrepreneurs first got into the business could be so similar. Paul Schiefer, for example, was an automotive enthusiast in the late 1930s who drove a typical "gow job" and regularly raced at the dry lakes. During World War II he volunteered to serve in the United States Navy, and upon his honorable discharge he returned to Southern California and his beloved hobby. Slightly older and a bit wiser than he had been four years earlier, Schiefer decided to open an automotive repair shop in his native San Diego in 1947. Word of mouth began to attract young rodders to his shop, and by the early 1950s he had begun to specialize in high-performance engine rebuilds. Over time his business slowly morphed into a speed shop that retailed manifolds, headers, cams, and other high-performance products manufactured 125 miles to the north. Soon, however, Schiefer discovered what he considered to be a gaping hole in the aftermarket's offerings, and by the end of the period in question, the Schiefer Manufacturing Company was one of the largest producers of lightened flywheels and clutch components in the industry.[13]

Schiefer's story is somewhat exceptional in that his business was based in San Diego rather than Los Angeles, but still, a few substitutions here and there would transform his tale into that of any of a number of other postwar manufacturers. Replace "Navy" with "Army Air Forces" and "lightened flywheels" with "camshafts," and you've essentially got the story of Ed Iskenderian.[14] Substitute "Army Air Forces" for "Navy" and "manifolds and heads" for "lightened flywheels," and you're talking about Barney Navarro.[15] Jack Engle, Alex Xydias, Roy Richter, and a number of others share similar origin stories, for it was common for returning veterans to successfully transform their prewar passions into postwar profits.[16]

For most of those involved in the high-performance industry of the late 1940s and early 1950s, those profits largely derived from the production and sale of add-on parts for Ford and Mercury flathead V8s.[17] So good was the flathead business, in fact, that some firms never felt compelled to diversify their operations. Barney Navarro manufactured high-compression cylinder heads, manifolds, and other high-performance parts exclusively for the flathead mill throughout the period in question (fig. 15), as did Eddie Meyer and Earl Evans.[18] However, most did look to expand their incomes by adding speed equipment for other makes to their catalogs. Weiand added a line of Studebaker products to his growing inventory of Ford components in 1948; and Iskenderian, Offenhauser, Harman & Collins, and many others did the same with boutique lines for everything from six-cylinder Chevrolets to Chrysler Firepower V8s.[19] For some, in fact, these oddball products actually accounted for the majority of their sales. Frank McGurk skipped the flathead market entirely and manufactured manifolds and other parts for Chevrolets. Likewise, Wayne Horning focused on GMC and Chevrolet cylinder heads,

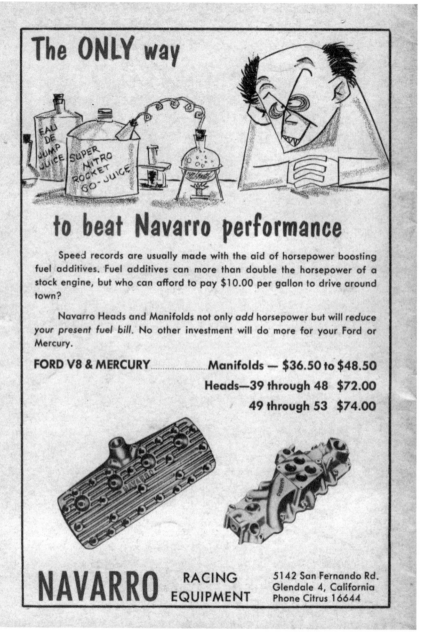

Figure 15. By the early 1950s, many dragsters and lakesters ran on alcohol, nitromethane, or several other expensive "special blends" of fuel. However, this Navarro advertisement from 1953 correctly points out that on the streets, aftermarket cylinder heads, manifolds, and other speed equipment could deliver a noticeable boost in power with simple and abundant pump gasoline. (Navarro advertisement, *Rod and Custom*, December 1953, back cover. Reproduced courtesy of Kevin Lee, editor of the Rod and Restoration Group of Primedia publications that now includes *Rod and Custom*.)

Frank Morgan on Studebaker products, and Nicholas Brajevich on speed equipment for the tiny Crosley engine. On the other hand, Vic Edelbrock manufactured plenty of flathead V8 components in the late 1940s, but parts for V8-60 midget racers were his staples in the very early 1950s.[20]

Thus, although Ford and Mercury V8s were by far the most important targets of the industry's research, development, and manufacturing, there was considerable variety. Given the expansion and diversification of hot rodding at the time, the melange of products available by the early 1950s is hardly surprising. However, aftermarket product diversity was almost nonexistent in the days of Laurel and Rajo, and the fact that postwar companies like Weiand and Offenhauser managed to juggle numerous niches small and large with such success is therefore worthy of note. Perhaps it speaks more to the ways in which the domestic automobile market itself had changed since the late 1920s—by the 1940s, variety was as important to the OEMs as volume—but it also suggests that there was something very different about the postwar Southern California industry's approach to the speed equipment business.

And indeed there was: the vast majority of postwar high-performance entrepreneurs were not simply racers who decided to trade their gloves and goggles for a business suit, as were many prominent prewar manufacturers like Jaegersberger and Roof. Neither, for that matter, were they elite racers and racing-engine designers who profited on the side from the sale of aftermarket parts that often bore no resemblance to those that they generated for big-time races like the Indianapolis 500, as to a large extent were Arthur and Louis Chevrolet. Instead, individuals who drove and raced hot rods themselves spearheaded the postwar industry, and they remained committed enthusiasts throughout their manufacturing careers. Don Blair, Stuart Hilborn, Chuck Potvin, and Jack McAfee were speed-equipment pioneers whose performances at the dry lakes in the late 1940s were legendary.[21] Paul Schiefer, Barney Navarro, Howard Johansen, Fred Carrillo, and many others who ran successful aftermarket businesses also found the time to prepare and race their own cars at Bonneville in the early 1950s.[22] Manuel Ayulo and Vic Edelbrock participated in oval-track and midget events well into the 1950s, even as their businesses expanded rapidly.[23] Meanwhile, other manufacturers, like Ed Iskenderian, Dave Mitchell, and Dean Moon, built daily driven hot rods that were deemed sufficiently well-executed to grace the pages of Hot Rod.[24] Thus, what set these postwar manufacturers apart from many of their prewar predecessors was that they were run by men who were in every sense enthusiasts themselves—they and their competitors, that is, were precisely the sort to whom they sold their products. This was more than slightly advantageous, of course, for it meant that they were perfectly positioned to closely monitor the latest trends.

Almost without exception, these enthusiast-entrepreneurs took advantage of their insiders' advantage with very basic general-purpose machine tools: lathes, drill presses, table- and hand-grinders, and various cutting apparatuses (fig. 16). Engine and chassis dynamometers were common as well, as were precision scales and flow benches, but in-house foundries and heavy-duty forges, stamps, and presses were exceedingly rare.[25] On the whole, the tools of the trade were simple, small, and readily scalable: Wayne Horning used nothing but a general-purpose lathe, a drill-press, and a hand-held grinder to finish his exotic 12-port GMC heads in the early 1950s; and the methods at Bell, Edelbrock, and other period companies were similar.[26] Piston manufacturers like Speedomotive often relied on more specialized tools like piston lathes, while cam companies like Iskenderian, Harman & Collins, and Clay Smith commonly used commercial camshaft duplicating machinery to simplify their operations.[27] But even among these firms, dedicated single-purpose machinery was virtually nonexistent: camshaft duplicators could be used to grind bumpsticks of all types, and piston lathes could be used to finish slugs of all sizes.

In a number of cases, the tools on hand at a given company had more to do with what was available on the used market at rock-bottom prices than with what the owner believed to be optimal.[28] Nevertheless, the general-purpose machinery prevalent among the manufacturers of the period was in fact a strategic blessing. For none of these firms was a mass producer; hot rodding was far too fickle for that. Instead, these companies carefully managed warehoused inventories of their many different product lines and adjusted their purchasing and production schedules as the market warranted.[29] The flexibility of their general-purpose machinery was absolutely crucial in this regard, for it allowed them to mix small batches of low-demand items, such as Studebaker single-intake manifolds, with much larger runs of higher-volume parts, like dual-plane manifolds for flathead V8s. If the warehoused inventory of any one of their products fell to low levels, further batches of raw castings could be ordered from the foundry and finished as needed to maintain an appropriate level of ready-to-ship stock. The typical aftermarket company of the period was therefore well-positioned to manage the uncertainties that serving the ever-shifting demands of hot rodders entailed.[30]

Quite a few of these firms also performed special-order work as part of their everyday operations. The Egge Machine Company stocked aluminum pistons in a variety of sizes but also offered custom-made slugs to those with special needs (and deep pockets). So too did Venolia and Speedomotive. Howard Johansen, on the other hand, dedicated a corner of his camshaft facility to the assembly of special-order racing engines, as did Barney Navarro, Clay Smith, and countless others at their shops.[31] High-performance muffler and exhaust-header manufacturers typically performed in-house standard and custom-application installations on a regular basis as well. Custom camshafts could also be finagled from most grinders, as could oddball stroker cranks and hand-contoured heads from others. The rule of thumb, even among the largest firms, was that anything was possible, provided that the price was right.

To advertise their custom-order and standard-stock items, everyone involved with the industry in the early 1950s knew that a spot in one of the nationally circulated enthusiast magazines was by far the most effective way to reach prospective customers. Before 1950, however, there were very few of these magazines on the market. A number of firms therefore took out regular advertisements and classified listings in less-well-targeted periodicals like *Popular Mechanics*; others, particularly those in the Los Angeles area, also advertised in SCTA, Bell, and Russetta event programs. By the late 1940s, car sponsorships were common as well, as svelte streamliners and minimalist dragsters emblazoned with Edelbrock, So-Cal, Clymer, and other prominent names proved to be useful not only as rolling demonstrators but also as solid income tax write-offs.[32] Throughout the period in question, word of mouth remained essential too. Navarro, Evans, Edelbrock, and Weiand earned positive reputations among enthusiasts very early on that served their businesses well; others, such as Edmunds, Thickstun, and Kogel, were less well-thought-of and ultimately less successful.[33]

But if the question of how to advertise one's products was easy for equipment companies to definitively answer by the 1950s, that of how best to distribute and sell their products was increasingly vexing. Here the problem was that along with the traditional manufacturer-consumer and manufacturer-retailer-consumer sales channels, new twists—regional distribution centers, for example, and national mail-order houses—complicated matters considerably. Throughout the period in question, most manufacturers continued to retail their products directly to consumers as they had for many years. But at the same time, most also contracted with jobbers, traveling salesmen who sold a company's products to speed shops and auto-parts retailers within a given region.[34] Meanwhile, a number of extant firms—especially larger speed shops like Bell Auto and Newhouse Automotive— began to act as regional distributors for certain manufacturers, selling Evans heads, say, or Weiand manifolds wholesale to other retail outlets while continuing to sell them out of their own storefronts too.[35] Further complicating matters, some manufacturers began to act as local and regional distributors for *other* manufacturers.[36] In addition, more than a few speed shops began to manufacture their own products, which they retailed directly from their storefronts *and* sold wholesale to other speed shops, jobbers, distributors, and mail-order centers. Some companies were active on nearly every one of these many sales fronts. Bell Auto, for

Figure 16. Two views of Phil Weiand's shop, circa 1955. In both shots, stacks of high-compression flatheads are clearly visible, as are stacks of high-performance intake mani-, folds for various makes and models of overhead-valve V8s. (Photos by Lee Blaisdell, reproduced courtesy of the Don Garlits Museum of Drag Racing.)

example, was not only a speed shop but also a manufacturer, a mail-order center, and a regional distributor. Throughout the late 1940s and early 1950s, the result of all of this was especially confusing for consumers, for it meant that a certain part that sold for $75 from the manufacturer might also retail for $80 from the mail-order houses and $95 through independent speed shops.[37] Not until the late 1960s would this muddle be resolved.[38]

Nevertheless, in spite of—or perhaps because of—the tangled web of competing distribution outlets that evolved in the late 1940s and early 1950s, the industry grew at a phenomenal pace. By 1952 the top twelve high-performance firms alone did $50 million a year in combined sales, and scores of smaller firms pulled in enough to warrant expansion.[39] In fact, facilities expansions were so common within the Los Angeles industry of the period that very few of the companies that survived for longer than a year or two failed to move into a larger plant at one point or another, and some did so more than once.[40] Thus, by the mid-1950s, production facilities of 5,000, 10,000, and in some cases 20,000 or more square feet were increasingly common.[41]

But for most aftermarket firms, achieving a high sales volume and filling their ever-larger facilities during the late 1940s and early 1950s rarely necessitated the full-time employment of more than a handful of machinists and other laborers. Most had less than a half-dozen employees, in fact, and even the largest rarely maintained a staff of more than ten or twenty. Edelbrock, the largest manufacturer in the country, still employed only ten to twelve machinists as late as the early 1960s, and Engle maintained a staff of less than six well into the 1970s.[42] There were, of course, exceptions, including Harman & Collins, which peaked in the late 1950s with an annual volume of $600,000 and a staff of thirty.[43] But for the most part, postwar California firms relied on small staffs as they cranked out their ever-larger runs of ever-more-diverse performance products.

Flexible, diverse, open-ended, enthusiast-based, and small (but growing fast)— these traits characterized the postwar Southern California speed equipment industry. However, companies scattered throughout the United States during the same period could very well have exhibited these same features, and indeed, many did. So what, exactly, made these L.A. firms unique? Consider the clues that the following profiles afford.

Offenhauser

Shortly after Harry Miller's partnership with George Schofield began to fall apart in 1930,[44] one of Miller's longtime employees, Fred Offenhauser, jumped ship. A machinist by trade and Miller's engine-shop supervisor for more than twenty years, Offenhauser quickly launched a business of his own at which he and a small crew hand-assembled double-overhead-camshaft four-cylinder racing engines similar to those for which Miller was famous. By 1936, Offenhauser's motors powered nearly half of the field at Indianapolis, and from the late 1930s through the early 1970s, the "Offy" was *the* big-stage American racing engine to beat.[45]

Offenhauser's nephew, also named Fred, went to work for his uncle in the early 1930s, while he was still in high school. Through the late 1930s and into the early 1940s, the younger Offenhauser worked at his uncle's shop, where he learned the tricks of the trade and was frequently told that the prosperous business would one day be his. In 1946, however, while he was abroad with the navy, his uncle sold the entire business—shop, tools, patterns, and inventory—to Louis Meyer and Dale Drake. Meyer and Drake, already well-known for their prewar on-track exploits, went on to produce competitive Offy engines for more than thirty years.[46]

When he returned from the service, the younger Offenhauser left the business he no longer stood to inherit. Because he drove a souped-up prewar roadster, he was well aware of the burgeoning hot rodding craze, and he soon decided to try his hand at the manufacture of aftermarket parts. To do so, he teamed up with another young enthusiast, Fran Hernandez, and in 1947 they founded the Offenhauser Equipment Company in Los Angeles. A trained machinist and fellow veteran of the Offenhauser racing-engine firm, Hernandez worked with Fred for about a year and a half before leaving to manage part of Vic Edelbrock's enterprise. Later he would go on to work (and race) for Mercury in conjunction with its motorsports promotions of the late 1950s and early 1960s.[47]

For his part, the younger Fred continued to manage the Offenhauser Equipment Company with the help of his brother Carl and an experienced racing-engine builder by the name of Ollie Morris.[48] Their high-compression flatheads and well-executed intake manifolds for flathead Ford and Mercury V8s sold well among postwar rodders, and their business boomed. In 1950, Fred moved his company to a larger facility across town, and in the early 1950s, he began to diversify as well, adding manifolds, cylinder heads, and other parts for domestic overhead-valve V8 and straight-six engines to his inventory. Ten years later, Fred would play a role in establishing the aftermarket's first industrial organization, and in the early 1970s, his assistance in the performance industry's efforts to work with federal and state environmental regulators would prove to be significant.[49] In the early 1950s, though, Offenhauser was but one of many who managed, with the help of a few close friends, to turn an interest in hot rods into a business that depended on them.

Wayne / Horning

Just after World War II, a Southern California enthusiast named Wayne Horning secured a loan from an older racer, Marvin Lee, for the development and construction of twelve-port aftermarket cylinder heads for GMC and Chevrolet six-
cylinder applications. Cast of iron, Horning's heads featured six intake ports on
one side and six exhaust ports on the other, a cross-flowing design that produced
dramatic results when bolted in place of the standard GMC or Chevrolet units.[50]
However, very few of these cylinder heads were made before Horning decided to
cash out of the operation in 1949. At that point, he sold the patterns and tooling
for his twelve-port Chevrolet unit (the "Wayne") to an enthusiast named Harry
Warner, who established a company to manufacture it. But his patterns, tooling,
and rights to the GMC head (the "Horning") went to another young enthusiast,
Fred Fisher, who worked at a small Long Beach engine-machine shop called the
Electronic Balancing Company.[51]

Fisher, who was once a flathead-V8 devotee, claimed to have "seen the light"
in the late 1940s after riding in a friend's hopped-up Chevrolet. Thereafter, he
devoted most of his time and energy to that company's overhead-valve straight
six.[52] Fisher, who often went by the name of "California Bill," published numerous technical and how-to articles on the Chevrolet and GMC engines in the
early 1950s and also put out a full-length book on the subject in 1951.[53] When
he bought the patterns for the cast-iron "Horning" head in 1949, he contracted
with a local foundry for a batch of rough aluminum castings that he then machined, assembled, and marketed as the "Horning-Fisher" GMC head. Combining Horning's sound cross-flowing architecture with the superior heat-dissipating
and detonation-suppressing characteristics of cast aluminum, the Fisher-Horning
head was a smashing success, even though it retailed for more than $230.[54]

For his part, Wayne Horning remained an active member of the high-performance industry, assembling complete top-notch six-cylinder racing engines
and even a few special-order, racing-spec twelve-port heads in his Los Angeles
workshop well into the 1950s. Chet Herbert, an area camshaft grinder and valvetrain specialist, produced the tappets and related components for Horning's
special-order competition heads, while Offenhauser supplied auxiliary devices
like high-pressure oil and water pumps for Horning's racing engines. Venolia
pistons were also standard-issue in these mills, and Horning fitted quite a few of
them with Hilborn fuel injectors, too.[55]

In the early 1950s, therefore, cylinder heads designed by Wayne Horning were
available from three different sources: Harry Warner supplied street- and mixed-

use cast-iron Chevrolet units; Fred Fisher built cast-aluminum, mixed-use GMC units; and Horning himself produced all-out competition units for GMC-based engines. In the end, however, all would disappear by the end of the 1950s, victims of the emerging dominance of Chevrolet's new V8.[56]

Eddie Meyer Engineering

In the late 1930s Eddie Meyer, a Los Angeles rodder, began to produce dual-carburetor intake manifolds, high-compression flatheads, and converted ignitions for the L-head Ford V8, selling them to fellow lakes- and street-racing enthusiasts to finance his own hot rodding endeavors.[57] With the help of his brothers, Louie and Bud, not to mention a healthy dose of advice from Ed Winfield, Eddie soon established a solid reputation among area racers, and his business had just begun to pick up steam when the events of December 7, 1941, forced him to turn his attention elsewhere. Four years later, though, Meyer's West Hollywood shop picked up right where it had left off.[58]

During the late 1940s, sales expanded, as did Meyer's line of speed equipment. Active on the dry lakes and within speedboat-racing circles, Meyer remained firmly rooted in the world of the enthusiast, refining his products as his customers' needs evolved.[59] But Eddie Meyer was a flathead V8 Ford enthusiast through and through, and in the early 1950s he was one of only a handful of manufacturers that did not tap into the expanding market for overhead-valve V8s. His sales therefore leveled off in the mid-1950s, and by the end of the decade, the Meyer name was on the wane. According to his brother Bud, Eddie soon decided that "the speed equipment business [had become] a rat race"; in the early 1960s, the brothers therefore closed their hot rod shop and "went into the specialized world of exotic foreign car repair."[60]

Meyer manufactured a lot of flathead speed equipment for hot rods and racing boats during the 1950s, but his ultimate significance has very little to do with any one item in particular, nor does it stem from the sheer variety of flathead gear that he produced. Instead, Eddie Meyer Engineering was a critical aftermarket firm because of the people Bud and Eddie Meyer hired and worked with—or rather, because of the people they trained in their shop and what they went on to do within the high-performance industry.

The first was their brother Louie, who left Eddie Meyer Engineering in 1946 and purchased a share of the Offenhauser racing-engine facility.[61] The second was a young enthusiast by the name of Ray "Racer" Brown. A gifted engine builder and a talented racer, Ray got his start in Eddie Meyer's shop shortly after Louie Meyer left. Four years later, he went off on his own, and during the 1950s, *Car Craft*, *Hot Rod*, and other enthusiast publications often tapped his expertise for technical, how-to, and feature articles.[62] Likewise, an enthusiast named Lou Senter went to work for Eddie Meyer just after World War II, assembling engines and manning various pieces of machinery for about a year and a half before teaming up with another rodder, Jack Andrews, to found Ansen Automotive in Los Angeles.[63] Ed Pink, famous for his racing engines in the late 1950s, 60s, and beyond, also started off in Eddie Meyer's shop just after the Korean War. There he cut his teeth in the engine assembly business, learning the tricks of the trade before venturing off on his own.[64]

Brown, Ansen, and Pink—in the long run, they were the most significant legacies of Eddie Meyer's two-and-a-half decade reign among the speed-equipment-manufacturing elite.

Iskenderian

As a teenager in the late 1930s, Ed "Isky" Iskenderian was an apprentice machinist who spent his spare time hanging around with rodders in Los Angeles. Over the course of 1939 and 1940, he built a Model T roadster with a flathead V8 that was equipped with a rare set of Maxi overhead-valve heads. Soon he decided to add a high-performance camshaft as well, so he got in touch with Ed Winfield to negotiate the purchase of one of his famous bumpsticks. These cams were so popular at the time that Winfield was struggling to keep up with demand. Isky's order was among many that were backordered, and after waiting for it for a few weeks, he decided to visit Winfield's shop to get a better handle on the likely date of delivery. Winfield took a liking to the young man, welcoming him into his shop and showing him some of the tricks of the trade. Soon Ed began to wonder whether he too might have a future in the camshaft market, especially given the backlog at Winfield. But before his plans gelled, the United States entered World War II.[65]

During the war, Iskenderian served as a B-24 tail gunner in the Army Air Forces, and upon his return he immediately got in touch with Ed Winfield. Further tours of Winfield's shop and hours spent conversing with the "reclusive genius" convinced Iskenderian that he should indeed begin to grind his own cams. In 1947 he secured the use of a small corner of a friend's machine shop, where he set up a cylindrical grinder that he had obtained for next to nothing, adapted it to the task at hand, and launched Iskenderian Racing Cams. The operation was simple: Ed's only employee, Norris Baronian, would rough-grind five or six camshafts during the day, and Ed would come in at night to finish them off. Sales among rodders were disappointing, however, and he had to turn to the East Coast stock-car racing scene in order to generate some sales.[66]

Isky's fortunes among West Coast enthusiasts changed when an old friend of his, Vic Edelbrock, stepped in to help him expand his sales—and improve his standing—within the hot rodding community. Edelbrock, whose manufacturing business was fast becoming the largest in the industry, had quite a bit of influence among distributors and speed shops, and he helped Ed muscle his way into retail inventories coast to coast. As a result, Iskenderian's operation quickly grew in the early 1950s, moving twice as Ed sought sufficient machining and warehousing space for what was quickly developing into one of the largest camshaft outfits in the country.[67]

An innovator and an aggressive salesman, Iskenderian remained atop the camshaft field for many decades to come, but in the late 1940s, it was his friendships with Ed Winfield and Vic Edelbrock that helped to get his struggling enterprise off the ground.

ENTHUSIAST-ENTREPRENEURS

People, and the ways they flowed between companies, established spin-offs, and ultimately came to form thick webs of contact and experience, were among the most significant aspects of the speed equipment industry's postwar concentration in the greater Los Angeles area. Offenhauser learned the ropes from his uncle, who had in turn learned what he knew about racing engine design from the famous Harry Miller. Likewise, Fran Hernandez started off at the elder Offenhauser's plant before splitting off with the younger in 1947, joining forces with Edelbrock in 1949, and ultimately leaving the speed equipment industry altogether to help direct the motorsports agenda of Ford's Mercury Division. Another Miller protégé, Ed Winfield, went on to influence the methods and strategies of a number of aftermarket entrepreneurs, including George Riley, Vic Edelbrock, and Eddie Meyer. Meyer, in turn, brought the likes of Ray Brown, Lou Senter, and Ed Pink into the industry; while his brother Louie went off with Dale Drake to forge a future for the Offenhauser racing engine. Meanwhile, Edelbrock and Winfield helped propel a young Ed Iskenderian into a career in the camshaft business, and Wayne Horning, Harry Warner, and "California Bill" managed to spin a single cylinder-head design into three distinct and successful early 1950s businesses.

Other examples abound. Barney Navarro got his start at Phil Weiand's shop just before World War II. Clay Smith learned the ins and outs of the camshaft trade from Pete Bertrand in the late 1930s, as did Karl Orr. Similarly, Bill Spalding briefly worked for Harman & Collins before going into the camshaft business for himself, and Howard Johansen's top cam grinder during the 1950s, hot rodder Al Barnes, spun off toward the end of the decade to found his own racing-engine company. Likewise, one of Sandy Belond's top deputies at the Southern California Muffler Company in the 1950s, Bob Hedman, ultimately left to start his own exhaust-header firm in the early 1960s. Finally, prewar cylinder-head manufacturer Art Sparks teamed up with Fred Carrillo in the early 1950s to develop a line of virtually indestructible connecting rods.[68] These were but the tip of the iceberg: many other spin-offs and episodes of personnel cross-fertilization occurred as the 1950s wore on, and subsequent periods witnessed similar developments.

Ultimately, these webs of contacts and inter-firm relationships that were so prevalent within the Los Angeles speed equipment industry—both during the period in question and in some cases for decades prior—suggest at least a partial explanation for the region's postwar dominance. Consider Philip Scranton's concept of *networked specialists*, "clusters of smaller companies in urban industrial districts [that] offered diverse finished goods to households and enterprises, relying on thick webs of contact and affiliation to organize production and sales."[69] One of three distinct types of specialty manufacturers of the late nineteenth and early twentieth centuries described in Scranton's *Endless Novelty*, these networked specialists were in many ways analogous to the firms that constituted the Los Angeles hot rod industry of the late 1940s and the early 1950s: aftermarket companies were small, they were clustered in or near the city, they produced and sold a diverse array of products, and their business practices resulted in dense networks of contacts among them. Perhaps the industry's concentration within the Los Angeles area in the postwar period was merely a symptom, that is, of the way its firms conducted their affairs.

Perhaps. But the fundamental question remains: Why Los Angeles? Scranton's model suggests that regional clusters developed in part because of the presence in a particular area of large pools of skilled, often immigrant laborers that specialist firms could tap as needed. In addition, the presence of other companies that he calls "specialist auxiliaries" (foundries, general machine shops, and other firms that furnished rough castings and other producer goods and services on demand) within a particular region was also likely to attract specialist manufacturers.[70] Los Angeles certainly had both of these in the late 1940s and the early 1950s: a large, government-contract-based aircraft industry ensured not only that the region was filled with skilled machinists of varying levels of experience but also that a number of smaller foundries and machine shops thrived there. Its economic climate thus was highly conducive to precisely the sorts of specialty-manufacturing practices that the speed equipment industry embraced.

Nevertheless, a critical causal link is still missing. Why Los Angeles, that is, and not Detroit, which had not only a number of small foundries and a large pool of experienced machinists but also factories and personnel familiar with automotive parts production? Or Chicago, where in addition to these manufacturing prerequisites one would also have enjoyed the benefits of a centralized location better situated to cater to rodders nationwide?

The answer actually has very little to do with Los Angeles itself. Instead, it has to do with who was there at the time—who the speed equipment industry's employees were, who its employers were, and who its customers were. For with very few exceptions, they were all hot rodders. They all raced at the lakes, the drags, or at Bonneville. They all built hopped-up or customized cars. They all went to the shows at the Los Angeles Armory. They all read *Hot Rod, Hop Up,* and the rest of the periodicals. In short, they were all integral members of a community of enthusiasts, one that had deep roots in the prewar lakes- and street-racing scenes of the Southern California region. The speed equipment industry of the late 1940s and the early 1950s emerged from within this community, and in seeking to organize their production and marketing strategies, enthusiast-entrepreneurs looked not simply to an undifferentiated mass of skilled machinists, but rather to their friends and associates from the lakes, the drags, the shows, and the local bench-racing hot spots. Networked specialists they were, in other words, but their networks included the requisite contacts not only with other manufacturers but also with their employees and their end-use customers.[71]

In a broader sense, then, the Los Angeles speed equipment industry of the 1940s and the early 1950s was more than just a clustering of "independent craftsmen."[72] Instead, it was part of a larger productive network based in Southern California, a network whose "product" was nothing less than the American hot rodding phenomenon itself. As Bruno Latour explains in his analysis of the scientific community, "The word *network* indicates that resources are concentrated in a few places—the knots and nodes—which are connected with one another—the links and the mesh: these connections transform the scattered resources into a net that may seem to extend everywhere."[73] Latour's "knots and nodes" are scientists and scientific institutions, of course, whose interaction within a network known as the scientific community results in the production of scientific knowledge. In the Southern California hot rodding community, or "network" of the 1940s and 50s, however, individual speed equipment companies and their factories certainly were among the critical "knots and nodes." But so too were dry lake beds, dragstrips, clubhouses, bench-racing hot spots, and ordinary rodders' garages. Binding these nodes into a productive network—a vibrant community that "produced" the hot rodding phenomenon that seemed to "extend everywhere"—were the enthusiasts themselves.

Nowhere else did such a network of enthusiasts, entrepreneurs, and enthusiast-entrepreneurs exist in the postwar period, and although the Southern California industry's share of the pie began to shrink perceptibly by the early 1950s, nowhere else did more than a handful of speed equipment manufacturers ever operate at the same time during the period in question. Nevertheless, by the spring of 1955, some within the Los Angeles hot rod industry had begun to fear that a threat to their cohesion and prosperity was gathering steam back East. But it was not Ed Almquist's Milford, Pennsylvania, speed equipment factory that they feared. Nor was it Gotha in Harvey, Illinois, or Crane in Hallandale, Florida. Instead, it was the OEMs and their escalating 1950s horsepower ratings that kept them up at night.

For as Barney Navarro lamented in *Rod and Custom* that March, "Chevrolet can [now] supply a 4 throat carburetor and manifold plus a dual exhaust system for a much lower price than any speed equipment manufacturer will ever find possible." Factory "power packs," that is, were infiltrating the street-use after-market to a degree that was disconcerting to folks like himself whose livelihood depended on bread-and-butter sales of bolt-on manifolds, headers, and similar items.[74] Pessimistic in the face of this competitive challenge, Navarro could not help but wonder aloud whether the speed equipment industry would be able to continue to play a role in the evolution of American high-performance motoring in the years to come. Perhaps, that is, the aftermarket's fifteen minutes had already come and gone.

■■■

CHAPTER FIVE

FACTORY MUSCLE,
1955–1970

What Navarro witnessed in 1955 was the onset of the "horsepower race." Using tuning tricks long common among hot rodders, mainstream American manufacturers like Chevrolet, Ford, and Chrysler began to compete for performance-minded customers by tweaking their overhead-valve V8s for more horsepower. By the end of the decade, passenger cars' horsepower ratings had reached an all-time high, often exceeding 300 even in those models *not* equipped with optional, performance-enhancing "power packs." And this was only the beginning. During the 1960s, fierce competition within the all-new, youth-oriented muscle car segment quickly pushed the horsepower race into territory many considered irresponsible: by 1970, 400-plus horsepower "factory hot rods" were selling like hotcakes to lead-footed baby boomers coast to coast.

Popular legend holds that this OEM horsepower race—particularly its 1960s muscle car phase—all but wiped out shadetree hot rodding. In his delightful biography of a 1950s and 60s Ohio hot rod club, Ron Roberson records that once the muscle car boom began, "there was no [longer any] need to build a hot rod when you could buy one off the showroom floor and make monthly payments." "Muscle cars," he concludes, "replaced hot rods."[1] In their respective histories, Ed Almquist, Pat Ganahl, and Gary and Marilyn Meadors agree: OEM muscle may well have been good for performance enthusiasm, but certainly not for rodding. Perhaps more to the point, Tom Wolfe has argued that the horsepower race represented an attempt on the part of the OEMs to capitalize on hot rodding by appropriating its grassroots methodology and incorporating it into their mass-produced designs. For his part, H. F. Moorhouse rightfully emphasizes the frustration that leading members of the hot rod fraternity often expressed regarding this unwelcome OEM incursion onto their turf.[2]

But if this was the case—*if*, that is, the OEM horsepower race truly *was* responsible for a noticeable decline in hot rodding during the late 1950s and the 1960s—surely, one would expect the record to overflow with tales of once-prosperous speed equipment manufacturers forced to concede defeat as these factory hot rods rendered their businesses obsolete. However, apart from a few offhand and unsupported remarks that the aftermarket as a whole was outclassed, outmatched, and out-engineered by the OEMs during the 1960s, the secondary literature is all but silent on the matter.[3] So too are period sources. During the 1960s, a few outspoken critics occasionally rehashed Navarro's 1955 concerns, but on the whole very few involved in the design, manufacture, distribution, or retail sale of aftermarket speed equipment during the late 1950s and the 1960s ever again spoke of impending doom at the hands of the power pack and the muscle car. To a man, in fact, they were far more concerned about the implications of the emergence of federal and state automotive regulations than with yearly sales of Pontiac GTOs or SS Chevelles. For as it happened, the decade and a half that followed on the heels of Navarro's ominous warning proved to be a period of unprecedented growth and prosperity for the speed equipment industry.

This was especially true of the 1960s. According to one conservative estimate, modest annual growth in the mid- to late 1950s resulted in an industry worth in excess of $36 million annually by 1961. Five years later, however, speed equipment manufacturers recorded $148 million in annual sales, and by 1970 the figure stood at $1.168 billion (to put this in perspective, the smallest of the Big Four, American Motors, reported total sales of $1.1 billion in 1970; Chrysler, by contrast, reported just under $7 billion; Ford, approximately $15 billion; and General Motors, $18.75 billion).[4] The fact that aggregate growth of this magnitude occurred at all is remarkable; that it occurred precisely during those years in which the OEMs' muscle cars and other high-performance packages reportedly quashed hot rodding is all the more striking—even puzzling.

The truth is that the popular chroniclers have it wrong. Hot rodding did *not* decline in the 1960s, and it certainly did not suffer at the hands of the OEMs and their high-performance offerings. Enthusiasts did build fewer prewar roadsters in the 1960s than they did in the 1940s and 50s, but this was nothing new: traditional hot rods of this sort had been dwindling in number since the early 1950s. Hot rodding, on the other hand, soldiered on throughout the 1950s as enthusiasts turned their creative energy toward newer and slightly less orthodox sedans, coupes, specialized dragsters, lakesters, customs, and even family cars. This drawn-out process of fragmentation or diversification within hot rodding continued unabated well into the 1960s as hot rodders began to transform run-of-the-mill imports, domestic compacts, and sports cars into their own unique, high-performance machines. And for the parts to do so, they looked, as they had for years, to the speed equipment industry.

But this is only half of the story. Aftermarket growth during the late 1950s and the 1960s did not occur simply—or even *mostly*—because of the popularity among enthusiasts of Corvairs, Volkswagens, Tempests, MGs, and the like. Instead, it largely stemmed from booming sales of more traditional high-performance cams, cylinder heads, intake manifolds, and exhaust headers for large-displacement, American-made V8s. And especially during the 1960s, a sizeable portion of these parts went, *not* to the owners of 1932 roadsters, 1940 coupes, or 1955 Chevys, but rather to those with late-model Mustangs, Barracudas, Chevelles, and Camaros—to those, that is, who already owned an OEM-built high-performance car. Aftermarket manufacturers therefore did not simply cede the high-performance market to the OEMs in the late 1950s and the 1960s. Instead, they met the challenge head-on, successfully competing with Detroit's talented engineers, colossal factories, and cubic research and development budgets for their share of the spoils.

As they did so, though, their outlook changed dramatically. For never again could the likes of a Vic Edelbrock hope to succeed merely by outsmarting his aftermarket competition—the Phil Weiands and Fred Offenhausers of the world. Instead, by the 1960s, he also had to contend with General Motors, Chrysler, and Ford—and, as the 1970s dawned, with state and federal regulatory agencies as well.[5] The 1960s was indeed a time of plenty, but it was also a time in which the industry's carefree days gradually came to an end.

FACTORY HOT RODS

Prior to the early 1950s, the American automobile industry showed very little interest in high-performance motoring. During the 1930s, Ford briefly dabbled in Indianapolis 500 competition, and from the very beginning of the automobile age, dealerships often sponsored oval-track cars as well.[6] For the most part, though, Detroit focused the overwhelming majority of its engineering and marketing efforts on its core enterprise: selling ordinary cars to ordinary people. And this was precisely what Oldsmobile, Chrysler, Ford, and the rest of the OEMs had in mind when they unveiled their new overhead-valve V8s in the late 1940s and early 1950s. Modern and efficient, these new mills were capable of shuffling around the hefty postwar sedans in which they were installed, but they were by no means high-performance motors—not as delivered.[7]

By the mid-1950s, however, these relatively tame late-model engines had begun to undergo a radical transformation. Oldsmobile's "Rocket 88" V8, for example, generated 135 horsepower from its 303 cubic inches when it was introduced in 1949, but its output rose to 160 by 1953, 202 by 1955, and more than 300 by 1957.[8] Similarly, Chrysler's "Firepower" V8, rated at 180 horsepower when introduced in 1951, topped the 300 mark by 1958.[9] Concurrent developments at Ford and Chevrolet—most notably, the introduction of the 265-cubic-inch, small-block Chevrolet V8 in 1955—confirmed that Detroit had warmed to high performance.

What exactly caused this change of heart remains unclear. During the 1960s, some within the aftermarket looked back with an almost palpable sense of nostalgia and claimed that General Motors's decision to hire erstwhile hot rod parts manufacturer Zora Arkus-Duntov to head a small team of powertrain engineers in 1953 was the turning point. Others have argued that rising compression ratios, cubic-inch displacements, and horsepower ratings were merely the result of the industry's need to power its two-ton "insolent chariots."[10] Period sources, however, suggest that at least part of the industry's inspiration came directly from the fenderless coupes and roadsters that enthusiasts were building.

In the fall of 1951, for example, the City of Detroit hosted a hot rod exhibition with the support of the mainstream automobile companies, each of which dispatched a team of engineers to examine the reconfigured cars on display.[11] Within months, Chrysler disclosed that its engineers had squeezed 310 horsepower out of a run-of-the-mill Hemi V8 simply by mimicking the tried-and-true methods of the rodder: fitting bigger valves, enlarging the intake and exhaust ports, installing a special intake manifold with multiple carburetors, modifying the camshaft for optimal valve timing, raising the static compression ratio, and installing a free-flowing exhaust.[12] In the mid-1950s, when these performance-tuning tricks became available to the general car-buying public as optional power packs, many within the hot rodding fraternity bragged that "it was a happy day for motorists in general when the big wheels who design the automobiles most of us drive got the 'go' fever not so long ago and borrowed some of the hot rodders' sacred devices to make them stock equipment on their formerly dull machines," or that "during the past few years Detroit has been openly paying left-handed compliments to hot rodders and soup-up artists . . . by offering as optional equipment . . . soup-up additions [that] were developed and used by hot rodders many years before."[13] Contemporary and secondary sources confirm that this was the case—that Chevrolet, Chrysler, and Ford did indeed learn their 1950s horsepower tricks from hot rodders by observing their activities (and, in some cases, by hiring them as consultants and advisors, a critical point addressed later in this chapter).

However, although they were interested in the ways in which hot rodders squeezed additional power out of production engines, none of the mainstream manufacturers ever seriously contemplated going after the do-it-yourself, straight-line-speed-obsessed enthusiast's dollar—not in the mid- to late 1950s, at least. Instead, the OEMs were enamored of stock-car racing. For back in the 1950s, when the "stock" in "stock car" actually meant what it implies, the bragging rights (and sales) that accrued to those companies whose vehicles proved their mettle on the track were simply too tempting to pass up. Fortunately for the OEMs, NASCAR[14] rules allowed competing cars to be modified appropriately, but there was a catch: the parts used to modify them had to have an OEM part number. All of the special camshafts, manifolds, pistons, and so forth that went into stock-car Fords therefore had to be marked with a Ford part number, and they also had to be available through Ford's parts-supply channels.

These high-performance factory parts did not necessarily have to be *widely* available in order to qualify, however. Consequently, the OEMs quickly took to the practice of offering limited numbers of these add-on components through their official supply system, either as "export-only" or as "police cruiser" parts. Would-be racers could get them if they wanted to—and Ford, Chrysler, Chevrolet, Hudson, and the rest certainly made sure that serious teams racing with their brands got them—but the OEMs never advertised them to the general public.[15] Instead, your average Joe who witnessed triumphant Chevrolet victory after triumphant Chevrolet victory at the stock-car races could go over to his local bow-tie dealership and order a brand-new V8 Bel-Air equipped with a power pack consisting of high-performance parts and accessories more appropriate for ordinary, spirited street driving. Naturally, the competition on and off the track was fierce, and year after year standard and optional factory horsepower ratings rose to new heights. By 1958, in fact, the *lowest* advertised rating of *any* OEM V8 stood at 215, while the highest topped 400.[16] Hitched as it was to NASCAR's rising star, the horsepower race of the mid- to late 1950s swiftly carried passenger-car performance into territory only the most optimistic of rodders would have dreamed of ten years earlier.

Significantly, though, the cars in which these newer engines and power packs were available were not high-performance vehicles per se. That is, they were not the sorts of cars that were likely to command the attention and respect of dyed-in-the-wool hot rodders. For these were *ordinary cars*—four-door sedans, station wagons, and the like. Their overhead-valve V8s were of course interesting, but only insofar as enterprising rodders could envision pulling them from low-mileage, wrecked behemoths and dropping them into lightweight coupes, roadsters, and quarter-mile dragsters. What's more, the high-performance packages available for these 1950s family cars weren't over-the-counter, do-it-your-self kits, but rather factory-installed options available only in conjunction with a new-car purchase. In other words, the OEMs were not yet seriously attempting to lure die-hard enthusiasts from their prewar coupes and roadsters, and they were not yet making any concerted effort to bring them to their own parts counters either. Instead, the 1950s phase of the horsepower race aimed to woo the ordinary middle-class consumer with the promise of space-age speed and vicarious oval-track mastery.

And at this, it was a phenomenal success. So much so, in fact, that it began to provoke a backlash both within the industry itself and among a number of its critics. According to the popular historian Roger Huntington, "Ford and Chevy were spending millions on their NASCAR programs, really butting heads. Safety authorities were upset about the emphasis on speed and acceleration in advertis-ing and sales promotion. Insurance people were scrutinizing rates on sports and high-horsepower models. And worst of all, government officials in the regulation and anti-trust areas were watching the automakers closely."[17]

As a result, rumors began to circulate as early as the summer of 1957 that a pre-mature end to the horsepower race was imminent. "If current Detroit whisperings are accurate," *Motor Life* reported that July, "auto makers are on the verge of a 'tacit agreement' to discontinue emphasis on horsepower and top speed," which "would likely mean the end, or at least the curtailment, of factory participation in racing."[18] The whisperings were indeed accurate: that fall, the Automobile Manufacturers Association (AMA) enacted a resolution that "not only banned direct factory participation in racing, but [also] gently suggested the companies should stop emphasizing horsepower and performance in their advertising."[19] Seemingly, this is precisely what General Motors, Ford, Chrysler, Hudson, and the others did: their accountants severed their ties to NASCAR racing, and their advertising no longer bragged of victory laps or cubic-inch advantages, but rather of comfort, convenience, and, as the 1950s came to a close, the economic merits of their new compact cars.

In the long run, however, the AMA's so-called "racing ban" didn't stand a chance. For even as the OEMs proudly paraded their smaller and ostensibly more responsible Corvairs, Falcons, and Valiants before the public and the press, behind-the-scenes developments confirmed their ongoing commitment to triple-digit cruisers. Oldsmobile and Ford dropped their optional power packs in 1958, for example, but both continued to sell unadvertised "export" and "police" equip-ment to oval-track stars in 1958, 1959, and 1960. Pontiac, Chevrolet, and Chrysler, on the other hand, continued to offer not only limited numbers of these add-on components but also their factory-installed power packs.[20] And, thanks to a new generation of high-octane leaded fuels that the oil industry introduced at the time, static compression ratios on all domestic automobiles were on the rise as well.[21] What's more, Dodge began to dabble in drag-racing competition on an unofficial basis shortly after the AMA's racing ban went into effect, winking in assent as a group of its engineers formed the well-funded Ramchargers racing club.[22] Meanwhile, Ford unveiled an all-new line of larger, more powerful overhead-valve V8s in 1958, as did Chevrolet.[23] In theory, then, the OEMs may well have intended to curtail high-performance marketing with their resolution of 1957, but in practice, they never did.

Not altogether surprisingly, the racing ban sham was short-lived. It came to an end in 1960, when Ford upped the displacement of its big-block series of motors to 352 cubic inches and released a 360-horsepower power pack option.[24] Other manufacturers quickly followed suit. By 1962, Chevrolet's small-block V8 of 1955 had grown to 327 cubic inches, and its big-block V8, introduced at 348 cubic inches in 1958, had grown to 409. Buick also entered the fray with a 401-cubic-inch V8, and Oldsmobile's Rocket 88 swelled to 394. Not to be outdone, Ford brought out engines of 390 and 406 cubic inches by 1962, and Chrysler began to rework its Hemi V8 for racing applications.[25] Each of these companies also introduced improved power packs for their road-going engines, and several also jumped headlong into factory-sponsored racing.

According to Roger Huntington, however, this 1960s horsepower-race revival differed from its 1950s predecessor in at least three respects. The first involved the types of racing with which it was associated. In the 1960s, manufacturers like Pontiac actively supported and sponsored not only NASCAR-style stock-car racing, as they had in the 1950s, but also organized, NHRA-style drag racing.[26] In part, this was due to several changes that had taken place within drag-racing circles in the late 1950s, most notably the reemergence of the dual-use, high-performance car as enthusiasts began to race their power pack–equipped sedans at quarter-mile strips on the weekends while continuing to use them for their daily errands during the week. By the end of the decade, the NHRA had christened several special racing classes for these late-model "stock" and "super stock" bombs, classes that were tremendously popular.[27] Thus, when the OEMs openly returned to racing in the early 1960s, they no longer shunned the strips in favor of the ovals. Instead, they poured money into both.

The second way in which the 1960s horsepower race differed from that of the 1950s, Huntington reports, was that these new early-1960s OEM motors were "literally re-engineered for performance." In other words, whereas the high-output engines of the 1950s were essentially run-of-the-mill V8s that the engineers at Chrysler, Ford, and General Motors improved through the addition of bolt-on, hot-rod-style parts and accessories, the engines of the 1960s were actually designed from scratch as high-performance mills.[28] To an extent, Huntington is right, for high-compression pistons, fully-machined combustion chambers, reinforced connecting rods, counterweighted crankshafts, and thick-webbed heavy-duty engine blocks—all standard fare on 1960s performance engines—certainly weren't in the same league as the bolt-on manifolds, carburetors, and ignitions of the 1950s power packs. However, Huntington's implication that these re-engineered mills demonstrated the superiority of the "college-trained engineers" vis-à-vis the "intuitive designers who had come up through the ranks via the hot rodding sport" is more than a bit off the mark.[29] For indeed, whether they were engineered into these new mills from the get-go or not, every one of these performance enhancements had been standard practice among hot rodders for decades.[30] That the OEMs used them in their new-car engines was indeed a breakthrough, that is, but one of application, not conception.

Huntington's third point is right on the money though: in the early 1960s, the OEMs realized—and attempted to exploit—the potential of the market for do-it-yourself, add-on high-performance parts.[31] For above and beyond their factory-installed power packs, they began to advertise and sell over-the-counter, bolt-on performance-enhancing products for their engines. Dodge had a number of do-it-yourself components available by 1961, including special-ratio gears, ram-induction manifolds, high-lift camshafts, and high-tension valve springs.[32] Ford also had an extensive line of add-on parts available by the mid-1960s, including seventeen distinct packaged kits of bolt-on parts for all of its V8s.[33] In addition, some manufacturers' 1960s mills left room for parts-swapping with older models. Thus the owner of a late-1950s, 283-cubic-inch Chevrolet V8 optioned out with all of that company's power pack components could actually take them along, so to speak, if he chose to upgrade to an early-1960s, 327 V8 model, for all of the parts and accessories designed for the former would bolt directly to the latter without a fuss.[34] On the other hand, the owner of an early-1960s 327 V8 Chevrolet could easily install a 350 V8 crankshaft into his 327 block.[35] Between their do-it-yourself, bolt-on high-performance accessories and the inherent interchangeability of their 1950s and 60s V8s, the OEMs therefore began to compete with the speed equipment industry directly, luring traditional speed shop and mail-order devotees to *their* official parts departments.

In the mid-1960s, the mainstream industry also began to mass-produce a new type of automobile that was conceptually similar to the traditional hot rod itself. Compact, lightweight, and equipped with monstrous V8 engines, the muscle cars of the 1960s were specifically designed to appeal to those among the up-and-coming baby boomers who otherwise might have considered building more traditional hot rods. The first of these muscle cars appeared in 1964, when Pontiac dropped a 389-cubic-inch, 325 horsepower V8 into its compact Tempest, a combination the company marketed as the GTO (fig. 17).[36] Within another year, Plymouth and Dodge followed suit with their V8 Valiants and Darts, respectively, and by the late 1960s all of the major manufacturers had joined in as well. 426 Hemi Barracudas, 455 SS Chevelles, and 427 Fords soon ruled the roads: by 1970 there were no less than thirty-six such models available.[37]

The story of their evolution over the course of the 1960s is often gripping,[38] but for our purposes, what warrants emphasis is that the muscle car was the OEMs' long-overdue answer to the hot rod. Complete with warranties and easy financing, they were designed to win over would-be build-it-yourselfers with the promise of instant tire-smoking gratification. Consider the following pitch, which appeared in a full-page ad for the Dodge Charger R/T in the July 1969 issue of *Hot Rod*:

> There you are with your plug wrench clenched in a set of badly battered knuckles, wiping the other paw on the back of your jeans, when this black maw of a grille attached to a wingless Mach 2 jet throbs up. "Ha," you scoff. "Bet he has to beat it with a whip to even get it out of the garage." With a snick that can only mean close-coupled four-speed and a howl that says 440 cubes of mean, it disappears. Charger R/T just arrived. End of the road for the do-it-yourself kit, Charlie.[39]

The message was crystal clear: why waste your time, energy, and blood hopping up your present ride when you could obtain a brand-new and considerably faster Charger R/T from your local dealership? Not everyone bought into this sort of logic, of course, but hundreds of thousands of speed-obsessed young boomers did.[40]

During the 1950s and 60s, the OEMs began to compete with hot rodders, hot rodding, and hot rod parts manufacturers in a number of significant ways. Nevertheless, neither their large-displacement standard engines, their association with organized drag racing, their brand-name parts and accessories, their factory-built muscle cars, nor their easy financing and new-car warranties actually marked the end of the "do-it-yourself kit." Instead, in nearly every one of the OEMs' high-performance forays of the period, speed equipment manufacturers saw new opportunities for profits and prestige.

THE SPEED EQUIPMENT INDUSTRY AND THE HORSEPOWER RACE

By the end of 1953, when it was discontinued after several million examples and more than twenty years, Ford's flathead V8 had served as the technological foundation of American hot rodding and speed equipment manufacturing for more

Figure 17. A 1965 Pontiac GTO, equipped with a 335-horsepower, 389-cubic-inch V8. (From the collections of The Henry Ford.)

than fifteen years. In 1954, some enthusiasts and aftermarket companies quickly embraced its overhead-valve "Y-block" replacement,[41] and back in the late 1940s and the very early 50s, many others had already abandoned the flathead engine in favor of the postwar overhead-valve V8s available from other manufacturers. Still, heading into 1955, the long-out-of-date and now officially obsolete flathead remained the dominant powerplant within the hot rodding fraternity, and a clear successor had yet to emerge. For the first time since the early 1930s—and only the second since the late 1920s—the speed equipment industry therefore relied for the majority of its aggregate sales on the lingering popularity of a powerplant that was altogether out of production. In the short term, this was no great cause for concern: the flathead market remained strong, and especially in 1954 and 1955, no one whose business depended on the sale of high-performance parts for the old Ford V8 had any trouble paying the bills. In the long term, though, the industry needed to find a replacement—an affordable, hot new engine available in new-model cars. Those within the industry therefore kept their eyes peeled for the "next flathead," hoping not only that they'd recognize it if and when it came along but also that they'd be prepared to work with it.

Fortunately for them, the wait was brief. In 1955 Chevrolet brought out a 265-cubic-inch, overhead-valve V8 that developed 162 horsepower. Powerful, lightweight, and compact, the new mill seemed to have potential, and dozens of aftermarket manufacturers immediately set themselves to the task of unlocking it.

But it was not yet clear to anyone that this new powerplant would ultimately prove to be the standout among the postwar crop of overhead-valve mills. Not, that is, until the following January, when Vic Edelbrock and his son, Vic Jr., appeared on the cover of *Hot Rod* with a Chevrolet V8 that they had worked over to the tune of an astonishing 229 horsepower. That same issue of *Hot Rod* also carried a feature that waxed enthusiastic regarding the compact dimensions of the Chevy powerplant and its tremendous engine-swapping potential.[42] Within another six months, high-performance parts for the Chevrolet V8 were available from no less than forty individual firms, and rodders and equipment manufacturers alike were certain that they had found the "next flathead." They were right: to this day, the small-block Chevrolet V8 and its many derivatives continue to dominate traditional, V8-oriented hot rodding in the United States.

Not every aftermarket company enthusiastically embraced the new mill. Barney Navarro's firm continued to specialize in flathead gear, for example, as did those of Jim Kurten, Earl Evans, and several others. But in time their businesses declined as the demand for flathead components slipped.[43] The market for flathead V8 parts did not collapse altogether, of course, and Edelbrock, Weiand, Offenhauser, and a number of other companies continued to profitably produce L-head components for years to come.[44] But by the early 1960s, it was no longer possible to base an aftermarket company entirely on the old Ford mill, as Navarro, Kurten, and Evans ultimately learned.

For every Southern California company that did not make the switch to the new Chevrolet V8, however, there were at least five or six that did. Thus, for the first time in the brief history of aftermarket manufacturing in the United States, the overwhelming majority of extant firms survived a major, OEM-level transition intact. This had not been the case in 1927, when Ford dropped the Model T in favor of the Model A, and neither had it been the case with the early 1930s advent of the V8. But in 1955 and 1956, the transition was smooth and swift, and the Southern California companies that had risen to dominance during the late 1930s and the 1940s remained on top.

Simply joining the small-block revolution was not enough, though. For the mainstream industry's ongoing horsepower race still threatened, with its optional power packs and "police" equipment, to leave the aftermarket in its dust. What faced the speed equipment industry by the beginning of 1956 was therefore no longer the challenge of withstanding the flathead's decline, but rather that of meeting the OEMs head-on. Doing so—remaining competitive vis-à-vis the mainstream automobile manufacturers of the 1950s—required a renewed focus among speed equipment companies on their core competency: finding a way to wring more horsepower and torque out of standard production engines. This is precisely what Edelbrock and his son did in the winter of 1955–56 with their first V8 Chevrolet build-up. After all, the 229 horsepower they achieved greatly exceeded not only the standard engine's 162-horsepower rating, but also that of Chevrolet's optional 180-horsepower power pack. And as the mainstream industry's horsepower race escalated, speed equipment manufacturers continued to offer similar performance gains to power-hungry motorists, even when OEM horsepower levels approached 300. For indeed, as Louis Hochman observed in a popular how-to manual in 1958, "no matter how good an engine is, there's always room for improvement."[45] Thus, whether through dual exhausts, triple-carb manifolds, long-duration camshafts, or reworked heads, the key for speed equipment manufacturers was to locate and tap the additional power hidden within each of the OEMs' hot mills. That they were able to do so in the 1950s proved that the OEMs had yet to fully decipher the hot rodding fraternity's many secrets.

But in the 1960s, the mainstream industry began to do just that. Larger, better-engineered high-performance engines with an array of optional, factory-installed performance upgrades and even hotter do-it-yourself add-ons threatened to diminish aftermarket opportunities and win away would-be rodders. Most speed equipment manufacturers were undaunted, however, and they managed to stay on message: *there's always room for improvement*. Writing in 1963, for example, *Popular Hot Rodding*'s Jerry McGuire insisted to his audience of ordinary rodders that "factory assembly methods will never be able to produce engines with as much brute horsepower and torque—cubic inch for cubic inch—as you can build up by careful hand work in your own garage, using special speed equipment from the many commercial hot rod suppliers."[46] Mass-produced compromise would always be the hallmark of the OEMs, that is, and thus the independent speed equipment industry would always be able to produce bolt-on improvements. Dennis Pierce of Petersen Publishing's *Hot Rod Industry News* struck a similar chord while preaching to the choir in 1967, reminding aftermarket leaders that their industry "is made up of men who have a desire to improvise, improve, invent and individualize on an existing product" in the belief that "Detroit did a nice job, but they didn't quite finish it."[47]

Others, perhaps less optimistic that aftermarket products would continue to trump OEM muscle in the long run, nevertheless maintained that theirs was certain to be a prosperous future. "As Detroit has borrowed from our bag of tricks and transformed their ugly ducklings into beautiful swans," Ray Brock wrote in the April 1967 issue of *Hot Rod Industry News*, "we have been carried along by their momentum and have had more products to create and ultimately more items to sell to an ever-increasing clientele."[48] In other words, the pie was now larger: OEM performance advertising had cultivated a growing interest in performance motoring, particularly among the up-and-coming baby boomers, and this in turn had helped to *increase* equipment manufacturers' opportunities, sales, and profits.[49] By the end of the 1960s, all of these justifications, explanations, and words of encouragement coalesced into something of a standard aftermarket rallying cry, heard whenever errant naysayers within the industry criticized the OEMs for horning in on "their" turf: the mainstream industry's involvement in the high-performance sector amounted to free publicity for the speed equipment industry, and there would always be a place for specialty manufacturers and speed shops, because no matter how high Chevrolet, Ford, and Chrysler might eventually raise the bar, there would always be room for improvements.[50]

Fortunately for Edelbrock, Iskenderian, and the rest, locating opportunities for improvement was usually easy. For every brand-new muscle car and high-performance-optioned sedan they moved in the 1960s, the mainstream industry sold at least four or five plain-Jane models and one or two low-performance, muscle-car "look-alikes."[51] Often, these cars were but an intake manifold, carburetor, camshaft, and exhaust system away from achieving a level of performance comparable with that of their more expensive brethren. Chevrolet's expensive, top-of-the-line Z-28 Camaro, for example, could be duplicated by enthusiasts of ordinary means by purchasing a run-of-the-mill Camaro and buying separately the parts and accessories needed to transform it into a Z-28 clone—if, that is, said enthusiast avoided the pricey factory parts books and obtained his high-performance add-ons from the more reasonable speed-shop counter or mail-order aftermarket catalog.[52]

Although rising insurance premiums began to price many a would-be owner out of the muscle-car market entirely toward the end of the 1960s, aftermarket add-ons could serve as rate-dodging loopholes. "Get the low-horsepower V8 and skinny tires," Ray Brock advised *Hot Rod's* readers in September 1970, "so that the insurance companies will set their rates on your driving record rather than what you are driving." Then, "after you are insured, . . . go ahead and put on the mag wheels and fat tires, big four-barrel carburetor—and whatever else you had in mind in the first place."[53] Often precisely because of the appeal of out-of-reach OEM supercars, owners of low-priced, insurance-friendly "look-alikes" and ordinary sedans thus were readymade sales for 1960s speed equipment companies.

However, aftermarket manufacturers were by no means limited to those the OEMs had failed to sell or had otherwise been unable to accommodate, for the owners of high-performance sedans and muscle cars—even those with top-notch 4-4-2 Oldsmobiles, Shelby Mustangs, and Z-28 Camaros—were potential customers as well. Finding ways to improve these high-horsepower beasts could be difficult, but more often than not, talented speed equipment gurus were able to unearth a trick or two that the OEMs had missed. The key, in most cases, was to have a keen eye either for minute detail or for the unusual; successful manufacturers often had both. Don Alderson of Milodon Engineering, for example, came up with a unique angled oil pickup tube in 1968 that swiveled through a 100-degree arc, virtually eliminating engine oil starvation by enabling its intake end to remain in contact with the oil pan's supply even when the car was cornering at the limit or accelerating hard.[54] Likewise, Fred Offenhauser developed an unusual reconfigurable intake manifold in 1969 that allowed owners of late-model, dual-use Chevrolet V8s to switch between an efficient single-carburetor street setup and a more-powerful multiple-carburetor strip combination in a matter of minutes.[55] For its part, B&M Automotive pioneered the art of tuning automatic transmissions, elevating the much-maligned slushbox into a state of relative parity, performance-wise, with the three- and four-speed manuals of the period.[56] Others devised transistor ignitions, gear-driven camshaft assemblies, super-lightweight rockers, precision exhaust headers, fuel-injection systems, and other enhancements in the 1960s, each and every one of which addressed a particular area the OEMs had missed or otherwise neglected on their top-notch models.[57]

In addition to having a fundamentally sound idea, speed equipment manufacturers also needed to be quick on the draw during the 1960s in order to compete with the OEMs. This was especially true in fields like performance camshaft and intake manifold manufacturing, for if a company that specialized in either of these areas were to be slow to bring their new designs to market, they would lose a lot of sales to dealership parts counters. After all, why should the owner of an all-new Ford or Chevrolet wait for an aftermarket cam or manifold for his new car when he could get an OEM one right away? The problem for aftermarket companies, of course, was that new-part research and development not only took a great deal of time, but it also required access to the engines and/or vehicles for which the new parts were intended. Since very little could be done about the latter—only very rarely did the OEMs make their new mills available for advance aftermarket research and development[58]—most speed equipment manufacturers focused instead on streamlining their new-product programs so that at the very least their cams, manifolds, exhaust systems, and the like would reach the market in a reasonably timely fashion. Iskenderian devised a system of OEM-powerplant evaluation that devoted part of every working day to new-mill dynamometer testing, for example, so that Ed and his technicians would have all of the information necessary to produce performance-enhancing camshafts for any new OEM engine that proved to be popular among enthusiasts.[59] Other companies hired additional machinists and engineers, setting them to work in dedicated research and development facilities so that they would be prepared, at a moment's notice, to supply performance parts for whatever might turn out to be the next big thing.

By 1970 most of them had it down to a science, their new-product releases often coinciding with the OEMs' own new-car launches.[60]

Another key to aftermarket success in the 1960s was for individual speed equipment manufacturers to focus on a particular niche. Niche marketing was by no means universal, of course, for a number of companies were by then well diversified. Neither was it altogether new, for Iskenderian had always focused exclusively on camshafts and valvetrain components, for example, and Offenhauser on intake manifolding, Venolia on pistons, the Crankshaft Company on stroker cranks, and so forth. But in a broader sense, niche marketing within the high-performance industry of the 1960s centered increasingly, not on the *type* of product one produced, but rather on its *application*. Consider those that manufactured parts strictly for all-out racing. Some, such as Stuart Hilborn's Fuel Injection Engineering, were established aftermarket companies that had always focused on the racing-only market.[61] Many more, such as Schiefer Manufacturing and the Crankshaft Company, were older firms that had originally produced parts for a variety of applications but had begun, by the 1960s, to home in on the racing-only scene.[62] Others, including Ed Pink, Donovan, Tubular Automotive, Lakewood Chassis, and Simpson Safety Equipment, were altogether new firms founded in the 1960s that focused on the racing-only market from the outset.[63] Other niches, such as off-roading and V8 street performance, were equally strong and equally well-served by aftermarket specialists.

One final key to success in the speed equipment industry during the period in question is what we might call the "personal touch"—the active promotion of one's business as one that offered genuine one-on-one service and advice. This was especially true for smaller speed shops, who in the 1960s found themselves in the unenviable position of having to compete not only with large mail-order houses and distribution chains but also with the OEMs' own high-performance counters.[64] For many equipment manufacturers, the personal touch was no less important. Norris Baronian built his tiny camshaft business on his reputation as a custom "cam man" for whom quality and personal service trumped outright growth and dollar volume.[65] The same was true of Jack Engle, who deliberately controlled his company's growth during the 1960s in order to remain flexible and custom-oriented.[66] Even Edelbrock, an enormous enterprise, maintained a number of custom operations in order to stay true to its tradition of personalized, hands-on craftsmanship, as did many others.[67] To be sure, there were at least a few aftermarket companies that sought to secure a competitive advantage by adopting the mass-production dogma of the OEMs, but they were the exception: most aftermarket manufacturers clung tenaciously to their reputation as specialists.[68]

By emphasizing their specialty status, focusing on their particular niche, streamlining their research and development programs, and falling back on what they had always done well—finding ways to improve mass-produced automobiles, no matter how hot—the speed equipment manufacturers of the late 1950s and 1960s more than held their own. Indeed, they thrived as never before, and the more the OEMs advertised their own high-performance models, the better the aftermarket did. In fact, the "free publicity" of the period actually stoked the buying public's interest in high-performance automobility to such an extent that it opened a number of altogether new markets for speed equipment. Thus, serious aftermarket high-performance research, development, and marketing for automobiles that were neither sporty nor equipped with large-displacement V8 engines began in earnest.

Consider the domestic compacts of the late 1950s and early 1960s. First introduced during the AMA's racing ban, these smaller automobiles were predominantly aimed at the consumer less interested in garish ornamentation than in value. Equipped for the most part with inline four- or six-cylinder engines, these Tempests, Valiants, Falcons, and Darts were built for economy, not speed. Nevertheless, shortly after their introduction, a number of hot rodders began to experiment with them, and it wasn't long before equipment manufacturers offered camshafts, intake manifolds, ignitions, and exhaust headers for them.[69] But in the mid-1960s, the OEMs began to use these vehicles as the basis of their new muscle cars. And as Tempests, Falcons, and Valiants gave way to GTOs, Mustangs, and Chargers, the bottom largely dropped out of the high-performance compact aftermarket.

Chevrolet's Corvair was the exception. Introduced in 1959, the Corvair was an unusual automobile that featured a rear-mounted, air-cooled, six-cylinder engine built for utility and economy. Because of its unconventional design, the Corvair was not well suited to the "shoehorn-in-a-large-V8" strategy that transformed other compacts into muscle cars, although Chevrolet did produce a turbocharged version to satisfy those buyers who craved a few more horses.[70] But for the most part, the Corvair remained a popular economy car well into the 1960s, Ralph Nader's reservations notwithstanding. It also remained popular among hot rodders and speed equipment manufacturers.[71]

The Corvair's motor wasn't particularly easy to work with, though. Its intake manifolds were cast as single units integral with its cylinder heads, which necessitated complex cut-and-weld procedures even for something as simple as fitting additional carburetors. Similarly, clearances within its engine case were close, complicating the addition of a stroker crank or larger-diameter pistons. Nevertheless, a variety of aftermarket components were available for the engine during the 1960s. Camshafts could be procured from most of the major manufacturers, as could mufflers, ignitions, and rocker assemblies. Stroker cranks were also available from adventurous firms like Weber, and several companies also offered turbos and/or superchargers for the air-cooled mill.[72] For in spite of its disdain for convention, or perhaps precisely because of it, the Corvair was a profitable market niche for many within the speed equipment industry throughout the 1960s and well into the 70s.

In fact, apart from domestic V8s and muscle cars, the Corvair's popularity among performance enthusiasts and equipment manufacturers was eclipsed only by that of another unlikely candidate for high-performance tuning, the Volkswagen. Every bit as unusual as the Corvair both mechanically and aesthetically, the lowly two-door Volkswagen Beetle had risen from the ashes of the war-torn German state of Lower Saxony to become the leading imported car in the American market by the end of the 1950s. Like the Corvair, the Volkswagen featured a rear-mounted, air-cooled engine, but in the Beetle, it was a tiny, single-carbureted, four-cylinder unit that displaced well under 100 cubic inches and developed, at its evolutionary peak, but 60 horsepower.[73] Designed with longevity and fuel economy in mind, the Volkswagen was about as unassuming a vehicle as could be imagined. Yet by 1957 the little car had attracted a handful of American boosters who promoted it as a sound and reasonable basis for a high-performance build-up.[74] And within another ten to twelve years, in spite of the German company's strict policy forbidding its modification on pain of new-car warranty forfeiture, the Volkswagen became just that for thousands of American performance enthusiasts.[75] Looking back on the car's phenomenal rise as a high-performance player from the vantage point of 1969, however, Lee Kelley noted in *Popular Hot Rodding* that the Beetle's popularity was due, not to the efforts of boosters or equipment manufacturers, but rather to the inherent appeal of its tiny motor. Cheap, plentiful, easy to work on, interchangeable year-to-year, and responsive to simple bolt-ons, the flat four, Kelly maintained, was not unlike the popular small-block Chevrolet V8—conceptually, at least.[76]

Indeed, its short-stroke, big-bore configuration gave the Volkswagen's motor tremendous top-end potential, but as delivered, its undersized carburetor, circuitous intake manifold, small valves, low compression, conservative cam, restrictive exhaust, and low displacement limited its real-world capabilities— and this was where the speed equipment industry stepped in. Many established manufacturers offered high-lift camshafts, stroker cranks, large-diameter piston and cylinder kits, and exhaust headers for the Beetle's engine in the 1960s.[77] In addition, a number of altogether new companies joined the industry during the late 1950s and the 1960s explicitly as Volkswagen specialists. European Motor Parts Incorporated (EMPI) of Riverside, California, sold dual carburetor kits, suspension components, exhaust systems, and even complete, highly modified

Beetles.[78] Shoemaker of Long Beach focused on the manufacture of performance exhaust headers and mufflers for the Volkswagen, while Revmaster in Riverside and Deano Dyno-Soars of Orange County cultivated esteemed reputations in the Beetle performance-engine business.[79] Another Southern California firm, Scat, sold dual carburetors and exhaust systems as well, but they also manufactured stroker cranks, camshafts, and piston and cylinder kits.[80] Several European companies, including Okrasa of West Germany and British Racing Motors (BRM) of England, also produced performance parts for the little car in the late 1950s and 1960s, though both sold them in the United States through domestic companies rather than through North American bases of their own.[81]

What exactly could an owner expect if he plunked down the requisite cash, broke out his hand-tools, and modified his Volkswagen? Anywhere from 25 to 100 percent more power, according to period sources.[82] Moreover, for those willing to remove their engines and send them, say, to Revmaster or to Deano for a complete performance rebuild, horsepower gains in excess of 200 percent were possible. Hence the tremendous popularity of the Volkswagen among enthusiasts, particularly those on a budget and those looking to learn the wrench-turning tricks of the trade on a relatively simple "starter car."

❦

Together with its top-end, late-model V8 speed equipment and its racing-only parts and accessories, sales of add-on components for VWs, domestic compacts, imports, off-roaders, sports cars, and other new markets helped ensure continued growth and prosperity for the high-performance aftermarket during the 1960s. In the end, that is, it didn't really matter to the Vic Edelbrocks and Fred Offenhausers of the world that the OEMs had gotten into the high-performance field for themselves. For indeed, at the very worst, speed equipment manufacturers lost a tiny portion of their V8 sales to the Javelins, Camaros, and Barracudas of the period, while the OEMs' relentless performance-oriented advertising helped create a more generalized and widespread interest in high-performance automobility that ultimately brought more customers and sales to specialty firms across the United States. Convinced that there would always be room for improvement no matter what the OEMs produced, aftermarket companies simply embraced the opportunities that came their way, whether they involved 425-horsepower domestic V8s or 53-horsepower German flat fours. And in so doing, they more than held their own.

However, if we dwell for too long on the ways in which the aftermarket *responded* to the challenge of OEM competition during this period, we run the risk of overlooking a critical element of the story. For adversarial though their relationship often was during the late 1950s and the 1960s, speed equipment manufacturers and mainstream automotive firms weren't always at odds. Throughout the period in question, dozens of aftermarket companies and personalities lent their expertise to the OEMs in a number of ways. Zora Arkus-Duntov, who was hired out of the speed equipment business in 1953 to head General Motors' fledgling high-performance team, was but the most famous of the many individuals that the OEMs either hired away from the aftermarket or tapped on a periodic basis as consultants. In the late 1940s, Ford hired Los Angeles camshaft grinder Clay Smith to assist with some of its racing-oriented special projects. And in 1952, Ford also tapped the expertise of Bill Spalding, who traded his small Southern California camshaft business for what turned out to be a twenty-three-year career at the Dearborn firm. Later, Ed Iskenderian and Ed Winfield consulted with several mainstream companies, helping Chevrolet to develop its first stock-car-racing camshafts, for example.[83] Right up to the onset of the AMA "racing ban" in 1957, these and other aftermarket specialists contributed to the OEM horsepower race.

According to Roger Huntington, however, the mainstream industry no longer required the assistance of these "intuitive designers" once the horsepower race resumed in the early 1960s. For with their teams of college-educated engineers and their cutting-edge testing gadgetry, he explains, their enhanced research and development capabilities rendered obsolete the advice of these old-school rodders.[84] But this is simply not the case. Ford hired Ak Miller as a "performance advisor" in the mid-1960s, for example, tapping his expertise to help develop high-performance systems for its smaller, six-cylinder mills. In addition, Fran Hernandez of Edelbrock joined Ford's high-performance team in 1966, and his influence within the company won Ed Pink a number of racing-engine contracts with Ford during the late 1960s and early 1970s. Ford was not alone: Chrysler hired Keith Black to head its West Coast marine division, while American Motors secured the help of Barney Navarro for its late-1960s Indy program.[85] Above and beyond these direct hires and consultancies, every American OEM hired aftermarket companies to perform high-performance research and development, to produce add-on parts and accessories, or to join in the marketing and sale of performance-oriented products during the 1960s muscle-car boom.

Indeed, Huntington admits as much. Edelbrock, he reports, not only sold its hi-riser ram-effect intake manifolds to Chrysler for it to use in its optional 440 Magnum package in the mid-1960s, but it also supplied add-on manifolds stamped with American Motors parts numbers to that firm in 1970. Vic Jr. and his employees also worked with American Motors on its Rambler racer in 1968.[86] But the phenomenon wasn't limited to Edelbrock. American Motors also teamed up with Grant Piston Rings in 1967, delegating the task of assembling a supercharged funny car to that Southern California firm.[87] In addition, American Motors entered into production and marketing agreements with Iskenderian, Doug's Headers, and Offenhauser in the late 1960s, as did Mickey Thompson with Buick and Pontiac, Sig Erson with Buick, and George Hurst with Oldsmobile.[88] For its part, Ford hired Carroll Shelby, an aftermarket tuner who joined the industry with the assistance of Dean Moon in the early 1960s, as its official performance-parts manufacturer during the muscle car era.[89] On the research and development front, Bob and Don Spar of B&M Automotive worked with Chevrolet on an on-again, off-again basis during the 1960s, as did Gary Hooker of Hooker Headers (with Chrysler) and Joe Mondello of Mondello's Porting Service (with Oldsmobile).[90]

Both by competing with the OEMs and by cooperating with them, the speed equipment industry helped bring high-performance motoring into the American mainstream in the late 1950s and the 1960s, prospering as never before. However, speed equipment manufacturing during these years wasn't all about the Southern California industry's relationship(s) with the OEMs, adversarial, cooperative, or otherwise. For as the aftermarket closed in on the $1-billion-a-year mark, a number of important internal developments unfolded as well, changes in the ways in which the industry conducted its affairs that were every bit as consequential in the long run as were power packs, muscle cars, and factory racing programs.

CHAPTER SIX

BOLT-ON POWER, 1955-1970

In 1955, prior to the OEMs' initial foray into the high-performance market, there were fewer than 200 speed equipment manufacturers in the United States. Fifteen years later, there were exactly 750, many of which were located outside of California.[1] Writing in 1969, the publisher of *Hot Rod*, Ray Brock, reported, "Eighty percent of the nation's population is east of the Rockies and that's where we sell 80% of our magazines. The same with speed equipment manufacturers."[2] By the end of the 1960s, that is, most hot rodders lived back East, and that's where *Hot Rod* sent most of its issues and the aftermarket most of its products. More than 75 percent of the speed shops, retail outlets, and mail order houses that dealt in performance-oriented products were based in the East too.[3] However, although the geographic distribution of the *retail* end of the industry came to more closely match that of performance enthusiasts during the 1950s and 60s, the *manufacturing* end did not.

To be sure, California's share of the industry had declined steadily since the war, falling from more than 80 percent in 1948 to less than 50 by the end of the 1960s.[4] But even in 1970, California's dominance was undeniable: home to 46 percent of America's speed equipment manufacturers, California trumped second-ranked Illinois by nearly 40 percent. In fact, against the shares of its six closest rivals combined, California still held a commanding advantage: Illinois, Ohio, Michigan, New York, New Jersey, and Pennsylvania together accounted for only 31 percent.[5] Moreover, if we narrow "California" down to the Los Angeles area alone, we find that in 1970, 38 percent of the entire American speed equipment industry was based in Southern California.[6] By the end of the 1960s, in other words, performance enthusiasm and performance-parts retailing both had spread throughout the United States so thoroughly that their geographic distribution closely matched that of the overall population, but the same was simply not the case for those that *manufactured* high-performance products.

Within Southern California, the industry did spread out during the late 1950s and the 1960s. In 1954, there were only a handful of companies in greater Los Angeles that were not within a twenty- to twenty-five-mile radius of the city's center. By 1970, though, in addition to the 235 companies located in the immediate vicinity of L.A.—defined here, however loosely, as the area encompassing Los Angeles, southeastern Ventura, and western San Bernardino counties—there were 37 more in suburban Orange County and 12 in Riverside.[7] Some of the more outlying firms were new, such as Revmaster and EMPI of Riverside and B.F. Meyers and Deano Dyno-Soars of central Orange County. Others were older, established firms whose leaders sought either to expand their manufacturing facilities by taking advantage of the cheaper industrial real estate then in abundance in more distant areas or simply to escape the crowded core. Weber, for example, an anchor on Whiteside Avenue in Los Angeles for nearly fifteen years, moved to Orange County in 1960 so that it could expand into larger quarters.[8] During the 1960s, several others did the same. Stuart Hilborn, on the other hand, moved his Fuel Injection Engineering company from Los Angeles to secluded Laguna Beach in Southern Orange County in 1964 specifically to trade smoggy skies for ocean views without having to sacrifice the advantages—particularly in terms of shipping and receiving—that relative proximity to L.A. held.[9]

That these companies fled to outlying areas during the 1960s—or, more generally, that an ever-greater portion of the region's speed equipment manufacturers, new and old, were based in Orange and Riverside Counties at the time—ought to come as no surprise. For in the 1950s and 60s, suburban towns that once served as bedroom communities for those who worked in downtown L.A. gradually morphed into large, decentralized "post-suburban" zones largely independent from the urban core itself.[10] And as they did, these outlying areas attracted manufacturers like Fuel Injection Engineering and Weber.

Whether located in downtown Los Angeles, in (post)suburban Orange County, or in a midwestern city like Chicago, high-performance manufacturers of the late 1950s and the 1960s grew at a phenomenal pace not only in terms of their collective numerical strength and aggregate annual sales but also individually. Period reports teemed with announcements to the effect that company A had moved into larger quarters or that company B had annexed a neighboring building. Not all of them expanded during the period in question, of course, but most did—in fact, many did so two, three, and even four times in the span of a few short years.

Significantly, too, not a single report of floor-space contraction or company down-sizing appeared at any time during the late 1950s and the 1960s.

In 1965, for example, the six-year-old Hooker Headers company of Ontario, California, burned to the ground, providing owner Gary Hooker and his partner Bill Casler with an opportunity to expand as they rebuilt.[11] Within two years, the company had already outgrown its new facility, forcing Hooker and Casler to move into a brand-new, 22,000-square-foot building across the street.[12] By the end of 1968, continuing growth necessitated the addition of 29,000 square feet to its still-new plant, but even this was insufficient. Less than a year later, construction was underway again at Hooker, as Gary and Bill oversaw the addition of 17,000 more square feet.[13] Prosperity also blessed Lakewood Industries of Ohio. In 1959 the company began as a basement operation in the town of Lakewood, but strong demand quickly prompted its founder, Joe Schubeck, to move into a larger two-car garage nearby. Steady growth soon enabled Schubeck and his partner, Bill Steiskal, to move into a 3,500-square-foot facility in Lakewood, where several in-novations in the design and manufacture of racing bellhousings led to explosive growth for the rest of the 1960s.[14] In 1971 Lakewood moved into a new plant nearly ten times larger than the one it had occupied nine years earlier, a 30,000-square-foot monster in Cleveland.[15] Likewise, Crower, a San Diego manufacturer new to the industry in the early 1950s, moved out of a tiny workshop and into an 8,500-square-foot facility in 1962, only to find it necessary to add 10,000 more four years later. Rapid growth during the late 1960s (and into the early 1970s) neces-sitated several more additions of a more haphazard nature, and by 1975 the com-pany had added on here and there so frequently that no one at the plant could say for sure just how large the place actually was.[16] Many others had similar 1960s plant-expansion experiences.

In nearly every case, however, the addition of manufacturing, research and development, and warehousing space at these companies in the 1950s and 60s amounted to an increase in the *scale* of their operations; only very rarely were these moves part of a more comprehensive change in *strategy*. In other words, equipment manufacturers across the country responded to the rapidly rising de-mand for their products during the period in question, not by moving toward mass production, but rather by adding more floor space and filling it with more general-purpose lathes, drill presses, boring machines, planers, hand grinders, and skilled machinists.[17] There were a handful of exceptions, though. Fenton's 90,000-square-foot manufacturing center used a number of automatic welders, tube-benders, drill presses, polishing machines, and unskilled attendants to turn out massive quantities of its run-of-the-mill wares during the 1960s, as did Cy-clone's 40,000-square-foot plant.[18] But for most, flexibility remained the rule.

Machining and assembly work was often done by hand, but in some cases multiple drill presses, double boring machines, and other more specialized—but not single-purpose—tools enabled larger batches of popular items to be produced more efficiently.[19] The trick in either case was to be prepared for the unexpected—to be prepared, that is, for the sorts of dramatic shifts in demand that might well make today's low-volume, custom-oriented products tomorrow's big sellers. And although few within the industry ever really mastered the art of long-term forecasting in a market as volatile as theirs, their general-purpose tools and skilled staffs allowed the unpredictable and the unforeseeable to be managed.

This is not to say that process refinement within the industry ground to a halt. At Iskenderian, for example, servicing the growing number of camshaft-grinders in use had, by the mid-1960s, become an onerous and time-consuming task. Even with regular fluid changes, the machines were designed to recycle the same coolant over and over again, which meant that after the first cam was ground on a freshly serviced machine, the recycled fluid would be contaminated with metal shavings. Isky and his technicians therefore came up with a novel solution: they installed a maze of interconnected pipes that linked each machine to a large, centralized coolant tank and filter. This ensured that Isky's machines would last longer and that his camshafts were ground in pure, fresh coolant every time.[20]

Process refinements at Schiefer, Crower, and Lakewood had less to do with machine-tool longevity than with quality control. In the late 1950s, Paul Schiefer developed (and later patented) a hard-facing process for his clutches and flywheels that reduced in-use friction, thereby lowering operating temperatures and all but eliminating the risk of warpage. Wear and tear was also on Bruce Crower's mind when he adopted a patented camshaft-facing process known as "Tuff-Tiding" in the late 1960s. But at Lakewood Industries, quality control was all about customer safety. During the 1960s, Lakewood became the first within the industry to apply a process known as "hydroforming" to the production of its metal bellhousings, which reduced the risk of catastrophic failure by ensuring that the housings were of a uniform thickness. Later, Lakewood also built an explosion test lab at its Cleveland plant in which its technicians spot-checked sample bellhousings from the assembly area in a controlled environment.[21]

For the most part, whether developed with safety, quality control, or machine-tool longevity in mind, these refinements found their way onto shop floors still firmly committed to flexibility. They enabled better-quality goods to be produced in a range of volumes, but they did not revolutionize the ways in which most of the firms within the industry conducted their affairs. Also, they typically were not used to produce new *types* of products: camshafts, crankshafts, headers, mufflers, manifolds, pistons, and ignitions remained the aftermarket's stock in trade. Thus, the industry's role—producing add-on parts for high-performance tuning—remained unchanged throughout the period in question. And since the late 1950s and the 1960s witnessed very little in the way of fundamental changes in automotive design at the OEM level, the same was true, by extension, at the level of the aftermarket. There were refinements, of course—360-degree intake manifolds, transistor ignitions, and roller-bearing crankshafts, to name but a few—but most of the products of the period were decidedly old-school.

Noteworthy exceptions included high-performance automatic transmissions, turbochargers, and fuel-injection systems. A more complete discussion of the first of these appears later in this chapter. As for the latter two, both illustrate the ways in which the aftermarket often led the OEMs in the development of new automotive technologies. A turbocharger, also known as a "turbo-supercharger," or simply a "turbo," is a type of supercharger that uses an engine's exhaust gases to drive an impeller that feeds additional air to the motor. As with standard belt-driven superchargers, turbos enable—or, more accurately, *force*—internal combustion engines to produce more power by *forcing* them to consume more air and fuel. However, whereas mechanical superchargers provide linear boosts in engine performance, turbos instead produce an exponential power curve. Theoretically, that is, the more horsepower a turbocharged engine makes, the more it is able to make.

First applied en masse to airplane motors during the Second World War, turbos only became common on land vehicles in the 1950s, when diesel truck companies began to use them to enhance the poor breathing of compression-ignition engines.[22] In 1960, however, the ever-resourceful Barney Navarro published an article in *Motor Life* about an experiment he conducted with a diesel turbocharger and a Chevrolet Corvair. The uniquely modified engine that resulted from his efforts made a lot more power, and Navarro boldly prophesized that the turbocharger "promises to be tomorrow's most highly touted hop-up accessory."[23] He was right. Two years later, Chevrolet and Oldsmobile both introduced turbocharged automobiles, the Corvair Spyder and the Cutlass F-85; but not until the 1980s would the OEMs apply this technology in a more widespread, routine manner.[24] In the meantime, aftermarket companies like Bell Auto, Rajay, and AiResearch filled the gap, spearheading turbo research and development and making turbocharger kits available to rodders who were after something a little bit different.[25]

The story of the emergence of automotive fuel injection systems, aftermarket or otherwise, is considerably more complex. This is because definitive links in the web of ideas, patents, and engineering influences that led to its widespread application in the 1970s and 80s are virtually impossible to establish. Fuel injection is a process that uses one or another means of pressurization to deliver vaporized fuel to an engine, rather than relying (as with carburetors) on intake vacuum to accomplish the task. In 1939 and 1946, respectively, aftermarket gurus Ed Winfield and Stuart Hilborn both received U.S. patents for pressurized injection processes, but only Hilborn was able to translate his idea into a marketable system.[26] During the late 1940s and the 1950s, Hilborn's racing-only "constant flow" injection set-up was a huge success, powering many dry-lake streamliners and quarter-mile dragsters and bringing fame and fortune to Hilborn and his company, Fuel Injection Engineering.

During the early 1950s, however, a number of OEM suppliers both in the United States and abroad developed fuel injection systems more appropriate for everyday street use. Then, in 1957, Chevrolet became the first American OEM to bring the new technology to market with its optional Rochester system.[27] Inspired, a number of speed equipment firms soon joined Hilborn in the fuel injection aftermarket, and a flurry of activity within the field during the early 1960s brought workable, if not affordable, street-use systems to the American market.[28] Meanwhile, Mercedes-Benz and Volkswagen worked with Bosch to bring out more advanced electronically controlled injection systems toward the end of the 1960s.[29] By the 1980s, computer-controlled systems were common, and by the 1990s, they were universal among American-market automobiles.

Aftermarket companies and OEMs both contributed to the development of automotive fuel injection, but the relative importance of their contributions has long been the source of much debate. We know, for example, that Ford learned a great deal by assigning Ak Miller to the task of reverse-engineering Mercedes-Benz's six-cylinder Bosch injection system in the 1960s.[30] Equally apparent is the fact that Chevrolet's introduction of the Rochester system in 1957 led to a considerable amount of fuel injection research and development in Southern California. But what, if anything, did American giants like General Motors, Ford, and Chrysler learn from their diminutive counterparts? The fact that Chevrolet's Rochester system hit the market one year after Ed Winfield's 1939 patent expired and just a few years before Stuart Hilborn's was set to do the same led some at the time and many more since to charge that Chevrolet did little more than apply the basic elements of one or another of these rodders' systems to their production engines. Writing in 1959, for example, Hot Rod's Bob Pendergast explained that the technology behind Chevrolet's 1957 system surprised no one in the rodding field, since "rods had [long] been the testing device for a form of fuel injection that had proven the practicability of the basic principle used in the Rochester," namely, Stuart Hilborn's constant-flow idea.[31] Ed Almquist agrees, charging in his collection of biographical sketches that "Chevrolet . . . copied the Hilborn system—

adding only electronic controls."[32] Tom Medley, on the other hand, maintains in the first of his two volumes on hot rodding that it was Winfield's more complex design that actually influenced the Rochester.[33]

Period evidence from Chevrolet suggests that Pendergast and Almquist are correct—that Hilborn's basic principle was indeed what the company applied to its V8s in 1957. Unlike electronic systems, which use electric pulses to activate tiny solenoids in manifold- or port-mounted valves, Hilborn's system used a set of high-pressure lines to carry fuel from a central pump and deliver it, in a constant flow, to each intake port.[34] This is precisely what the Rochester injector did, "offer[ing] constant-flow port injection," in Chevrolet's own words.[35] However, whereas Hilborn's system could only adjust the pressure of the fuel delivered according to the position(s) of its eight individual throttle valves, Chevrolet's Rochester unit could also adjust the flow according to engine temperature and vacuum. Hence, Hilborn's system was less well-suited for street use than was Chevrolet's.[36] In short, GM may well have "borrowed" Hilborn's constant-flow idea, but its system was by no means simply a copy of his.

Nevertheless, as fuel injection evolved during the 1950s and 60s, aftermarket manufacturers were certainly in the vanguard, just as they were with turbochargers. However, the vast majority of their efforts during this period went into the production of a more diverse array of traditional performance products. And as demand for their add-on parts increased, especially during the mid- to late 1960s, speed equipment companies from coast to coast expanded their facilities, bought new equipment, and hired new machinists. But as they did so, their collective capacity overwhelmed the existing distribution system, which relied on a combination of direct mail orders and speed shop sales. Some performance-aftermarket companies therefore began to sell their products through larger chains like J.C. Penney and Sears, while middlemen—manufacturers' representatives, independent jobbers, and wholesale distributors (WDs)—handled a growing percentage of their manufacturer-level sales.[37]

All of this worried many independent speed shop owners, for whom the chain stores and the new middlemen were larger-than-life competitors against whom the "little guy" was all but powerless. Hot Rod Industry News took up their cause, running numerous articles in the late 1960s designed to educate speed shop owners on the many ways in which their status as speed equipment *specialists* could, with an adjustment here and a bit of effort there, enable them to stay afloat and even prosper in spite of their ostensibly unfavorable marketing circumstances. And indeed, many speed shops did continue to do well, even if their grumbling never quite ceased.[38] On the other hand, apart from some minor complaints about the pricing policies of certain WDs and chain stores, speed equipment manufacturers largely welcomed these new outlets because they broadened the scope of their market and enabled them to better serve their customers, particularly those outside of California. But because they did not wish to alienate any of their retail outlets, especially the tradition-steeped speed shop, most equipment companies adopted a neutral public stance on the matter.[39]

Besides, many 1960s manufacturers were far too busy squabbling among themselves to have the time or the inclination to get involved in the brouhaha between the speed shops and the WDs. Competition within the industry had always been intense, but in the late 1950s and the 1960s, it grew fierce. As a result, manufacturers often traded barbs publicly in widely circulated enthusiast periodicals, using precious advertising space to charge their rivals with everything from patent and copyright infringement to deliberate product misrepresentation. In 1960, for example, the California Equipment Company of Seattle noted, in an advertisement in *Drag News*, that some of its competitors were copying its popular floor-shift conversions and warned that it was prepared "to avail itself of the protected clauses provided by the U. S. Patent law and to proceed against anyone" that did so.[40] California Equipment was not alone in its frustration with the knockoff problem, but it was one of the few to ever threaten legal action over it. For as Delores Berg of Gene Berg Enterprises later explained, the very nature of the industry's products—parts and systems that modify other parts and systems—made it very difficult to prove infringement in most cases.[41] Consequently, few among them ever tried, and even fewer bothered to file patents in the first place.

Far more common, though equally difficult to prove, were charges that one or another of a given company's competitors lied about the critical matter of racing and motorsports affiliations. When Venolia claimed as its own several of Mickey Thompson's sponsored drag-racing champions in 1966, for example, Thompson called them on it. In an open letter published in *Drag News*, Thompson challenged Venolia's leaders to a mediated sit-down to sort out which racers actually used which company's pistons. When representatives from his competitor failed to show up at the designated time and place, Thompson wasted very little time. Within a month, he responded to Venolia's silence by securing a clever full-page spread in *Drag News*: the first half of the page repeated Thompson's claims, and the second half was left blank apart from a tiny notice that the space had, courtesy of Mickey Thompson Enterprises, been "reserved for [the] Venolia Piston Company."[42]

Similar charges were exchanged between a number of camshaft manufacturers in the 1950s and 60s, though only rarely were they as civil as Thompson and Venolia's phantom exchange. Crane Cams of Hallandale, Florida, and Iskenderian Racing Cams of Southern California seldom held back in their frequent

1960s tiffs, for example. Never one to err on the side of caution when making racing-related claims in his advertisements, Ed Iskenderian nevertheless was taken aback when Crane declared in *Drag News* in the summer of 1966 that it had surpassed his firm as the leading manufacturer of racing camshafts. Crane, borrowing from Isky's own bag of advertising tricks, based its claim on the number of entries and winners in select NHRA classes; according to this measure, Crane was in fact "number one." Iskenderian responded swiftly, charging Crane with "mud-slinging," while re-asserting its traditional claim that across the board, in all associations and all classes, Iskenderian cams were actually "number one." Back and forth the two firms went in what must have been an entertaining exchange for *Drag News* readers.[43] But for Iskenderian and Crane, none of this was trivial: in the competitive high-performance aftermarket, racing triumphs mattered. And as these and other exchanges in the 1960s "ad wars" demonstrated, manufacturers were often willing to stretch the truth—and lash out at their competitors—in order to claim plausible motorsports advantages here and there.[44]

Lending a measure of urgency to at least some of these claims and counter-claims was the fact that a number of aftermarket companies were being swallowed in a massive wave of mergers and acquisitions at the time. In some cases, particularly toward the end of the 1960s and into the 1970s, corporate umbrellas were the ones doing the buying.[45] But for the most part, these buyouts were peer-to-peer transactions. For example, Moon purchased Potvin Cams from its founder, Chuck Potvin, in 1962. That same year, Schiefer obtained Harman & Collins's roller camshaft and magneto divisions. Meanwhile, Frank McGurk sold his company in chunks over the course of 1968 and 1969: his rocker lines went to Crane Cams, while his camshaft operations went to Iskenderian.[46] And the list goes on. Some of these transactions were amicable, voluntary, and mutually beneficial, but many of them were not. Either way, though, these moves kept competitive pressures within the industry high. At best, those that remained unaffected by them quickly found themselves facing larger, more powerful rivals; at worst, they knew that they could be the next to lose their independence.[47]

Whether because of mergers and acquisitions or simply because of the unit and dollar-value growth the industry experienced during the period in question, speed equipment companies were getting bigger, and increasingly, this lent a corporate feel to many firms. Though their enthusiast-founders were for the most part still at the helm, that is, growth forced many firms to adopt structural characteristics more in line with giants like Chrysler and General Motors than with the sole proprietorships of the 1930s and 1940s hot rod industry. Chief among these was their adoption of independently operating functional divisions of precisely the sort that Alfred Chandler has described at GM.[48] For Chandler, of course, these sorts of operating divisions were necessary for large corporate entities, but the basic idea proved no less effective for the much smaller companies that manufactured speed equipment. Thus, Ed Iskenderian elected to establish McGurk Camshafts as an independent operating division of Iskenderian Racing Cams when he bought the McGurk enterprise in 1969.[49] Similarly, when Mickey Thompson bought Autotronic Balancing in 1963, he set up its erstwhile owner, Bill Hitchcock, as the head of a new Mickey Thompson Balancing Division. Thompson, in fact, spent much of the early 1960s reorganizing his entire operation along these lines: his balancing, light-metals foundry, piston manufacturing, engine assembling, and other departments all were reestablished as independent operating divisions by the end of 1963.[50] Other equipment manufacturers took the corporate model a step further, establishing boards of directors in some cases and going public with stock offerings in others.[51] We should not overstate the case, because for every publicly traded company and for every division-oriented speed equipment enterprise in the 1960s, there were literally dozens of sole proprietorships. Still, as the industry matured during the 1950s and 60s, "aftermarket" no longer necessarily implied "small" or even "owner-enthusiast-controlled."

Thus, when Ford, General Motors, and Chrysler decided to enter the high-performance market to win would-be rodders away from speed shops and aftermarket manufacturers, what they found among their Southern California competitors were not the stereotypically tiny, garage-based hot rod companies of the 1930s and 40s, but rather a collection of prosperous performance specialists willing to tackle whatever technological challenges or marketing ploys might come their way. And take them on they did: to date, the aftermarket has outlived the OEM challenge of the 1950s and 60s by more than four decades. But before we get too far ahead of ourselves, let's take a moment to further flesh out some of the developments that took place within the industry during this period by briefly examining the experiences of Crane, B&M, and Gene Berg Enterprises.

Crane Cams

By 1953, Harvey Crane had had enough. An enthusiastic young rodder with a 1932 Ford, Crane found that none of the commercially available high-performance camshafts of the time consistently produced the results he was after. But because he was a resident of Southern Florida, Crane was unable to simply swing by one or another of the Los Angeles camshaft shops for tips, advice, or custom work. Frustrated, he examined the various bumpsticks he had tried, hoping to find a flaw common to them all that might point him in the direction of a homespun solution. To his surprise, he found that all of the cams he examined were inconsistently machined, lobe to lobe. That is, the difficulties Crane experienced whenever he tried a reground cam in his flathead mill were largely due to sloppy finishing and poor quality control. After fixing one of these camshafts by carefully regrinding each of its lobes to match the others, the proverbial lightbulb lit up. Convinced that he could do a better job manufacturing camshafts than the West Coast boys, he began to work on a handful of cam profiles of his own. Toward the end of 1953, he struck a deal with his father to rent a portion of his general machine shop, and with the purchase of a Storm-Vulcan camshaft grinder, the Crane Engineering Company was born.[52]

The Storm-Vulcan unit wasn't cheap, however, and Crane struggled at first to make the payments. But there was a method to Harvey's madness: the Storm-Vulcan machine was a high-end tool that enabled him to grind his cams with a lobe-to-lobe precision the established California grinders had yet to achieve. But before he was really able to get going, Uncle Sam called, and Crane was forced to put his plans on hold and serve with the armed forces in Korea. When he returned, Crane got straight back to business, and his precision-ground camshafts fast became a hot-selling item among area rodders and drag racers. Soon his success enabled him to purchase an even better grinding machine and to move into his own 3,500-square-foot facility in Hallandale.[53] His sales were mostly local, though, until a loyal Crane customer, Pete Robinson, won the NHRA Nationals drag racing event in 1961.[54] Orders soon flowed into the Hallandale shop from all over the United States, and by decade's end, the company employed more than eighty engineers, technicians, machinists, and other personnel in a sprawling complex totaling more than 50,000 square feet.[55] By then, Crane was one of the largest camshaft companies in the business.

Yet Crane Cams never really lost its "upstart" character—not in the 1960s, at least. For Harvey Crane relished the role of the underdog, promoting himself as the East Coast spoiler who had crashed the West Coast grinders' exclusive party. Period advertising for the company often bragged of how the little Florida company had grown to prominence by stealing customers from the likes of Iskenderian and Howard's, established West Coast firms too busy and perhaps too arrogant to notice the Hallandale firm until it was too late.[56] Though his aggressive, chest-thumping ads embroiled his company in squabbles that appear silly in hindsight—the 1966 "ad war" between Mr. Crane and Mr. Iskenderian, for example, over which of their firms was actually "number one"—the tactics worked. Relative to the more established firms, that is, Crane grew at a phenomenal clip during the 1960s. Also, Harvey Crane never lost an opportunity to remind whoever was listening (or reading) that his was an *East Coast* firm, for he was something of an informal spokesman within the industry for all of the

East Coast manufacturers, racers, and ordinary enthusiasts who were by then fed up with the activity's lingering obsession with Southern California and the West Coast.[57] Legendary East Coast racers like Don Garlits had by the end of the 1950s proven that California didn't have all the champions, of course, but what Crane brought to the spirited 1960s East-West rivalry was the forcefully argued claim that California didn't have all the technical expertise either.[58]

In fact, expertise was precisely what Crane claimed as his *advantage* vis-à-vis the California competition. And to a large extent, he was right. Other camshaft firms—Crane's archrival Iskenderian, most notably—adopted new testing devices, machine tools, and manufacturing processes during the 1950s and 60s, transforming the camshaft business into one of the most advanced segments of the industry. But Crane took things a step or two further. Content neither with his current means for product evaluation and quality control nor with the prospect of outsourcing the tasks, Crane therefore arranged to purchase the Camcheck Company of Ypsilanti, Michigan, in the winter of 1966–67, moving that company's computerized camshaft proofing equipment and highly skilled staff to his Hallandale headquarters. With this move, Crane could legitimately claim to have the most advanced quality-control system in the business.[59] In addition, toward the end of 1967, the company broke ground on a cutting-edge, 10,000-square-foot research and development facility adjacent to its plant. Equipped with a custom-made dynamometer specifically designed to test the output of drag-racing engines within a brief five-second window, this new facility enabled Crane's engineers to evaluate their designs without subjecting their sensitive quarter-mile racing engines to the strains of extended dynamometer testing—and without sending an actual dragster (and an actual driver) down the strip with an unproven grind.[60] At the same time, Crane also went to great lengths to ensure that his manufacturing space was filled with the best machinery available, including computer-controlled tools and design programs that would eventually be common within the industry but were exceedingly rare in the mid- to late 1960s.[61] For Crane, then, what mattered was precision, and he was never afraid to try new methods and new tools to guarantee that his cams were spot-on.

Unfortunately, Crane's proclivity to do things a little bit differently ultimately cost him his company. In 1989, the privately held firm's board of directors voted to fire Harvey Crane, and today Crane Engineering is employee-owned and operated. For Crane distributed a lot of company stock over the years to his machinists, engineers, and support staff in an effort to retain his best employees; but in the end, when they discovered they could oust their demanding boss, they did. In his thirty-six years in the business, however, especially during Crane Engineering's formative years,

Harvey Crane did prove that an upstart, innovative, East Coast firm could succeed in an industry overwhelmingly made up of ensconced, old-school, West Coast rodders.[62]

B&M Automotive

In 1951, while in his senior year of high school in Hollywood, California, Bob Spar enrolled in what was known as the "four-four plan," a vocational training program that allowed students to spend four hours a day at school and four hours working at an outside job. Spar, a lifelong automobile enthusiast, went to work for an automotive repair shop just across the street from his school. As it happened, one of that shop's regular customers was a colorful used-car dealer based in Hollywood known as "Madman" Muntz, who had recently begun to produce his own low-volume sports car. A roadster with a Cadillac V8 and a GM Hydra-matic automatic transmission, Muntz's car was based on tooling and designs purchased from Kurtis Kraft, a Glendale company that specialized in custom sports cars built around standard OEM running gear. Assembled in Chicago, Muntz's cars arrived at his Hollywood dealership not yet fully prepped for sale, and he regularly left the necessary final adjustments and tuning to the tiny shop at which Spar worked. In the spring of 1952, several months after graduating from high school, Spar went to work for Muntz full-time, turning wrenches at a brand-new dealership that Muntz established on Sunset Boulevard.[63]

For the most part, Spar spent his days at this dealership doing routine new-car prep work, tune-ups, and run-of-the-mill repairs. From time to time, however, he and the others in the shop did custom work too. Then, in mid-1952, GM cut off Muntz's supply of Cadillac V8s, almost certainly because GM's Chevrolet division was at the time preparing to launch its own sports car, the Corvette. This forced Muntz to reconfigure his design to accept the considerably less desirable (and less powerful) Lincoln V8. Consequently, Spar and his associates began to spend a great deal of their time swapping heads, installing multiple carburetors, and the like—hot rodding Lincoln engines, that is, for those among Muntz's customers who weren't satisfied with its performance. In addition, Spar performed a number of engine swaps for those who did not want a Lincoln mill at all, replacing the stodgy motor with surreptitiously sourced Cadillac and other GM V8s. Finally, some of Muntz's customers were also dissatisfied with their Hydra-matic transmissions, so, according to Spar, "we . . . came up with little tricks we could do here, and tricks we could do there" to boost the automatic's performance.[64] His confidence and experience growing, Spar began to wonder if the time had come for him to leave the Muntz operation and strike out on his own.

In September of 1953, with a friend of his from the Muntz dealership, Mort Shuman, Spar opened a small garage of his own in Van Nuys. Inspired by then-famous hot rod shops like Ray Brown Automotive and C&T Automotive, Spar and Shuman decided to call their new enterprise "B&M Automotive." But they were not exactly in the hot rod business—not yet, at least. Indeed, much of what they did in their first year or so of operation wasn't all that different from what they had done down in Hollywood: general repairs, for the most part, with a modification job or two thrown in on occasion. Muntz himself was their first repeat customer, in fact, for he often hired Spar and Shuman to fix a particularly problematic car or to perform an especially grueling engine swap. However, by the time the Muntz sports car bowed out in 1954, Spar and Shuman had an established, general-repair customer base of their own that more than paid the bills.[65]

But Spar in particular continued to hold out the hope that B&M might some-day morph into a full-time hot rod shop à la Ray Brown or C&T. So on evenings and weekends, he did custom engine and drivetrain work for area enthusiasts. Rather quickly, though, Spar realized that if B&M was to become a reputable hot rod shop, it would need a signature niche. From the work he was doing on the side, Spar knew that the standard transmissions fitted to flathead V8 Fords were especially prone to early failure when coupled with hopped-up engines. He also knew that there weren't yet any hot rod shops that specialized in this oft-overlooked but increasingly crucial area of automotive modification. So bit by bit he began to focus on gearboxes, and before long, area enthusiasts regarded B&M as "the transmission guys."[66]

Still, B&M Automotive remained a general repair shop, and though Spar spent a lot of time repairing and modifying manual Ford transmissions, he felt that this was not a niche area of expertise sufficiently distinct to enable B&M to break into the hot rod field full-time. Then, early in 1955, Bob's younger brother Don wrapped B&M's rolling test bed (a 1940 Ford) around a telephone pole. Don was fine, but the manual-gearbox, flathead-V8 hot rod was a total loss, and Spar decided to replace it with something a little bit different: a 1949 Oldsmobile with a Hydra-matic automatic transmission. From his time spent working on the Muntz cars, Spar knew that a tweak here and an adjustment there could transform a run-of-the-mill Hydra-matic into a capable transmission well suited to hot rod applications. With this new car, Spar, his brother Don, and his partner Shuman began to experiment with the automatic, developing tricks and testing them out at the local dragstrip and, when the coast was clear, on their local streets. By the fall of 1955, they felt they had developed their modifications to the point where it was time to test the market, so they started heading to the strips on the weekends. There they ran Spar's Oldsmobile in open competition, hoping that a few key victories might spark some interest in what they were up to. And indeed, when Spar beat the local hero, Dick Harryman, at the Saugus strip in early 1956, Harryman himself became one of their first Hydra-matic customers.[67]

Shortly thereafter, however, Spar was drafted by the army for a two-year stint. In his absence, his brother Don and his partner Shuman managed to hold things together, but by the time Spar returned from the service in 1958, Shuman was thoroughly fed up with the smaller paychecks that accompanied the shop's move into the high-performance transmission business. Don Spar therefore purchased Shuman's interest, and B&M became a brother-brother operation. With Bob's return, sales began to pick up, and B&M, with its Hydra-matic niche, was soon on its way to becoming the all-out hot rod shop that Bob Spar had long desired. By the end of the decade, B&M had three different levels or "stages" of tune to which they built their transmissions, depending on the intended application. The mildest was known as "Street and Strip," the middle as "Competition," and the hottest as "Blown Competition." Thus, B&M built and sold transmissions for street-use, mixed-use, and all-out drag racing. And with a number of loyal racing customers, particularly among the growing ranks of those with blown Willys "gassers," Bob and Don's business boomed.[68]

In 1962, however, General Motors introduced an improved automatic known as the Turbo Hydra-matic, or Turbo 400. Chrysler, meanwhile, brought out a unit it called the Torque-Flite, and the following year, Ford followed suit with its C-6. Part of the resurgent horsepower race of the early 1960s, these new transmissions from three of the four leading OEMs were superior to the standard Hydra-matics of the 1950s in every way. They were also stronger and quicker-shifting than even the modified Hydra-matics B&M built, as Bob and Don Spar learned to their dismay. However, when lead-footed rodders began to break these new OEM units, the search was on at B&M for ways to hop them up. By 1964, Bob and Don had abandoned the Hydra-matic altogether in favor of the newer Turbo 400, to which they were able to successfully apply their tried-and-true tweaks. Like their counterparts in the engine-modification business, in other words, Bob and Don Spar found a way to continue to prosper in the face of the OEMs' performance-oriented push of the 1960s.[69] To this day, B&M remains among the premier manufacturers of high-performance transmissions, torque converters, and related accessories.

"Manufacturing" isn't quite what B&M was up to, however—not in the 1950s and 60s, at least. The company did produce a handful of shifting accessories at the time—brackets and levers that Bob, Don, and their employees finished and assembled after obtaining the rough aluminum castings and stampings from outside foundries. For the most part, though, Bob and Don sold OEM-built transmissions and torque converters that they modified. Sourced, in the case of the

Turbo 400, from a division of General Motors known as End Products, these transmissions arrived at the B&M plant in stock trim, individually boxed and prepped for replacement duty. Bob and Don then disassembled these brand-new units, applied their tuning tricks, reassembled them, and sold them as B&M-spec "Street and Strip," "Competition," or "Blown Competition" automatics. Theirs, in other words, was a manufacturing operation only in the loosest sense. Nevertheless, Bob and Don were remarkably innovative in their production techniques. By the end of the 1960s, the company had a number of automatic machines on the floor, and the brothers had set for themselves the goal of using the best available computer-controlled testing equipment to increase their volume of modified transmissions and torque converters without compromising quality or dramatically increasing their overhead. Given the nature of their signature products, that they were able to do so was an extraordinary feat.[70]

Extraordinary as well were the testing procedures and devices they developed at their Van Nuys shop. Chief among these was the automatic transmission dynamometer, a unique device that earned the brothers a lengthy feature article—and the cover shot—in the December 1969 issue of *Hot Rod*. Engine dynamometers had been in use within the industry for decades, providing even the smallest of speed equipment manufacturers with critical data regarding the performance-enhancing potential of their products. But for B&M, a transmission specialist, engine dynamometer testing was of limited value, and as a result, Bob and Don spent a considerable amount of time over the years testing their products in actual cars. But because real-world testing is exceedingly difficult to control—everything from relative humidity and temperature to altitude and pavement conditions can affect the results—Bob and Don began to work on a lab-based transmission dynamometer in the mid-1960s that would allow them to simulate real-world street and racing conditions in the controlled comfort of their shop. This work led to the celebrated "automatic dyno" of 1969. Capable of holding any automatic transmission (and of coupling it to any engine), B&M's device used an adjustable-weight flywheel to simulate the conditions and stresses the transmission would encounter in the intended application. Linked to a control panel with an oscillograph that recorded variables from multiple sensors, B&M's automatic dyno was a versatile piece of equipment that gave the firm a critical edge. Bob and Don were so well-respected for their testing capabilities, in fact, that by the end of the 1960s they were spending a considerable amount of their time testing transmissions for General Motors and Chrysler.[71]

In spite of the success of their slushbox business, by 1970 neither Bob nor Don felt entirely comfortable pinning their futures on such a narrow niche. Casting about for a way to diversify their operation, they hit upon the idea of establishing a new division under the B&M umbrella to manufacture recreational vehicles (RVs). The idea wasn't altogether new: back in 1964, when President Johnson first began to escalate the war in Vietnam, B&M saw its youth-oriented business take a terrifying, albeit temporary, nose-dive; and to fill the income gap, Bob and Don had sold modified Turbo 400s to RV owners fed up with having to shift their vehicles manually. Confident that an RV division would complement their transmission business nicely, the brothers therefore sold one-third of their hitherto privately held shares of B&M stock to a group of outside investors in 1970, using the money to establish the Sportscoach Corporation of America.[72] But among automotive enthusiasts, B&M would remain synonymous with high-performance transmissions, its core business.

Gene Berg Enterprises

Located in the City of Orange in central Orange County, California, Gene Berg Enterprises was not formally established as a high-performance business until December 15, 1969, the very end of the period in question. However, Berg's personal involvement with performance tuning and the speed equipment industry actually dated back to the mid-1950s. More specifically, during the thirteen years prior to the issuance of his official California business permit in the fall of 1969, Gene Berg played an active role in the emergence of the specialized segment of the performance aftermarket focused on air-cooled Volkswagens. Significant though his later accomplishments within this field would be, the pre-history of his company during the 1950s and 60s therefore warrants a closer look.

Gene Berg was a resident of suburban Seattle and a lifelong performance enthusiast who spent much of his youth in the early 1950s tooling around in flathead-powered hot rods. In 1956, however, he spent some time behind the wheel of a Volkswagen while visiting a relative in Montana. Impressed with the way the little car drove, Gene placed an order for one at the local dealership when he returned home. Eight-and-a-half months later, his turn on the waiting list came up, and he took delivery of a brand-new 1957 Beetle. Equipped with a 36-horsepower engine, the car, while faster than the one he had driven the year before, nevertheless proved to be intolerably slow to Gene after only a few short weeks. Convinced of the car's potential—and determined, as a dyed-in-the-wool hot rodder, to unleash it—he therefore pulled the still-fresh engine from the still-new car, tore it down, and closely inspected its parts. After making a tweak here and an adjustment there, he carefully reassembled the mill and put it back in the vehicle. The results were astounding: the otherwise ordinary car "would run 80 or 85 mph when other VWs were running 70 mph."[73]

Pleased with his accomplishment, he tooled around town in his hot VW for a couple of weeks before a problem with the front suspension prompted him to take it to the dealership for some routine adjustments he was certain would be covered by his new-car warranty. He was right, and the dealer set one of its technicians to work on the car. When he returned from the routine post-service test run, however, the technician was baffled. Subsequent test drives by the service advisor, the service manager, the sales manager, and the owner of the dealership resulted in a consensus opinion that although the suspension problem had been corrected satisfactorily, the car was simply way too fast. The group descended on the unsuspecting Berg, demanding to know what he had done to the car to make it run so fast. Well aware of Volkswagen's official pronouncements prohibiting the modification of its cars on pain of warranty forfeiture, Gene adamantly denied that he had modified the car—after all, he didn't want to wind up footing the bill for the front suspension fix. In the end, the dealer let him go, convinced that Berg had indeed fooled around with the car, but lacking the evidence to prove it. Within a few days, the car developed a misfire, but rather than risking his new-car warranty once again by returning to the dealership, Gene decided to repair the problem himself. On a whim, though, he called the dealer when he was done and reported not only that the car had been misfiring but also that he himself had fixed the problem. Impressed with Gene's mechanical prowess (and still convinced that Berg had modified his VW's engine), he offered the young man a position as a technician in his service department. Gene, a bus driver for the Seattle Transit Authority who had been itching for an opportunity to go to work on cars full-time, gratefully accepted the offer.[74]

Gene worked at the dealership for several years, and by 1962 he was its service advisor. Unhappy with its longstanding, unspoken, across-the-board policy of overcharging customers for routine maintenance and adjustments, however, Gene eventually told his boss that he didn't want to be a part of the scam any longer. His boss obliged, firing him on the spot. With a wife and three young children to support, Gene needed work, and he needed it fast. But instead of searching for another dealer job, he decided to strike out on his own. In 1962 he and his wife Delores (Dee) founded a small repair shop near their home. Gene's Service, as they called it, was a general Volkswagen repair shop, and a lot of folks who had been regular customers of his when he was turning wrenches at the dealership began to come to his independent shop instead. Word spread quickly, and before long Gene had a thirty- to forty-day waiting list for routine maintenance jobs alone. His regular business booming, Gene and his wife began to branch out a bit on the side, doing custom-basis VW modification jobs for a handful of customers in the evenings and on weekends. But Gene had never been a fan of the aftermarket

"California trash," as he called it, and the more he dabbled in Volkswagen engine and transmission modification, the more convinced he became that he could do a better job of building the special intake manifolds, rocker assemblies, and stroker crankshafts that went into his performance rebuilds.[75]

After checking out a couple of do-it-yourself books on casting from the local library, Gene and Dee decided to have a go at it. Their plan was to cast a magnesium intake manifold, but the molasses-and-sand molding they crafted for the purpose wouldn't hold, and even after repeated experimentation, they simply could not achieve the results the books promised. Determined to get into the parts-manufacturing business but convinced that they just were not cut out for the delicate casting work, they shopped around for a local foundry that would be willing to undertake a small-batch job or two. Once they found one, they began to obtain small runs of rough magnesium castings that Gene finished by hand in his shop and installed on his performance-oriented jobs—all the while, of course, continuing to perform routine maintenance and repair work for his regular mainstream customers. In his spare time, he also developed a small racing program, building a VW-powered rail dragster and a Beetle-based "stocker" that he campaigned at drag-racing meets in the Seattle area.[76]

Gene began to travel to Southern California on a regular basis to race his cars as well. There he established some firm contacts within the fledgling VW performance industry. For Gene, meeting and befriending Joe Vittone and Dean Lowry of EMPI—a booming firm that by then was building quality merchandise with which Gene was impressed—was an inspiring and transformative experience. The feeling was mutual: Dean and Joe were struck by the quality of the hand-built parts Gene used on his dragsters, and during the latter half of the 1960s, EMPI often brought to market items that Gene had pioneered in his Seattle shop, sometimes with and sometimes without his explicit permission. Nevertheless, Gene was genuinely fond of his California counterparts, and with every racing trip he took to Los Angeles during the latter half of the 1960s, Gene felt more and more a part of the growing VW performance business.[77]

Consequently, when Lowry asked him to come to California early in 1969 to join him and his brother in a spin-off Volkswagen performance firm, Deano Dyno-Soars, Gene Berg didn't hesitate a whit: he and Dee packed their things, closed their shop, and left for Southern California. But within a few short months of their arrival, things began to go awry. The company itself was doing well, and bolstered by a smattering of coverage in the hallowed pages of *Hot Rod*, its reputation as a first-rate Volkswagen engine firm was solidifying.[78] But Gene's partners in the enterprise were brothers, and their wives were sisters. In the behind-the-scenes decision making, Gene and Dee therefore felt as though they were often the odd men out, so to speak. Confident in his abilities as an engine builder and as a performance-parts manufacturer, Gene decided to sell his share of Deano Dyno-Soars to the Lowry brothers in December of 1969, using the cash to set up Gene Berg Enterprises.[79] After thirteen years of side jobs and garage-scale production, Gene Berg was now at the helm of a genuine high-performance company of his own. And in the quarter-century that followed, he built an unrivaled reputation for excellence within the VW aftermarket: among enthusiasts, his crankshafts, manifolds, close-ratio gears, five-speed conversions, and other signature products were regarded as the best that could be had at any price.[80]

However, the story of Gene Berg Enterprises, its innovations, and its many spin-offs and affiliated personnel is one that belongs to a later era. For the moment, though, consider the implications of the 1960s pre-history of Gene Berg's successful business. For here was a man who eased himself into the speed equipment industry over the course of many years, drawing on his experiences on the street and at the strip to guide him in the production of small quantities of hand-built Volkswagen parts in a manner that echoes the way the Edelbrocks, Weiands, and Spaldings of the 1930s backed into their roles as flathead-equipment manufacturers.

▼

The foregoing accounts illustrate many of the most critical developments and trends within the speed equipment industry of the late 1950s and the 1960s. Harvey Crane was an East Coast aftermarket partisan, a relative newcomer to the speed equipment industry whose company relied on the latest research and development, manufacturing, and testing techniques, coupled with an aggressive advertising strategy, to grow from obscurity to prominence within a few short years. B&M Automotive, on the other hand, muscled its way into the performance industry by seizing the initiative in what hitherto had never been regarded as a category of automotive technology worth the rodder's while: automatic transmissions. Furthermore, the ways in which B&M adapted to the better-performing OEM transmissions of the 1960s is illustrative of the responsive flexibility so common within the high-performance industry of the period. Finally, Gene Berg's gradual evolution as an aftermarket manufacturer during the 1960s demonstrates above all else that although much within the industry had changed substantially since the 1930s, the well-trodden path from hobbyist to manufacturer certainly had not. For even in 1969, the speed equipment industry was still dominated by enthusiast-entrepreneurs like Berg.

Heading into the 1970s, these enthusiast-entrepreneurs certainly seemed to have plenty of reasons to smile. Collectively, their industry was edging toward the $1-billion-a-year mark. Individually, their businesses were growing fast, and the tiny manufacturers of just a few short years before had evolved into diversified operations with tens of thousands of square feet of floor space, innovative production methods, and cutting-edge products. Millions of baby boomers coast to coast were buying their first cars—and countless thousands of them their first aftermarket gear. Through their muscle car sales and their performance-oriented advertising, Chrysler, General Motors, Ford, and American Motors were continuing to contribute to a culture of speed in the United States as well, much to the delight of everyone involved in the high-performance business. Indeed, as Dennis Pierce of *Hot Rod Industry News* remarked in January of 1970, the speed equipment industry "has grown at a fantastic rate" during the 1960s, "and all indications point towards even more growth in the 70's."[81]

Pierce was right: the high-performance aftermarket would continue to grow at a phenomenal rate in the 1970s, much as it had in the 1960s. The difference, though, was that the political and economic climate of the 1970s would prove to be far less welcoming to those who cherished power, torque, and speed.

CHAPTER SEVEN

THE SPEED EQUIPMENT MANUFACTURERS ASSOCIATION

"The Detroit super car is dead." So mourned Lee Kelley, the editor of *Popular Hot Rodding*, in October 1971 as the OEMs curtailed their production of high-performance automobiles for 1972. Gone were a number of performance icons, including Pontiac's GTO and Oldsmobile's 4-4-2.[1] Within another year, Ford's 351 HO Mustang was arguably all that remained of the muscle car era, and by the fall of 1973, it too was gone.[2] The demise of the American super car was swift: as the decade began, Ford, Chevrolet, Buick, Oldsmobile, and American Motors all introduced larger, more powerful engines to their lineups, and Dodge and Plymouth added altogether new muscle cars to theirs;[3] prospective buyers thus had dozens of high-performance models to choose from in 1970. Two years later they had but a handful, and within three they had virtually none.

Why were General Motors, Ford, Chrysler, and the rest of the OEMs so quick to abandon this lucrative segment of the new car market? Rising insurance rates seem to have had something to do with it, for during the mid- to late 1960s, several companies slapped surcharges of $400, $600, and even $800 or more—per year—to the premiums of those who tooled around in Barracudas, SS Chevelles, and other factory hot rods.[4] Another possible reason is that the muscle cars of the late 1960s were no longer the svelte and compact rockets that they once were. Longer, wider, and much heavier than their immediate predecessors, the super cars of 1969, 1970, and 1971 often struggled to haul their over-accessorized heft around with much authority, in spite of their larger and more powerful engines. However, muscle car insurance premiums and gross vehicle-weight ratings had been on the rise for years by 1970, but OEM performance-model sales had yet to level off. The rapidity of the genre's subsequent decline therefore suggests that something else was going on.

Indeed, the period and secondary evidence is nearly unanimous on this point: more than anything else, tightening federal motor vehicle safety standards and the "ecological war on pollution" were what brought the prosperous era of OEM performance to its abrupt halt.[5] For as *Popular Hot Rodding*'s Lee Kelley explained to what must have been a distraught readership in the fall of 1971, government regulations had begun to "tak[e] the super out of super cars": "It's one thing to pay $5000 for a muscle car that'll run in the 13s off the showroom floor and another to pay $5200 (that $200 extra is for added safety and smog equipment) [for a car] that can barely get into the 14s! It would take an extra $500 (at least) to get this new 'muscle' car to run with the old one, and to lots of people the effort just [isn't] worth it."[6] Hobbled by federal mandates, the muscle cars of the early 1970s no longer lived up to enthusiasts' expectations, and thus, although insurance woes and inept engineering and marketing decisions may well have doomed the muscle car by the end of the 1960s, the safety and emissions regulations of the early 1970s were in fact the final nails in its coffin.

Celebrated by insurers, safety advocates, and the environmental lobby and universally lamented by the sorts of folks who read *Hot Rod* each month, the passing of the OEM muscle car nevertheless did not mark the end of the affordable high-performance automobile in the United States, even in the short term. Neither, for that matter, had the so-called power packs of the 1950s and the GTOs of the 1960s marked its genesis. Rather, the era of the muscle car was but a single, relatively brief chapter within a much longer tradition of high-performance motoring in America, a tradition that began with the modified Model T "speedsters" of the 1910s and continues to thrive among the sport-compact "tuners" of today.

According to nearly every popular and academic history of hot rodding, though, the end of the muscle car era did not mark a return to normalcy within the hot rodding fraternity. Instead, it heralded the coming of a dark age, a period in which precisely the same sorts of regulations that brought an end to the 400-horsepower factory beasts of the 1960s also swiftly brought the average rodder to his knees. No longer free to tinker with his car as he saw fit, the sorrow of the ordinary enthusiast was eclipsed in the 1970s and the early 1980s only by that of the average speed equipment manufacturer, who was no longer free to make and sell whatever the marketplace demanded. As the "sexy sixties" gave way to the "sad seventies,"[7] in other words, desperate rodders and aftermarket leaders alike found themselves on the defensive, fighting off those who, they believed, were seeking to replace their '32 roadsters, '55 Chevys, and '69 Chargers with bland, "four-door shoe box[es] with a top speed of 40 mph."[8] Some enthusiasts searched for ways to circumvent the law, while others—including the editors of most of the popular periodicals—made use of the printed word to give voice to the rodder's cause. But in the end, all of the legal loopholes closed, and much of what was published in the editorial pages of *Hot Rod*, *Popular Hot Rodding*, and *Hot Rod Industry News* ultimately mattered not a whit. For in spite of their best efforts, the "hot rod apparatus," in the words of H. F. Moorhouse, "lost a lot of battles against the federal and local state."[9] The seventies, that is, were in fact quite sad.

Or so it would seem. To be sure, the fears that inspired the fiery editorials and the dreary predictions that found their way into the opening pages of the popular periodicals each month were indeed real: during the late 1960s and the 1970s, performance enthusiasts did openly fear government regulations and the implications that they bore for the future of their beloved pastime. So too were speed equipment manufacturers genuinely concerned for their businesses. However, the ways these parties expressed their concern in print represent but part of a larger and far more complex story. In other words, Moorhouse, Ed Almquist, and the rest of those who have written about the dire straits in which the hot rodding fraternity found itself during this time accurately capture the desperate tone of the period, but in so doing they have all but overlooked what actually took place. For the speed equipment industry enjoyed a string of sales-record-breaking years during the late 1960s and the 1970s and, language of doom notwithstanding, actually managed to win far more legislative battles—both for itself and for the average rodder—than it lost.[10] That it was able to do so under precisely the same legal and regulatory pressures that the OEMs faced during the same period—*and* that it was able to do so by relying on the very niche the OEMs had abandoned, the high-performance market—renders its experience all the more remarkable. So remarkable, in fact, that one cannot help but wonder how and why it has escaped the notice of so many for so long.

To be fair, Almquist does inform his readers that the speed equipment industry "fought back" in the 1970s. He also briefly mentions that an industrial organization known as the Speed Equipment Manufacturers Association played a leading role in its response. With the exception of a passing reference to "a new crop of environmentally-friendly products" that the industry developed in the 1970s, however, Almquist shies away from the details.[11] So does Moorhouse. A sociologist by training, Moorhouse explicitly frames *Driving Ambitions* as "a social analysis of the American hot rod enthusiasm."[12] Consequently, when it comes to the 1960s and 70s, Moorhouse focuses on the editorial content of the enthusiast publications and the industry's unofficial journal, *Hot Rod Industry News*, in an attempt to reconstruct the symbolic dimensions of this regulatory struggle. Although he acknowledges that the opinions published in these periodicals were often at odds, the bulk of his analysis emphasizes the rhetorical strategies that their editors shared: their attacks on the "uninformed bureaucrats" bent on eradicating the automobile; their appeals to the enthusiasm of the hot rodder and the specialty manufacturer; their doomsday prophesies of a world without supercharged Hemis, shadetree mechanics, and neighborhood speed shops; and, as the years wore on, their increasingly deflated and defeated tone. However, *Driving Ambitions* is often misleading on all of these points precisely because Moorhouse relies exclusively on the *editorial* material that appeared within the first few pages of *Hot Rod* and *Hot Rod Industry News* each month. Had he turned beyond page ten in any one of the many periodicals he consulted, that is, he would have found a veritable plethora of additional, behind-the-scenes details directly relevant to his analysis. He would have read about negotiations between the speed equipment industry and the National Highway Traffic Safety Administration, the Environmental Protection Agency, and the California Air Resources Board. He would have read about industry-sponsored testing and self-certification programs. He would have read about courtroom victories and legislative triumphs. In short, he would have read about the many ways in which the "hot rod apparatus" diligently worked during the course of the late 1960s and the 1970s to ensure that its own desperate forecasts never actually came to pass.

What follows here, in chapters seven, eight, and nine, is an attempt to fill in these details. Looking past the heated rhetoric and the wounded egos of the period, chapter eight examines the emergence of federal, state, and local automotive safety and noise-control legislation, tracing their evolution and elaboration during the 1960s and 70s and highlighting the many ways in which the hot rodding fraternity dealt with the unprecedented challenges that these new laws posed. Chapter nine does the same for automotive emissions regulations. Bear in mind, however, that it is only for the sake of narrative clarity that this story unfolds over the course of more than one chapter. For in the minds of those within the hot rodding fraternity, the many new federal, state, and local mandates of the 1960s and 70s were of a single piece — together, that is, they amounted to an undeclared and manifestly unwelcome "war on high performance."[13]

For anyone who has browsed through the automotive periodicals at Barnes & Noble or spent an afternoon watching *The Speed Channel*, the fact that the rodding apparatus ultimately won this war is readily apparent. After all, there wouldn't be low-slung Jettas, winged Civics, and periodicals like *Import Tuner* today if things had turned out otherwise. We must be careful, though, not to allow the benefit of nearly forty years of hindsight to skew our perspective. For in the late 1960s and the early 1970s, victories in this war — for enthusiasts and aftermarket manufacturers alike — were seldom certain, and they were almost always hard-won. Invariably, too, they involved the Speed Equipment Manufacturers Association, or SEMA, a performance-aftermarket organization of which I have thus far said very little. Our story therefore begins in this chapter with SEMA's origins and early evolution.

CREDIT MANAGERS AND LEGAL LIABILITIES

In the late 1950s wholesale distributors had yet to emerge as a significant link in the aftermarket distribution system, which meant that manufacturers still dealt directly with their mail-order and speed-shop customers.[14] While this arrangement worked well enough back in the days when *Hot Rod* only printed a few thousand copies each month, by the end of the 1950s, the task of keeping track of who ordered what—not to mention who *owed* what—was nearly overwhelming. Some equipment manufacturers toyed with the notion of eliminating credit-basis and COD sales entirely, but for most this was an unattractive option. After all, for every dishonest customer who bought on credit, there were dozens more who made their payments on time every time. What's more, these industry insiders knew that if they dropped their credit options altogether, their sales would drop. Most therefore continued to take their chances in an open-credit market, taking the good with the bad.

For a small minority, however, the risks associated with "business as usual" were no longer acceptable. They needed to retain their thirty-, sixty-, and ninety-day cycles, but they also needed to find a way to manage the risks involved. To do so, the heads of these select few companies set aside their secrecy and their type-A competitiveness and turned to an unlikely source for help: each other. So began the short-lived and informal "credit managers group," also known as the "Speed Equipment Manufacturers Credit Association."[15]

This group usually met at the home of Phil Weiand, one of the founding fathers of the Los Angeles speed equipment industry. There the likes of Els Lohn, Paul Schiefer, Bob Spar, and others discussed their credit problems in an open forum. Significantly, they never talked about their own credit strategies per se. Instead, they discussed actual, specific problems with actual, specific companies. Els Lohn, for example, might bring up the fact that a certain speed shop—which he would identify by name—was unreliable on a COD basis or that a certain mail-order house—again, identified by name—had fallen behind on its 60-day installments. Paul Schiefer might then volunteer that he had no problems with that particular speed shop but that he had in fact experienced similar difficulties with the aforementioned mail-order house. Round and round the talks then went until everyone had a chance to air their specific concerns, and then they would adjourn. Never once did they explicitly agree to blacklist certain companies or otherwise rig the market, however. In fact, Weiand always invited a lawyer to attend the meetings, whose job it was to prevent the group from venturing into illegal collusive territory. Armed with what they learned at these meetings, however, Weiand, Lohn, Schiefer, Spar, and the rest were able to more easily identify potentially problematic customers and deal with them accordingly. This went on, with meetings every few months, into the early 1960s.[16]

By the beginning of 1962, though, credit issues were no longer all that the group discussed. For during 1961 and 1962, the NHRA had rewritten its official rules in ways that troubled many of the manufacturers who showed up regularly at Weiand's house. Driver and spectator casualties were on the rise at the nation's many sanctioned drag strips by the early 1960s, and as a result the NHRA was having trouble making its liability insurance payments. In an effort to reduce its premiums, the association therefore adopted a new program for its technical inspections. Known as "NHRA-Approved," this program aimed to standardize the technical-inspection process each entry had to go through before every officially sanctioned meet. To accomplish this, the NHRA developed of a list of specific products — transmissions, superchargers, roll cages, seat belts, brakes, and the like — that were acceptable. And in order to more easily distinguish that which was approved from that which was not, the NHRA also issued an official list of authorized brand names to its technical inspectors. What was unclear, and what frightened those who regularly attended the credit managers group, was the question of how exactly the NHRA determined which brands made this list and which did not. Fearing the worst, attendees began to air conspiracy theories, wondering aloud whether back-room deals and outright bribery soon would be the order of the day. None of them seriously doubted that the NHRA meant well, but they were skeptical about the methods behind "NHRA-Approved." Consequently, they began to cast around for a means through which they, the manufacturers, might obtain a collective voice in the proceedings. Fortunately for them, regular attendee Bob Spar of B&M Automotive had an idea.[17]

Back at the 1960 NHRA Nationals in Indianapolis, several of Spar's customers were turned away during the mandatory technical inspection process because their B&M automatic transmissions were not equipped with scattershields.[18] Metal engine-bellhousing covers developed to contain clutch and flywheel fragments in the event of a catastrophic failure, scattershields were required for certain racing classes by the safety-conscious NHRA, and with good reason: when a flywheel or a clutch disintegrates at 6500 rpm, the shredded pieces literally become shrapnel, tiny shards of flying, red-hot metal that pose a serious hazard to drivers and spectators alike.[19] However, catastrophic failures of this sort were common among dragsters equipped with manual transmissions but almost entirely unheard of among vehicles with racing automatics. For Spar and his small band of customers, therefore, it seemed as though the NHRA was missing the point. After spending the better part of an entire day sitting in a makeshift waiting area in the NHRA's on-site Nationals headquarters, Spar's raised hand finally attracted the attention of a harried official who happened to pass by. After he explained the situation, the official, Jack Hart, told Spar that he would grant him a hearing in the presence of the technical inspectors the following morning. And at that meeting, the inspectors decided that the B&M cars should be allowed to run as they were, sans scattershields.[20]

Two weeks later, back in Southern California, Spar received a call from Hart, who invited him to a one-on-one meeting to discuss what happened at the Nationals. Spar agreed, and over lunch, Hart told Spar that a big part of what had won the day for B&M in Indianapolis was Spar's frank and informed approach to the bellhousing problem. Most of the time, rules-related disputes at NHRA meets were far less civil and informed, he explained. He also mentioned that Spar's perspective as a manufacturer was one the NHRA often lacked when contemplating technical and safety requirements. With this in mind, Spar and Hart began to meet for lunch a couple of times a month to talk about the NHRA and its approach to racing-safety standards. As 1961 came to a close, they also began to discuss Spar's concern over the ostensibly arbitrary nature of the new "NHRA-Approved" system. Hart's response was simple: if speed equipment manufacturers could find a way to band together *formally* so that they could come up with their own product-safety specifications, then there wouldn't be any need for the controversial "NHRA-Approved" program. In other words, as far as Hart was concerned, the NHRA was prepared to drop its program altogether if the manufacturers, as a group, would be willing to fill the resulting void. And *this*, Spar reported to his fellow credit managers, was the answer they'd been searching for. Why not form an actual association of manufacturers, a formal body through which their resources could be pooled in order to bring an end to the specter of the "NHRA-Approved" system and to establish, in its place, their *own* drag-racing parts specification and certification program?[21]

However, it was one thing to get a small group of speed equipment manufacturers to meet occasionally on an informal basis to discuss their credit problems, but it was quite another to try to set up a genuine industry-wide association. Consequently, though the subject often came up in the spring and summer of 1962, nothing actually happened. They talked about it favorably, but none of the regular attendees of the credit managers meetings seemed to know how to get the ball rolling.

Enter a man by the name of Henry Blankfort. A vice president at the Revell Model Car Company, Blankfort began to pop up here and there among the equipment companies during the summer and early fall of 1962. The Revell Company planned to launch a new line of plastic model kits based on championship dragsters and funny cars, and Blankfort's task—at which he was failing miserably—was to obtain permission to reproduce a number of aftermarket manufacturers' official logos and racing-sponsor stickers in miniature decal form so that the company's kits would be more authentic. Blankfort's problem, though, was that Revell's rival, AMT, had the same idea, and *their* point-man, Dick Day, always seemed to be a step or two ahead of him. It wasn't a question of timing necessarily, but rather of contacts: before accepting his position with AMT, Day had been a major player at the Petersen Publishing Company, so he already knew most of the key manufacturers personally. For Day, in other words, it was simply a matter of calling up old friends and asking for a favor. Blankfort, on the other hand, had only a couple of meaningful inside contacts. Accordingly, Day quickly won a number of exclusive licensing deals for AMT, while Blankfort struck out almost every time for Revell. However, through a friend of his, Ed Elliot of Elliot-McMullen Advertising, Blankfort was able to meet Dean Moon of Moon Equipment and Roy Richter of Cragar, neither of whom had signed with AMT. Although he was quick (and pleased) to seal a deal with these two companies, Blankfort was also dismayed when he discovered in the course of his negotiations with them that there was no blanket association of speed equipment manufacturers through which he could make a few more contacts—and perhaps a few more deals.[22]

Shortly thereafter, Blankfort placed a call to another of his woefully inadequate list of contacts, Els Lohn of Eelco. In addition to being an old friend of Blankfort's, Lohn was one of the original members of the credit managers group, and over the course of 1962, he, like Bob Spar, became a tireless advocate within that group for the establishment of a formal, industry-wide association. Blankfort needed to sign a few more licensing agreements, and Lohn needed a spark to get the organizational process rolling; together, the two men hatched a scheme. Blankfort's firm was an active member of a small trade group known as the American Model Association, and he offered to host a general meeting of speed equipment manufacturers at Revell's offices in Venice, California, so that he could tell them all about his association—how it worked, what exactly its formation had entailed, and, especially, the benefits it offered its members. For his part, Lohn was responsible for turning out the troops. On January 10, 1963, approximately two dozen manufacturers gathered at the Revell offices to listen to Blankfort's presentation, which ended, not altogether surprisingly, with an appeal to those who were gathered not to sign a deal with AMT.[23]

Blankfort got what he wanted: in the days that followed, he secured a few more decal deals. Lohn was happy too, for as the meeting adjourned, those in attendance agreed that it was a good idea to go ahead and form an organization. Two months later, on March 23, 1963, thirty-five people representing twenty-four different speed equipment companies showed up at Dean Moon's Gay Nineties Bar to formally establish an association. Blankfort was there as well, for he had offered the American Model Association's bylaws as a model for those of the new aftermarket body. After drawing up their rules and regulations, those in attendance voted unanimously to file a charter with the State of California, and two months later, on May 13, 1963, the legislators in Sacramento approved their request.[24]

By the time of its official launch that May, thirty-five Southern California companies had signed on to the idea of the Speed Equipment Manufacturers Association, or SEMA. However, none of those thirty-five was particularly well-known among the general public, so the association's members voted unanimously to invite Ed Iskenderian to join them as an additional charter member *and* to serve as their first president. Isky had not been a member of the credit managers group, and he had not attended any of Lohn, Blankfort, and Moon's organizational meetings. But Isky was a colorful character who was better known outside the rodding fraternity than the likes of Roy Richter, Dean Moon, or Harry Weber; much to their delight, he accepted their invitation.[25] Association finances were tight, but fortunately for the fledgling group, Ed Elliot agreed to serve as its executive director, and from his office at Elliot-McMullen, SEMA launched its first campaign.[26]

Over the winter of 1962–63, the Federal Trade Commission (FTC) had begun to investigate the business practices of the speed equipment industry, for it had received a number of complaints regarding pricing inconsistencies in manufacturers' and distributors' advertisements. With the backing of their newly chartered industry association, four of SEMA's members flew to Washington, D.C., to meet with FTC officials. The agency's complaint turned out to be relatively minor, and after a brief discussion, SEMA's representatives were able to secure its resolution.[27] Little did they know, as they boarded the plane for the long flight home, that within a few short years flights to D.C. on behalf of SEMA would be all too common.

Indeed, in the summer of 1963, federal automotive regulations were virtually unheard of, and several years would pass before they would even become a minor problem for the OEMs, let alone a major concern for the high-performance industry. Instead, what tended to trouble SEMA's charter members at the time wasn't any different from what had bothered many of Weiand's credit managers back in 1961 and 1962—namely, the new "NHRA Approved" program. Rather quickly, charter members Bob Spar and Holly Hedrick of Schiefer Manufacturing therefore moved to create a technical committee, volunteering their time and their companies' resources in an effort to develop an industry-based replacement for the NHRA's system.

With the full support of Wally Parks and Jack Hart of the NHRA, Spar and Hedrick's new committee got to work that fall. They aimed to develop a set of minimum performance-based specifications for each of the NHRA's many product categories; their goal, in other words, was to replace the NHRA's hated list of approved brand names with a list of categorized specifications. Under the new system, individual aftermarket companies would need to submit to the SEMA Technical Committee independent test results demonstrating that their products met the applicable standards in order for them to be approved for NHRA racing. Once their parts were approved, companies would be permitted to "self-certify," in their catalogs and on their product packaging, that this particular supercharger-drive kit, say, meets that particular set of specifications. Yearly reviews of the standards would ensure that the program kept up with the frantic pace of technological change within the racing world, and participating companies would need to submit further independent testing results whenever the committee's annual reviews revealed the need to modify an extant standard. For their part, all that the NHRA's technical inspection officials would need to worry about was the question of whether or not the aftermarket parts that were fitted to a given racing entry carried SEMA's seal of approval. Accordingly, from the point of view of the average weekend warrior, very little would actually change: would-be racers would still need to build their cars using only officially sanctioned parts and accessories. For the industry, however, the difference would be critical: *they*, not the NHRA, would set the relevant standards and police their enforcement.[28]

"They," of course, meant SEMA—specifically, SEMA's Technical Committee. And in order to ensure that each of its new standards would be less arbitrary than the NHRA's outgoing brand-name lists, Spar and Hedrick conceived of their new committee as one whose membership would change with each new specifications project. Spar and Hedrick would retain their permanent chairs, but the remainder of the committee would consist of other SEMA members considered to be experts in the particular product category then under review. At the behest of Jack Hart of the NHRA, for example, the Technical Committee's first project was to develop a set of standards for racing clutches and flywheels.[29] Accordingly, Spar and Hedrick called on Bill Hays of Hays Clutches, Paul Schiefer of Schiefer Manufacturing, and Harry Weber of Weber Speed Equipment—best known for their clutches, flywheels, and related products—to serve as Technical Committee experts for the clutch and flywheel project. Forty years later, Bob Spar vividly recalled the skepticism with which many SEMA members greeted the plan at the time. "You're never going to get Paul Schiefer and Bill Hays and Harry Weber in the same room at the same time," he was told on more than one occasion as he prepared for their first meeting. "You're just not—they *hate* each other."[30]

Spar and Hedrick's chosen experts were indeed fierce rivals. But on the appointed day, all of them showed up, shook hands, and together developed SEMA Spec 1-1. Subsequent iterations of the Technical Committee looked at everything from dragster-chassis construction to driver firesuits, and in every case, expert members of the group put aside their differences and worked together.[31] It was in their interest to do so, of course, for it was a chance for each of them to have a say in what would ultimately be required of them and their products. Besides, each of them knew that if they declined to participate, they might well be forced within a few short months to manufacture their racing-application parts to the SEMA standards developed by their participating rivals. A case in point is Bill Hays. Hays wasn't even a SEMA member when the plans for a set of clutch and flywheel standards were announced, but he was quick to answer Bob Spar's call to join the association—and its Technical Committee—so that Schiefer and Weber wouldn't be able to put him at a disadvantage in the racing-products market.[32]

It was never simply a matter of rival manufacturers sitting down and brokering a compromise, however. For in every case, the overriding concern, according to longtime Technical Committee member Carl Olson, was *liability exposure*. Olson, who managed the Trans-Dapt Company in the mid- to late 1960s while its owner, Willie Garner, served as SEMA's president, was a lifelong racing enthusiast who campaigned a rail dragster in his spare time. In 1966 the chair of the Technical Committee, Bob Spar, decided that the time had come to bring another permanent member on board, preferably someone with real-world racing experience. Garner recommended Olson, who reluctantly agreed to assume the additional responsibility. Olson therefore missed the Technical Committee's formative first few years, but what he witnessed during the late 1960s is telling: committee members, in everything they did, always assumed the worst. If, for example, they were talking about lightweight flywheels, they would frame their work in terms of disintegration and explosion—the worst in-use flywheel-failure scenarios imaginable. They would also assume end-user incompetence: incorrectly-torqued bolts, off-center installations, and even inappropriate applications. In so doing, they would ultimately arrive at a set of flywheel specifications that were designed to minimize the impact of the worst-case outcome that the least competent, most bone-headed user might inadvertently unleash on himself, his opponent, and the crowd in the course of a quarter-mile run. This is turn would minimize "Specs"-compliant manufacturers' product-liability exposure as well as that of dragstrip owners and the relevant sanctioning bodies.[33]

On the other hand, SEMA's liability exposure grew, for in the end, *it* was the organization that set the standards, and *its* was the official seal of approval that appeared on Specs-compliant product packaging.[34] Nevertheless, not until the early 1970s would the association's board of directors come to be at all concerned with the liability nightmare implicit in its Technical Committee's work. Instead, the board actually sought to broaden the applicability and the appeal of its committee's actions during the 1960s. In the fall of 1965, for example, SEMA arranged a meeting between its Technical Committee and a number of officials from the American Hot Rod Association (AHRA), the United Drag Racers Association (UDRA), and the NHRA both in order to obtain the input of these racing sanctioning bodies and in order to better position itself as *the* official and trustworthy source of racing-safety specifications.[35] Two years later, toward the end of 1967, SEMA's board of directors voted to formalize its Technical Committee's work, establishing the "SEMA Specs" program—also known as the "Meets SEMA Specs" program—as an independently managed functional division of SEMA.[36] By the early 1970s, the program had resulted in the establishment of dozens of separate racing-product specifications, but at the same time, it was no longer the unqualified success it had been when it first was launched. For as its lists of specifications continued to expand, verification became a serious problem.

After all, "SEMA Specs" was a program based on trust. SEMA trusted that participating companies would actually manufacture their off-the-shelf parts to the very same specifications as their initial, officially approved prototypes. And because the specifications SEMA issued were *performance-* rather than *design-* based, it wasn't always easy to determine whether a given firm's run-of-the-mill products actually met the applicable standards. Abuse was rare, but it was always a possibility.[37] As early as 1967, therefore, the Technical Committee began to develop a random testing process for its fledgling program, as we will shortly see.[38] But even in the early 1970s, the looming specter of widespread "Specs" abuse still troubled a number of SEMA members. By the middle of the decade, in fact, many of them wondered aloud whether the program might in fact be far more trouble than it was worth, at least in terms of the association's liability exposure. In 1978 SEMA therefore voted to further distance itself from the program, reestablishing it as an independent, wholly-owned subsidiary known as the SEMA Foundation, Incorporated. Ten years later, lingering concerns finally prompted SEMA's board of directors to sever its ties with this foundation entirely.[39]

But now we're twenty years ahead of ourselves. For back in 1968, the "Specs" program was an unqualified success for the association. In fact, it was in many respects SEMA's raison d'être. It was where the association spent the bulk of its dues, and it was by most accounts responsible for winning over countless erstwhile doubters—the many manufacturers who simply weren't interested in the notion of an industry-wide organization back in 1963—and swelling the association's ranks.[40] It also gave SEMA a taste of things to come, for although no one could have known it back in 1963, 1964, or even 1965, the "Specs" program would ultimately serve as a strategic prototype for SEMA's response to local, state, and federal regulations in the late 1960s, the 1970s, and beyond. Worst-case assumptions, performance-based specifications, round-table discussions, inter-firm cooperation, and a measure of self-restraint, that is, all of which were vital to SEMA's self-regulatory efforts of the early 1960s, would inspire and directly inform its approach to the challenges later posed by the National Highway Traffic Safety Administration, the Environmental Protection Agency, and the California Air Resources Board.

SEMA AND THE HIGH-PERFORMANCE INDUSTRY

During the early to mid-1960s, the Technical Committee's racing-specifications program was the Speed Equipment Manufacturers Association's most important project, but it was by no means all that the new group did. For in the early 1960s, "hot rods" and "hot rodders" were once again attracting negative publicity, particularly through television and newspaper exposés on illegal street racing and the alleged dangers of the quarter-mile strip.[41] Therefore, as a secondary associational cause, SEMA immediately began to try to nip this problem in the bud by hosting meetings with law enforcement agencies and community leaders, by encouraging professionalism among its members, and, it seems, by installing a well-known personality as its president.[42] For Isky was chosen largely for his million-dollar smile: he was seen as someone who could help the industry burnish its image.

For the most part, though, the association spent its first few years promoting a number of internal projects. First and foremost among these, of course, was its fledgling "Specs" program, but in addition, SEMA negotiated a group-rate health insurance contract for its members and their employees; hosted frequent seminars to deal with pressing issues such as manufacturer-wholesaler and wholesaler-dealer relations; and, perhaps most importantly, began to hold a few of its regular meetings in places like Indianapolis and New York in order to foster stronger ties between its East Coast and West Coast members.[43] As important as these efforts were, however, none of them, with the possible exception of the "Specs" program itself, held a candle to the group's pièce de résistance, the Annual High Performance and Custom Equipment Trade Show.

Better known as the SEMA Show, this yearly exposition remains the focal point of the aftermarket calendar to this day, and it is truly a sight to behold.

Hundreds of thousands of exhibitors, representatives, and journalists from all over the world gather every November in Las Vegas, Nevada, where the show has now been held for more than thirty years, to catch up with old friends, to show off their new products, and, of course, to conduct business. Though it is closed to the general public, the SEMA Show nevertheless ranks among the largest trade shows in the world today. In fact, the show fills the city's massive convention center on Paradise Road and spills over into nearly every parking lot and hotel conference room in the surrounding area. Steeped in tradition and long-enshrined in SEMA's mission statement, the Annual High Performance and Custom Equipment Trade Show[44] is largely taken for granted these days, especially among younger enthusiasts and aftermarket entrepreneurs. But in the mid-1960s, there was no annual high-performance exhibition of any kind. Consequently, speed equipment manufacturers who wished to showcase their products had to do so publicly, at major drag-racing meets or at large car shows like the annual Motorama in Los Angeles. Some also set up booths at closed-to-the-public events like the yearly Automotive Aftermarket Manufacturers Association (AAMA) Show, which alternated between New York and Chicago. Though the AAMA was a replacement-parts aftermarket association, its show was the best available closed-access option for high-performance parts manufacturers.[45] These venues were less than ideal, of course, but SEMA was still a very new association in the mid-1960s, a cash-strapped body without the means to launch a trade show of its own. Several of its members grumbled regularly that this ought to be among its top priorities, but on its own, there was really very little SEMA could have done.

Robert Petersen of the Petersen Publishing Company, on the other hand, was an interested party with the means—and, by 1966, the motivation—to see just such a project through. That September, Petersen launched a new closed-circulation magazine called *Hot Rod Industry News*, a trade publication sent free of charge each month to thousands of manufacturers, wholesale distributors, and speed shops across the United States.[46] Billed as "The Voice of the High Performance and Custom Industry," Petersen and his editors hoped that *Hot Rod Industry News* would help to unite the speed equipment industry in the interest of its long-term growth. This, of course, was also one of SEMA's goals, and from the very beginning, the editors of this new publication—men pulled directly from the ranks of the country's many speed equipment manufacturers and distributors—therefore gave considerable editorial space to SEMA and its tentative endeavors. As a result, Petersen and his editors knew that there were many within the new association who wanted a yearly trade show; and for their part, Petersen's editors believed that "a trade show was a 'must' activity for the[ir new] magazine,"[47] a sure-fire way to promote their fledgling publication.[48] That fall, Petersen and

SEMA's board of directors therefore hammered out a deal. *Hot Rod Industry News* and the Speed Equipment Manufacturers Association would co-sponsor an annual, closed-access, high-performance trade show, and the task of planning and organizing the event would fall to Petersen Publishing's Special Events Department—specifically, to a young man within that department by the name of Dick Wells.

Wells was a lifelong automobile enthusiast who moved to Southern California from Lincoln, Nebraska, in the late 1950s. Shortly after he arrived, he managed to secure a job at Petersen Publishing's *Hot Rod*. There he advanced through the ranks rapidly, earning a promotion into the company's Special Events Department in the mid-1960s—just in time to be the point man for the SEMA Show project. But he wasn't simply given a carte blanche. For as it happened, Petersen was a good friend of the owner of the Los Angeles Dodgers, Walter O'Malley, who had already managed, long before the project fell into Wells's lap, to convince Petersen that the covered area under the grandstands at Dodger Stadium would be the perfect venue for the show. "It was a terrible, terrible thing to do," Wells would later claim, for the area under the grandstands was cramped and dank.[49] Besides, SEMA wasn't even the only group slated to be at the facility on the days appointed for the show, January 10–12, 1967. The Soviet Union's men's soccer team was scheduled to be in town, and it had already arranged to hold practice sessions on the field at Dodger Stadium on those very days.[50] Nevertheless, in spite of the additional security and the somewhat less-than-ideal nature of the facilities, the show went off without a hitch, and by the standards of its day, it was a phenomenal success. Wells sold 98 booths for the three-day "pipe-and-drape" affair, at which 120 industry representatives played host to more than 3,000 distributors, dealers, and speed shop owners from across the United States and Canada.[51] Planning for the 1968 Show began almost as soon as the gates closed, and Wells and the Special Events crew at Petersen quickly secured the necessary space at a larger and far more appropriate facility, Anaheim's new Convention Center in Orange County.[52] Attendance in 1968 rose to 3,800, with more than double the number of booths, and from that point on, the show was a vital part of SEMA's calendar.[53]

Together with the "Specs" program, the Annual High Performance and Custom Equipment Trade Show helped establish SEMA as a legitimate, purpose-driven industrial organization. Long before the ink dried on its first Spec and the turnstiles began to rotate at Dodger Stadium, however, external events far beyond the immediate control of the Speed Equipment Manufacturers Association began to gather steam, events that would require a radical revision of the group's core mission. For in the early to mid-1960s, urban air pollution and automobile safety swiftly worked their way onto the national agenda. And as Congress and a number of state legislatures scratched and clawed for ways to solve these pressing issues, both the OEMs and the entire automotive aftermarket found themselves on the defensive, their practices and their products scrutinized as never before. And for those aftermarket companies that manufactured speed equipment, SEMA's actions quickly came to be the heart and soul of their collective defense.

CHAPTER EIGHT

"INK-HAPPY DO-GOODERS," 1960–1978

Back in 1960, American automobility was at its zenith. Car ownership was at an all-time high, the American automobile industry was prospering as never before, Eisenhower's Interstate Highway Project was well underway, and auto-centric architectural and land-use patterns had become the norm in metropolitan areas across the United States. Cheered by many Americans as evidence of postwar progress and prosperity, these developments had come at a steep price, however. Fifteen years earlier, just after World War II, residents of the booming Los Angeles area were puzzled by the appearance of a brownish haze that enveloped their city during the daylight hours. Subsequent research concluded that this haze, known as photochemical smog, was largely due to automobile exhaust emissions.[1] Meanwhile, as aggregate miles and average speeds climbed, the number of annual automobile-accident deaths rose, steadily creeping toward the 40,000-per-year mark in the late 1950s.[2] Together with widespread reports of dishonest practices at automobile dealerships, a general level of new-car fit and finish that was hit or miss at best, and the specter of neighborhoods and cities torn apart by endless ribbons of elevated superhighways, these growing problems had, in the words of James J. Flink, led "many Americans [to begin] to have critical second thoughts about the automobile industry and its product."[3] Many, that is, began to doubt the wisdom of Charles E. Wilson's famous dictum, for indeed, it was no longer crystal clear that what was good for General Motors actually *was* good for America.[4]

In spite of their concerns, Americans continued to purchase new cars at a record pace as the 1950s came to a close. And as they did, they expressed an un-ambiguous preference for luxury options such as radios, automatic transmissions, power windows, and large-displacement engines over those geared more toward safety or economy. Most of the OEMs offered seat belts as optional equipment by the mid-1950s, for example, and Ford even offered a comprehensive safety package. But very few new-car buyers were actually willing to pay for these extra-cost features.[5] Or so the OEMs claimed, for critics both at the time and since have argued that the industry's inability to sell safety during the 1950s and the early 1960s wasn't actually due to consumer preferences. Instead, they claim, the OEMs' collective emphasis on performance and style in its designs and advertising came at the expense of any real attempt to engineer and sell meaningful safety features.[6] Either way, in reality power, style, and convenience sold well, and safety did not.

Neither the automobile companies nor the federal government seemed to be particularly concerned about any of this. The OEMs did make fleeting gestures toward the goal of improved highway safety during the 1950s, enacting an industry-wide, AMA-backed racing ban in 1957 and establishing a formal body to develop automobile safety standards known as the Vehicle Equipment Safety Compact (VESC) in 1958.[7] For its part, the government did begin to fund initiatives during the 1950s to explore the root causes and the long-term implications of urban air pollution.[8] But as the 1960s dawned, all of this was ancient history. The OEMs never actually took their self-imposed "racing ban" very seriously, and by 1960 all of them once again emphasized horsepower and performance in their advertising and openly sponsored racing associations and teams.[9] In addition, the industry's underfunded, understaffed VESC only managed to produce a single new-car safety standard.[10] Moreover, although the California legislature did make the quantum leap from passive research to meaningful action when it passed its landmark Motor Vehicle Pollution Control Act in 1960, Congress did not follow suit.[11]

By 1963, however, the federal government was no longer able to ignore the fact that urban air pollution was getting worse, and not just in Los Angeles. More to the point, public opinion was shifting as well, for although few were willing to give up (or even curtail) their new auto-centered lifestyles, an ever-larger share of the voting public was equally unwilling to live with dirty air. Congress therefore passed the first of what would become a series of Clean Air Acts that year.[12] So began the era of the regulated automobile.

Not until the end of the 1960s did any of these new federal and state pollution-control acts begin to have a measurable impact on the speed equipment industry's affairs, however, and not until the early 1970s did they begin to pose a serious problem for the majority of ordinary rodders. Instead, what troubled enthusiasts and entrepreneurs alike, in the early to mid-1960s, was the growing furor over automotive safety.

By the late 1950s, the total number of annual highway fatalities in the United States had been on the rise for years, but the fatality *rate*—whether expressed as a function of total population, aggregate miles, or number of registered vehicles—had been on the wane for more than two decades.[13] During the early 1960s, however, the rate began to edge up marginally, rising from a seventeen-year low of 20.8 deaths per 100,000 people in 1961 to 27.1 five years later.[14] Some blamed this sudden turnaround on the carelessness and lack of proper training of individual drivers; others blamed it on organized motorsports, for the "competitive driving habits" that oval-track, drag, and road racing allegedly encouraged among American motorists.[15] But in the end, those that blamed it on the OEMs won the day. In 1965 a young attorney named Ralph Nader published a scathing little book, *Unsafe at Any Speed*, in which he claimed that the major American automobile manufacturers deliberately ignored their products' "designed-in dangers" for the sake of their bottom lines.[16] Flawed though his analysis was in many of its specifics, Nader's overarching, two-pronged argument—that something needed to be done about automotive safety, on the one hand, and that the automobile industry shouldn't be allowed to decide what that something ought to be, on the other—was nevertheless exceedingly influential.[17] One year after Nader's book first hit the shelves, Congress passed and President Johnson signed the National Traffic and Motor Vehicle Safety Act of 1966.

This act established a new agency, the National Highway Traffic Safety Administration (NHTSA), under the broader organizational umbrella of the U.S. Department of Transportation. Its primary task, at least at first, was to develop a set of automobile safety requirements, mandatory standards that would apply to all new cars sold in the United States after January 1, 1968. However, new-car model years do not begin on New Years Day, but rather several months earlier. Thus, by the time Johnson signed the act into law in September of 1966, the automakers had already released their 1967 cars, so NHTSA effectively had less than a year from the date of its inception to develop the necessary standards, issue them to the automakers, and devise a way to enforce them. Fortunately for its administrators, another federal agency, the General Services Administration (GSA), had already established a set of safety standards for all cars purchased by the government back in the early 1960s. NHTSA therefore borrowed liberally from the GSA's list, and by the end of 1966 it had released its own.[18]

NHTSA's list established basic, new-car safety-performance guidelines for twenty different categories of automotive components. Dashboards, for example, were to be padded; seatbacks were to have integral headrests; braking systems were to be of the fail-safe, dual-circuit variety; windshields were to be of safety glass; and so forth. The government was doing just what Nader wanted, holding the OEMs accountable for the carnage on the highways and seeking, through a set of independent and specific mandates, to ensure that in the future, drivers and passengers alike would be better protected.

However, the National Traffic and Motor Vehicle Safety Act of 1966 *also* charged NHTSA with the task of generating "used car" (or more accurately, "in-use") safety standards—standards, that is, that could be applied to cars that had already left the showroom floor. Vehicle owners would be required to replace their brake shoes, for example, when they wore down to a certain point. The same was true of tires, shocks, tie rods, control arms, and any other part of a car that normal wear and tear might ultimately render unsafe. This the agency declared in a list of in-use automotive safety standards published in the fall of 1968.[19] However, NHTSA never planned to enforce these in-use standards itself—after all, how on earth could its administrators possibly know for certain whether Mr. Smith's Ford in Miami was equipped with adequate brake shoes, say, or that Mrs. Jones's Cadillac in Seattle had a decent set of tires? With this in mind, Congress stipulated in the language of the 1966 act that each of the fifty states should develop an annual inspection program to enforce NHTSA's in-use mandates. Technically, the federal government could not *require* the states to do this, but it could threaten to withhold federal highway funding from those states that chose not to. Thus, the 1966 act declared that 10 percent of nonparticipating states' annual federal road-construction and maintenance subsidies would be withheld until their lawmakers got with the program. Some states already had an annual inspection system; for them, it was simply a matter of registering their extant programs with NHTSA. Many others had never had such a program, but by the end of 1968, forty-five states either had an annual inspection system in place or had one in the works.[20]

The ways in which the OEMs responded to the passage of the National Traffic and Motor Vehicle Safety Act in 1966 and NHTSA's subsequent mandates have been described at length elsewhere.[21] For our purposes, it will suffice to note that the OEMs were unhappy with the federal government and its requirements, but in the end they complied with the law. However, the 1966 act was a comprehensive piece of legislation, and in a number of ways its provisions applied to the likes of Bob Spar, Els Lohn, and Roy Richter too. First and foremost, the new law *directly* targeted the products of the automotive aftermarket: NHTSA's 1966 new-car standards applied to *any* company that produced *any* of the parts in *any* of its twenty different categories. A number of general-replacement and high-performance aftermarket firms manufactured seats, brakes, and bumpers, for example; these firms, just like Chevrolet and Ford, were now required to certify their designs with NHTSA. Certification testing promised to be neither straightforward nor cheap, and aftermarket leaders immediately began to worry that individual companies, particularly those that were small, might be unable to afford to certify their parts.[22]

Indirectly, however, the 1966 act was far more troubling, particularly for the high-performance end of the aftermarket. For its very passage provided more than a measure of legitimacy for those who sought to associate performance enthusiasm in general—and anything having to do with "speed," in particular—with highway fatalities. Soon NHTSA would float the idea of requiring top-speed limiters on new cars, insurance companies would begin to hike their muscle car premiums, and the hot rodding fraternity would find that it had once again been saddled with precisely the same sorts of negative associations against which it had fought so earnestly a decade and a half before.[23]

Even more troublesome for speed equipment firms was the fact that the new law delegated the task of enforcing its in-use standards to state and local agencies. And in so doing, the legislation granted considerable interpretive leeway to lesser officials. For the 1966 act charged the states with the task of making sure that local motorists maintained their cars' original safety-performance capabilities, but never once did it define "original," and never once did it declare NHTSA's word to be the final say. Consequently, many state and local agencies took NHTSA's list of in-use standards to be little more than a starting point. As they formalized their programs, that is, many of them felt compelled to go the extra mile in order to ensure that the vehicles under their immediate jurisdiction were maintained in accordance with their own particular understandings of "original performance." By the end of the 1960s, many local transportation departments and inspection officials across the United States therefore had, under the open-ended authority of the 1966 act, begun to rule out modified suspensions, for example, on the grounds that cars with stiffer springs, shorter shocks, and thicker anti-roll bars would no longer handle as they had when they were new. Similarly, custom wheels of smaller (or larger) diameter (or width) than stock were often judged to be problematic, as were aftermarket seats, auxiliary lighting schemes, and add-on airfoils. For in the eyes of a growing number of local officials, automotive parts that weren't stock simply weren't needed.[24]

Both for those with modified cars and for those who manufactured add-on parts and accessories, this was an ominous development. The notion of government automotive regulation wasn't altogether new in the mid- to late 1960s, of course, either for enthusiasts or for speed equipment companies. Back in 1951, California's rodders had learned to live with a new state law that required their roadsters to be equipped with fenders, and throughout the 1950s, motorists elsewhere had dealt successfully with local laws that covered everything from aftermarket dual exhausts to headlight height.[25] What *was* new, though, was the sheer breadth of these newer laws. Together with a wave of municipal- and state-level noise-control acts (not to mention federal and state emissions control requirements) that also emerged in the late 1960s and early 1970s, these automotive safety mandates therefore sparked a multifaceted response from ordinary rodders, well-placed journalists, and speed equipment manufacturers.

Enthusiasts and journalists swiftly launched a war of words. In May of 1966, *Hot Rod*'s Bob Greene fired off an editorial ridiculing the federal, state, and local officials who pioneered these new laws as "overenthusiastic or ink-happy do-gooders."[26] Later that same year, one enthusiast, John P. Keelan, suggested that the government's decision to regulate the automobile industry and hot rodding was but part of a larger and manifestly un-American socialist conspiracy. Keelan also portrayed legislators and regulatory bureaucrats as ignoramuses who were barking up the wrong tree: "In short, the same pack of idiots who can't patch a street properly are going to tell Detroit how to design and build cars! Never bothered much by facts, our politicos have disregarded an age-old truism: people cause accidents."[27] Throughout the late 1960s and into the 1970s, others often echoed Keelan's sentiments: regulation is un-American, it attacks the problem of highway safety from the wrong angle, and those within the government who support it are reckless morons.[28]

Likewise, anyone who dared to write letters to the popular periodicals in *defense* of these bureaucrats and their regulations were themselves vilified as treacherous halfwits. When letter-writer Marc Sheridan claimed in November 1972 that "the do-gooders are helping our health, our future, and our sport" with their regulations, *Popular Hot Rodding* lambasted his naiveté: "Just what are you talking about, tiddlywinks? Some of the proposed state and federal legislation would virtually wipe out hot rodding by imposing strict regulations on what an individual could do to his car . . . Can't you see the dangers here?"[29]

The overwhelming majority of letter-writers agreed with *Popular Hot Rodding*'s sentiments, and by the mid-1970s, letters and editorials characterizing governmental regulation as an unconstitutional assault on individual liberties, as an unjust attack on free enterprise, and even as incontrovertible evidence that Orwell's dystopian vision was beginning to come true were common fare in *Hot Rod*, *Popular Hot Rodding*, and the others.[30] In addition, the Petersen Publishing Company itself began to run a series of regular editorials in its periodicals in 1973, scathing two-page spreads that typically combined outrageous commentary with an equally audacious cartoon.

Letters that teemed with paranoid suggestions, editorials that embraced a numbing pessimism, and publisher's remarks that boiled over—these are what are used by those who have written about the 1960s and 70s to capture the essence of the hot rodding fraternity's response to the advent of government regulation.[31] Unfortunately, such an approach to the events of this period altogether overlooks the many other more creative and proactive ways in which enthusiasts, journalists, and speed equipment manufacturers sought to deal with the challenges they faced. They were mad, that is, but they were also motivated.

SEMA AND NHTSA

When President Johnson signed the National Traffic and Motor Vehicle Safety Act into law in the fall of 1966, the Speed Equipment Manufacturers Association was still a marginal group. This is not to say that its first three years were unproductive, however. Its fledgling "Specs" program had already begun to transform the ways the largest drag-racing associations in the United States handled the delicate matter of self-regulation. In addition, SEMA had begun to organize the first industry-wide trade show, scheduled to take place the following January. Finally, and perhaps most importantly, the association had by the fall of 1966 proven to most of the naysayers within the Los Angeles speed equipment industry that it was indeed possible for rival manufacturers to work together to promote their common interests. But SEMA was small. Its funds were limited; its staff was part-time; and, in the fall of 1966, its name was virtually unknown among outsiders and even among enthusiasts.[32]

Nevertheless, in the wake of the National Traffic and Motor Vehicle Safety Act's passage, the members of SEMA's governing board quickly realized that they were going to have to step up. Fearful of what they believed to be the ominous implications of the new act, they were convinced that hot rodding desperately needed an organized means through which to promote its cause. But they also knew that none of the extant, broad-based associations were likely to act decisively to defend on-road hot rodding. The National Hot Rod Association (NHRA), for example, gradually lost interest in the street scene during the late 1950s and early 1960s, and by 1966, its exclusive focus was the promotion of drag racing. The same was true of its main rivals, the United Drag Racers Association (UDRA) and the American Hot Rod Association (AHRA).[33] SEMA could not afford to write off on-road rodding, however, because its members' profits overwhelmingly derived from the sale of street-use products. Unwilling to sit idly by — and, for that matter, unwilling to stake their collective future on the power of the written word — SEMA's board of directors voted to act.

At the time, according to industry pioneer and SEMA member Willie Garner, their top priority was "to get [their] industry's side of the picture before the public and the lawmakers."[34] Consequently, their first move, in the winter of 1966–67, was to hire a Washington, D.C., representative. After interviewing several candidates, SEMA chose Earl Kintner, a partner in the D.C. law firm of Arent, Fox, Kintner, Plotkin, and Kahn.[35] SEMA hoped that Kintner would be able to open a dialogue between the association and the National Highway and Traffic Safety Administration, but Kintner, a seasoned veteran of the D.C. scene, suggested to the group that it ought to tweak its name a bit before sending him on his rounds. "A name change would assist greatly our representation," he explained to the board of directors in February of 1967, because "elderly bureaucrats are not likely to appreciate the 'swinging' generation's preoccupation with 'speed.'"[36] SEMA's members agreed, and by the time Earl Kintner placed his first phone call on their behalf in March, they had voted to amend their body's appellation: henceforward, SEMA was to stand for the more politically correct *Specialty* Equipment Manufacturers Association.[37]

By the end of the spring, Kintner had established a working relationship with Dr. William Haddon Jr., the head of NHTSA and its parent agency, the U.S. Department of Transportation. In a series of meetings, Kintner explained to Haddon that above all else, SEMA hoped to obtain a clarification of NHTSA's expectations vis-à-vis its "new car" standards. Were any of these standards open to further interpretive negotiation? How exactly would the certification process work for aftermarket firms? Would it involve a centralized program with official inspectors, or could it instead be something more along the lines of SEMA's own self-regulatory "Specs" program?

By the beginning of June, Kintner had his answers, and he reported them first to SEMA and then to the staff of *Hot Rod Industry News*. For starters, Haddon agreed to relieve the entire automotive aftermarket—its high-performance and its standard-duty sectors alike—of the responsibility for no less than four-fifths of the initial regulations. Aftermarket manufacturers, that is, would be required to demonstrate compliance through certification for only four of the original twenty categories of automotive parts: brake lines, glazing materials, seatbelt assemblies, and wheels. In addition, Haddon explained to Kintner that final certification procedures for the aftermarket had yet to be determined but that they would be open to negotiation; as a stopgap measure, he agreed to allow aftermarket companies to self-certify, a la the "SEMA Specs" program, that their products met the applicable standards. In return, Kintner explained to Dr. Haddon that the speed equipment industry supported the spirit and the goals of the 1966 act and that SEMA would do everything in its power to ensure that in the years to come, hot rods, hot rodders, and the hot rod industry would be part of the solution, not the problem.[38]

SEMA's leadership was pleased, but it also knew that Kintner's compromise was temporary: no one knew for certain when NHTSA's final certification procedures would be handed down, and no one knew exactly what they might entail. So during the fall of 1967, SEMA officials focused on *their* end of the bargain. For they read *Hot Rod* and *Popular Hot Rodding* each month too; therefore, they knew that there were plenty of rodders out there who were furious with the "do-gooders" in Washington, and they also knew that many of them still raised hell in their local communities by racing on the streets.[39] Thus, SEMA faced the challenge of reining in these overzealous rodders without alienating the *average* rodder. On the one hand, that is, the association needed to demonstrate that it remained firmly on the side of the authorities when it came to highway safety lest it jeopardize its working relationship with NHTSA. But on the other hand, SEMA also needed to make sure that ordinary enthusiasts understood that it too was adamantly opposed to regulations that would unduly restrict their freedom to tinker with their cars.

SEMA attacked the problem from two angles. First, in August, *Hot Rod Industry News* published a lengthy interview with SEMA's new president, Willie Garner, in which he urged dealers and speed shop owners to lead the charge. "However odd it may sound to some," he said, "one of the best courses we can take is to make a conscious effort to cooperate with local law enforcement agencies. . . . Currently, dealers would be wise to encourage their customers to obey all local laws. The worst thing to happen could be a rash of arrests on 'hot rodders.' That kind of adverse publicity can't help anybody, especially our industry."[40] Garner's charge was simple: ordinary speed shop owners needed to convince their die-hard customers that compliance in the short term served their long-term interests. And in *Hot Rod Industry News*, many others pushed the same speed-shop-based strategy for several years.[41]

Rodders were a fiercely independent bunch, however, and Garner's crew knew that many of them were unlikely to actually heed the legal counsel of their local countermen. Hence their second angle, a direct appeal to hot rodding enthusiasts to stand in an "equal partnership" with SEMA to pursue their common regulatory interests. To do so, SEMA called on a well-respected old-school rodder, Don Francisco, who published his appeal to the rodding fraternity that October in a hard-core racing tabloid, *Drag News*:

> The problem both SEMA and hot rodders have is to convince the legislators that safety and smog programs don't have to outlaw hot rodding. To do this, we will have to prove that we are willing to work with the legislators in any way possible, and demonstrate that parts and methods used by responsible rodders do not make a car more unsafe nor cause it to add to the smog problem. (In many instances the car is made safer and the amount of smog it produces is reduced.) This is a big and

important project. It will require the complete cooperation of everyone who is even remotely concerned with hot rodding.[42]

Francisco clearly hoped that enthusiasts and entrepreneurs would work together to promote responsible rodding, but whether or not his appeal (or Garner's, for that matter) actually swayed anyone is impossible to determine with any degree of certainty. But it did seem to give Earl Kintner an ounce of reassurance, as he continued to meet with NHTSA officials during the fall of 1967, that his clients were cleaning up their act. And whether or not Dr. Haddon ever saw these published appeals, he certainly had no reason to believe that SEMA was not living up to its word. In fact, he remained impressed with the association and its commitment to highway safety—so much so, in fact, that in November of 1967 he decided to formalize NHTSA's certification arrangements with it. Henceforward, Haddon announced, the stopgap measures he and Kintner agreed on earlier that year would be NHTSA's official aftermarket "new car" parts-certification procedures. Firms, that is, would be required to do their own testing and to self-certify, on their product packaging, that their brake lines, glazing materials, seatbelt assemblies, and custom wheels met all of the applicable federal safety standards.[43]

For SEMA, this was a major triumph. The speed equipment industry was going to be regulated, of course, but because of SEMA's intervention, the task of actually meeting the government's mandates was going to require no more effort or expense on the part of the typical manufacturer than that of meeting the industry's own self-regulatory "Specs" requirements. Nevertheless, the news of Kintner's success was greeted neither with the widespread popping of champagne corks nor with the celebratory smoke of parking-lot burnouts. Instead, SEMA's leaders were cautiously optimistic at best, for they were well aware that things could just as easily have turned out differently. Therefore, Garner and his allies closely studied what had actually happened over the past year, seeking to learn from their experiences with NHTSA.

They began by reevaluating their successful "Specs" program—or rather, by continuing to evaluate it. For during the summer and fall of 1967, when Kintner's deal with Haddon remained tentative, SEMA's board of directors instructed its Technical Committee to begin to look for ways to broaden the program's scope, streamline its specifications-development procedures, and strengthen its enforcement. As a result, the Technical Committee had already launched a pilot program for voluntary product testing by the time the final deal with NHTSA was announced.[44] Still, SEMA's board of directors wanted to be able to use its "Specs" program to demonstrate to anyone who cared to notice that SEMA was a dynamic organization committed to pushing the envelope on automotive safety.

Thus, during the course of 1968, 1969, and 1970, the program's home-brewed racing specifications grew stricter, and its product-testing procedures became far more comprehensive.[45] Perhaps most significantly, though, the board voted in the early months of 1970 to transform the program from a function performed by its Technical Committee into a stand-alone committee of its own known as "Meets SEMA Specs."[46]

In addition, SEMA's leaders continued to try to boost the association's membership in the late 1960s, convinced that when it came to legislative lobbying, "strength in numbers" mattered. SEMA's board of directors had, of course, worked to attract new members from the very beginning back in 1963, and in 1965 the board even decided to allow interested parties who were not manufacturers to join their association as nonvoting "sustaining members." As a result, SEMA had 141 dues payers by the fall of 1967.[47] During the late 1960s, though, SEMA's leaders felt that they needed to expand their recruitment efforts. In August of 1967, they ran a two-page spread in *Hot Rod Industry News*, claiming in part that "well-meaning, but misinformed, federal and state legislators are launching tireless attacks against the manufacturers and sellers of automobiles and automotive accessory equipment, all in the name of safety and cleaner air" and that "unless these lawmakers are made aware of the harmful effects their efforts could produce, we could all be in danger of losing our livelihoods. The Specialty Equipment Manufacturers Association is currently leading the battle to save our industry . . . we need your ammunition."[48] Less dire appeals from Willie Garner and others followed during 1968, 1969, and 1970, and in 1969 the board of directors voted once again to loosen SEMA's membership requirements.[49]

Much to their delight, their efforts paid off well: by the end of 1970, SEMA was collecting dues from more than 400 different manufacturers, distributors, retailers, and publishing houses.[50] And dues, in the end, were what SEMA *really* needed in the late 1960s and early 1970s. Washington representation was exceedingly expensive, as was the "Specs" program. Beginning with its second iteration in the winter of 1967–68, 20 percent of the profits from the Annual High Performance and Custom Equipment Trade Show therefore went to fund these efforts, which helped tremendously.[51] But ultimately, SEMA's leaders knew there simply was no substitute for steady organizational income in the form of dues. Hence their ongoing quest for "ammunition."

Strategically, the late 1960s and the very early 70s was also a vital period for SEMA. As we have seen, Willie Garner, Don Francisco, and other point men began to try to calm things down as early as 1967, urging average rodders, speed shop owners, and equipment manufacturers not to overreact to the new regulations, but rather to cooperate and comply with the laws as a sign of good faith. Compliance was only part of SEMA's budding regulatory strategy though. For none of the association's members were actually prepared to simply comply with anything and everything the folks in D.C. might dream up, and neither, for that matter, were any of the popular rodding journalists nor the millions of ordinary enthusiasts. SEMA's overarching strategy therefore combined its emphasis on self-control, cooperation, and compliance with a concerted effort to establish solid working relationships and ongoing dialogues with the relevant regulatory agencies. "We shall police ourselves, we shall work with the established governmental bodies, and we shall continue to be creative," declared SEMA's Eric Grant with confidence toward the end of 1970, and most within the industry agreed.[52] By the end of 1972, in fact, association members had enshrined these basic tactics in an official policy statement.[53] SEMA's hope was that through level-headed conversations with the authorities, its opinions, its concerns, and especially its members' years of accumulated hands-on automotive experience might be taken into account. Thus, cooperation, compliance, and constructive dialogue, much more than Hot Rod's inflamed editorials, actually formed the basis of the industry's collective response to regulation.

During the late 1960s, however, SEMA's cooperative and collaborative approach was still new and untried. It took a great deal more than Kintner's initial success with NHTSA in 1967, in other words, to actually convince dues-paying association members that this was indeed the proper way for them to proceed in the years to come. SEMA's leaders therefore continued to test and refine their tactics. In 1967, for example, the Department of Transportation and NHTSA began to toy with the idea of requiring mechanical governors on all new cars to limit their top speed to ninety-five mph.[54] The notion fizzled in the early months of 1968, but toward the end of 1970, NHTSA's new director, Douglas Toms, revived the idea by proposing a new federal motor vehicle safety standard known as "High Speed Warning and Control." Toms's idea was simple: new cars would be limited mechanically to ninety-five mph, and at eighty-five mph, their hazard lights would flash and their horns would sound a continuous tone to warn other drivers—and, presumably, the police—that the car was approaching its terminal velocity.[55] Toms solicited opinions on the idea from both the OEMs and the aftermarket in the winter of 1970–71,[56] and SEMA submitted its response in March.

The popular periodicals got wind of the proposal and began to kick and scream about this blatant example of "bureaucratic over-reaction,"[57] but SEMA nevertheless declared that it supported the idea—in theory, at least. Specifically, the association agreed with NHTSA that there was no conceivable reason for a car used on the street to have the capability to exceed ninety-five mph on a public highway, but it also argued that flashing lights and honking horns were no less hazardous than speeding cars. Furthermore, SEMA warned NHTSA of the dangers of mechanical governors—if, for example, the low-speed performance of a given car were in any way restricted as a byproduct of its ninety-five mph limiter, then it would in fact be much *less* safe because of its inability to deal with everyday situations like low-speed passing and freeway merging.[58] SEMA also pledged to work with NHTSA and the Department of Transportation to further refine the standard, if necessary, but in any event, resistance from the OEMs ultimately consigned the notion to the regulatory scrap heap.[59]

Still, SEMA's willingness to work with the authorities on this (and other) federal automotive safety proposals did not go unnoticed. In January of 1971, Toms himself accepted an invitation to address association members at their annual trade show in Anaheim; there he told those in attendance that if they would continue to work closely with the authorities in the future, as they were at the time with NHTSA on the question of the ninety-five mph limit, then their industry would continue to thrive. "You fellows have too many resources, too much talent and far too much inventive power to allow any other conclusion," he declared to thunderous applause as he wrapped up his remarks.[60] One year later, also at the SEMA trade show, Undersecretary of Transportation James M. Beggs echoed Toms's claims: "We're not regulators. We're not automotive dictators. We're not the engineering and manufacturing experts. You are. And we need your help and support. I'm very optimistic about the degree of vehicle safety that can be built into an automobile . . . This is not to say that original equipment must be maintained. Doug Toms tells me, in fact, that aftermarket equipment is often of a higher quality than the original. Your contributions toward producing high-quality equipment can actually increase safety performance."[61] SEMA's cooperative and collaborative approach was beginning to work to the association's advantage, winning it some key allies in Washington. In fact, Doug Toms was so impressed with SEMA that he made a practice of consulting closely with it, and SEMA, equally delighted with Toms, unanimously voted him its "Man of the Year" in 1972.[62]

We must be careful not to overstate the extent of SEMA's involvement with the authorities in Washington during these years, though. SEMA did maintain a working relationship with NHTSA, and it did continue to refine its regulatory strategy in the federal city. However, SEMA's leaders also realized during the spring of 1968 that Kintner's deal with Haddon had ushered in a period of relative calm on the federal level. They recognized, that is, that at least for the time being there was no longer any need for full-time legal counsel in Washington. Back in California, on the other hand, the industry faced stricter regulations from that state's Motor Vehicle Pollution Control Board.[63] Association finances in 1968 were tight: not for another couple of years would SEMA's ongoing recruitment efforts actually begin to pay off. Consequently, Willie Garner and SEMA's board of directors redirected their scarce resources to the place where they were most urgently required. And in the spring of 1968, that was Sacramento, California. Within the span of a few short months, SEMA therefore released Kintner and his D.C. law firm, set up an expanded office in California, hired an experienced industry insider named Dan Roulston to act as its first paid and full-time executive director, and lured a young lawyer from the staff of the MVPCB, Eric Grant, to serve as SEMA's counsel in the Golden State.[64] All of this was done to better enable the association to deal with the pressing issue of California emissions, and for several years, it proved to be a gamble more than worth its while.

By the end of 1971, however, SEMA's federal fortunes began to shift as the lull of the late 1960s came to an end. Specifically, the passage of the Clean Air Act of 1970 meant that smog control was soon to become as much if not more of a challenge for the speed equipment industry on the federal level as it was in the late 1960s in California.[65] This development alone was enough to change SEMA's regulatory priorities, but as it happened, it coincided with a renewed push within NHTSA for stricter federal oversight of aftermarket manufacturers.

In the fall of 1971, SEMA's newly inaugurated president, Vic Edelbrock Jr., received a phone call from NHTSA's Doug Toms. Toms told Edelbrock that he and his associates had begun to receive reports of quality-control problems among aftermarket custom wheels, and he warned Edelbrock that if SEMA could not whip its wheel producers into shape, then he would have no choice: stricter standards and a mandatory federally operated certification program would have to be developed.[66] Edelbrock's response was twofold. First, he enabled SEMA to reestablish a stronger presence in Washington by securing a substantial grant from Robert Petersen of the Petersen Publishing Company, which he used to hire a permanent Washington representative, Dale Hogue.[67] And second, Edelbrock sat down with members of the SEMA Wheel Committee to develop an appropriate approach to Toms's new challenge.

Formed in 1966 at the request of the Southern California Timing Association (SCTA) officials who ran the annual time trials at Bonneville, the SEMA Wheel Committee's work began in earnest in 1967. That year the committee, led by Roy Richter of Cragar Industries, began to develop a set of strict performance-based wheel specifications, a centralized testing program, and a rotary-fatigue test rig adequate for their purposes. The committee's task, in other words, was to develop the official "SEMA Spec" for custom wheels, and as a result, it worked alongside Spar and Hedrick's Technical Committee throughout the latter half of the 1960s.[68]

But in the fall of 1971, Edelbrock informed the group of NHTSA's sentiments and pleaded with them to redouble their efforts and further refine both their standards and their testing procedures. One year later, Richter's crew presented their new Spec to Edelbrock and SEMA's board, who reported it in turn to ordinary rodders in *Hot Rod*. SEMA's custom-wheel standard, known as the 5-1 Spec, now required not only that participating firms submit their testing data for initial Specs program approval but also that they then submit sample wheels to SEMA once a year for further testing, verification, and "requalification."[69]

Pleased with what the committee accomplished, Edelbrock dispatched Dale Hogue to NHTSA's offices, where he presented SEMA's revised 5-1 Spec to Douglas Toms. Toms was impressed, but he was not yet sold. Specifically, the quality-control provisions of 5-1 remained unsatisfactory to him; Toms, Hogue reported back to SEMA in the winter of 1972–73, wanted to see random sampling, independent testing, and greater organizational oversight.[70] In the summer of 1973, SEMA's Wheel Committee therefore tweaked its 5-1 Spec once more, incorporating provisions for random, off-the-shelf wheel testing and stricter annual renewal requirements.[71] In addition, SEMA's board voted to establish a permanent testing facility, mostly (though not entirely) to enable it to meet the demands of its new 5-1 guidelines.[72]

By the summer of 1973, however, Toms had resigned his post at the request of Richard Nixon, one of many within the president's administration who were shown the door as part of a second-term shake-up. SEMA's leaders mourned the loss of their close friend and ally in Washington, but on the bright side, at least in the eyes of folks like Richter, stricter aftermarket wheel requirements quickly vanished from the federal agenda in the wake of Toms's departure.[73] SEMA and its Wheel Committee were proud of what they had accomplished, though, and they went ahead and implemented the new 5-1 Spec. As an association spokesman explained in a press release that July, "If and when [NHTSA] decides to set wheel standards" down the road, "SEMA's quality control and qualification program will be of great value to them in the custom wheel category."[74]

And so it went with SEMA and NHTSA. The speed equipment industry maintained a working relationship with NHTSA and the Department of Transportation, dealing with the occasional new federal requirement with the same commitment to communication, cooperation, and compliance that it used when facing the initial challenges of the 1960s and early 1970s. Federal automotive safety requirements were more than just a temporary problem that the speed equipment industry solved, in other words. Instead, they became an ever-present constant, one that most aftermarket firms had simply learned to take in stride by the early 1970s.

SEMA AND THE STATES

Local automotive safety developments were much less constant and therefore far more problematic. Recall that under the terms of the National Traffic and Motor Vehicle Safety Act of 1966, state and local authorities were charged with the task of interpreting and enforcing NHTSA's in-use safety standards. Innumerable local variations sprang up as a result of this decentralized system: suspension modifications that were legal in one state (or even one jurisdiction) may have been illegal in others; and the same was true for exterior lighting arrangements, custom wheels, and certain types of body modifications. By the end of the 1960s, rodders, speed shop owners, and equipment manufacturers therefore faced a vexing array of local regulations.

Further complicating matters were a growing number of local automotive noise requirements. Not until 1972 would Congress authorize the Environmental Protection Agency to develop a national standard for the regulation of automobile exhaust-noise levels, and not until 1977 would that agency actually begin to work on the matter in earnest.[75] In the meantime, state and local authorities were free to deal with the problem as they saw fit. As a result, many rodders soon found that when they modified their cars, they had to deal not only with local automotive safety requirements but also with local noise-control ordinances. Many of them also found that even if their modifications were legal in their hometowns—and, accordingly, their vehicles were legally registered—neighboring towns, counties, and states weren't always willing to allow them to pass through *their* jurisdictions without citing them for equipment infractions.[76]

In short, SEMA faced a difficult task. Dealing with the uniform requirements of the National Traffic and Motor Vehicle Safety Act's new-car standards was one thing, but dealing with disparate local interpretations of the 1966 act's in-use requirements (not to mention differing city- and town-level noise ordinances) was quite another. Certain types of mufflers, say, that once were legal nationwide were now illegal in a number of jurisdictions. The same was true of lowered springs, exhaust headers, custom wheels, form-fitting seats, and countless other aftermarket staples. Thus, as disturbing as these local variations of the late 1960s and early 1970s were for period enthusiasts, they were an even greater hassle for an industry that had to worry about things like marketing and advertising, product distribution, and legal liability.

In an ideal world, SEMA's leaders would have simply dealt with each of these local issues on a case-by-case basis. And indeed, this is precisely how they dealt with problems of this sort in California during 1967, 1968, and 1969.[77] California was SEMA's home turf, however, so it was relatively easy for the association's leadership to keep up with local-level challenges that emerged there. Difficulties elsewhere were an entirely different matter, and what SEMA's leaders therefore needed was a way to monitor local- and state-level regulatory developments throughout the country as effectively as they could in their own backyard.

In the summer of 1968, Willie Garner and his crew hit upon a simple approach that they hoped would do the trick. That July, Garner published an article in *Hot Rod Industry News* in which he urged speed shop owners, wholesale distributors, and equipment manufacturers to "keep our SEMA Headquarters advised as to local developments affecting the automobile."[78] Local firms, that is, should serve as SEMA's eyes and ears. Well aware that this arrangement might be insufficient, Garner also issued an identical appeal to enthusiasts that summer in an open letter in *Hot Rod*.[79] Three years later, Eric Grant did the same in an editorial in the NHRA's *National Dragster*, asking ordinary rodders "to serve as part of a nationwide early warning system to alert SEMA of newly proposed restrictions on vehicle use."[80] During the late 1960s and early 1970s, *Hot Rod's* editors frequently reminded readers of SEMA's request for their assistance, as did the Petersen Publishing Company in its own *Hot Rod* spots.[81] However, the editors of *Popular Hot Rodding* never mentioned SEMA's call for enthusiasts and aftermarket manufacturers to act as its local eyes and ears, and they never echoed *Hot Rod's* calls for its readers to do the same. Instead, *Popular Hot Rodding* encouraged readers to join enthusiast-based lobbying groups such as the National Automotive Racing Enthusiasts Association.[82]

Nevertheless, SEMA's strategy worked: during the late 1960s and the early 1970s, the association regularly received word of problematic developments from enthusiasts and speed shop owners across the United States. This enabled SEMA to begin to dispatch representatives, send out letters, and otherwise organize and execute responses. Many of these local problems ultimately fell through the cracks, of course, for SEMA's California-based staff had neither the time nor the resources to address them all. Still, as a result of this informal watchdog system, the association began to get a handle on the problem of local variation.[83]

In Colorado, for example, state legislators passed a new requirement in 1971 that "outlawed any modifications to suspension or steering [systems] that would cause [them] to differ from original equipment." Colorado enthusiasts as well as a number of local aftermarket companies quickly passed word of this new development to SEMA, and in the fall, Eric Grant and SEMA member Ray Brown flew to Denver to meet with the administrator of Colorado's safety-inspection program. After a lengthy discussion, they reached a compromise: *extreme* modifications such as pavement-scraping lowering jobs and sky-high lift kits would be outlawed, but more ordinary modifications like custom wheels, heavy-duty shocks, and traction bars would remain legal.[84] By the end of 1973, similar compromises had been reached between representatives of the aftermarket industry and officials in Wisconsin, Minnesota, Washington, Oregon, and Virginia.[85]

Negotiations between SEMA representatives and state and local officials did not always proceed smoothly, though. Consider the so-called "Pennsylvania Crisis" of 1970–74. In the Keystone State, the motor vehicle code was amended in the wake of the passage of the National Traffic and Motor Vehicle Safety Act to read, in part, that any automobile "with an exhaust system which has been modified or altered in any way from that furnished by the vehicle manufacturer" would be denied registration. During the late 1960s, nobody within the Pennsylvania Department of Transportation (Penn-DOT) took the matter seriously, and local inspection centers rarely enforced the provision. Nevertheless, when Pennsylvania-based aftermarket manufacturers Jere Stahl of Stahl Headers and Herb Lipton of Kay Automotive Warehouse learned of the existence of the provision in 1970, they immediately contacted SEMA. They feared that if Penn-DOT ever chose to stage a crackdown, this obscure part of the code might well come into play, virtually eliminating the custom-exhaust business in Pennsylvania.

During 1970 and 1971, Stahl, Lipton, and SEMA's Eric Grant began to try to work with Pennsylvania lawmakers to officially purge the books of this restriction once and for all. But in the fall of 1971, their efforts were cut short when the state's governor, Milton J. Shapp, ordered Penn-DOT to begin to enforce the motor vehicle code more strictly, particularly its provisions for the prevention of unnecessary exhaust noise. Rather quickly, Stahl and Lipton's nightmare came true, as inspection officials denied registrations all across the state for infractions that were often as minor—and utterly meaningless in terms of exhaust noise—as the substitution of an exhaust clamp for a spot weld in replacement-muffler installations. In 1972 SEMA stepped up its efforts to open a meaningful dialogue with Penn-DOT on the matter, and local enthusiasts even staged a "rod run" to the capital city of Harrisburg in an attempt to reverse the troubling course of events. But by the winter of 1972, all of this had come to naught, and SEMA and the State of Pennsylvania had reached an impasse.[86] The association finally broke this deadlock in November of 1973, however, and the way it did so warrants closer scrutiny. For as it happened, the so-called "Pennsylvania Crisis" ultimately came to a close, not through the persistence of Stahl, Lipton, and Grant, but rather through the intervention of an influential character that SEMA officials Don Prieto and Dale Hogue first met at a Vehicle Equipment Safety Compact (VESC) meeting in the fall of 1973.

As we have seen, the VESC was an organization founded by the OEMs in the mid- to late 1950s and charged with the task of formulating voluntary automobile safety standards. The VESC's feeble efforts were rendered meaningless when Congress passed the National Traffic and Motor Vehicle Safety Act in 1966, but in the early 1970s, the VESC was reorganized as a loose alliance of industry representatives, state transportation officials, and police officers. This new VESC met periodically to develop safety-standard recommendations for those states that participated in NHTSA's in-use enforcement system. Forty-four states were affiliated with the revamped VESC, and although none of them were legally bound to adopt its recommendations, most did. The VESC was highly selective, however; it did not try to deal with minor interpretive differences that might emerge between its members, but only with the most difficult or the most pressing issues at hand. In other words, the VESC was never meant to rein in local variations altogether, but instead to handle those that seemed particularly troubling to the bureaucrats, cops, and industry representatives who were its members.[87] And by the fall of 1973, "reconstructed vehicles" had reached the top of their list.

In the eyes of VESC members, reconstructed vehicles included hot rods, street rods, 1950s-style customs, dune buggies, and any other type of automobile that was either entirely home-built or so extensively modified from its original configuration that NHTSA's in-use safety standards could not be applied to them. So, the VESC voted to hold a series of meetings on the issue in the fall of 1973 to come up with a set of uniform standards for these vehicles. Judiciously, the VESC's chairman thought it might be wise to invite interested parties from within the automotive aftermarket to join the group's discussions, and SEMA therefore dispatched Don Prieto and Dale Hogue to San Francisco to attend the first meeting on August 1, 1973.[88] By the spring of 1974, the series of meetings had run its course, and the VESC issued a set of preliminary recommendations for its members to consider. Prieto and Hogue presented these recommendations to the SEMA board in April, and within a few weeks, SEMA mailed its response to the VESC's chairman.[89] Meanwhile, Don Prieto published his thoughts on the recommendations in *Hot Rod* that May, urging readers not to overreact to them but instead to recognize that what the VESC *really* wanted from rodders was a measure of responsibility: "If a person wants to operate a street rod on the highways of America, there are certain responsibilities he must accept . . . certain considerations he must make towards others. . . . The responsible rodder will see the immediate necessity for seat belt use, the windshield made of safety glass, wipers that work, coverings for hot, exposed exhaust pipes, bumper protection, etc."[90] For Prieto, in other words, the VESC's reconstructed vehicle guidelines were sound and reasonable, and the average responsible rodder therefore oughtn't fear them.

In the fall of 1974, the VESC issued its final recommendations, and many of its member states immediately implemented them.[91] For SEMA, the VESC's reconstructed vehicle standards were of course less than ideal, but the association was nevertheless pleased with them — or rather, more precisely, it was pleased with the way they had been developed. For the VESC's meetings in the fall and winter of 1973–74 had enabled SEMA to work with a *single* body on a *single* set of safety standards for the first time in a number of years. Moreover, Prieto and Hogue had represented the association well in San Francisco, boosting the aftermarket's reputation among the authorities involved. Finally, the episode confirmed once again the wisdom of SEMA's overarching regulatory strategy: cooperate with the authorities, comply with the law, and whenever possible, present the positions of the speed equipment industry to the relevant legislators and regulators.

SEMA's experience with the VESC in 1973 and 1974 had two further implications. First (and in the short term, foremost), Prieto and Hogue's participation in the meetings led directly to a resolution of the lingering "Pennsylvania Crisis." For as it happened, the chairman of the VESC, a Mr. Brainard, once served as a high-ranking official at Penn-DOT. Hogue and Prieto therefore cornered Brainard during a coffee break at one of the meetings in the fall of 1973 and explained their difficulties to him. Brainard assured them that he would do whatever he could to try to get things rolling once again, and he kept his word: by the time the next VESC meeting was called to order, Brainard had pulled some strings in Harrisburg to arrange a meeting for them with the head of Penn-DOT. In October, Hogue and Prieto therefore flew to Harrisburg, met with Brainard's contact, and successfully jump-started the dialogue. Penn-DOT "indicated that they [were] in the process of revising the entire code and that they would welcome technical assistance from SEMA"; SEMA, in turn, not only promised to assist the agency but also offered its word that the aftermarket would continue to respect the law, regardless of the outcome of Penn-DOT's efforts. Pleased, Prieto and Hogue "left the meeting with strong feelings that this initial conference had been productive and that things were looking a little brighter in Pennsylvania."[92]

The second major consequence of SEMA's participation in the VESC's meetings in the fall of 1973 was that it brought SEMA's regulatory strategists into a closer working relationship with Dick Wells. Wells was certainly no stranger to the association: he was the point man within the Petersen Publishing Company's Special Events Department that organized the first SEMA Show at Dodger Stadium back in 1967, and during the late 1960s and early 1970s, he continued to produce the show for Petersen each year. Moreover, Wells was an active SEMA member in the early 1970s, serving alongside Els Lohn, Harry Weber, Lou Baney, Ray Brock, and Alex Xydias on its Special Events and Show Committee.[93] But this was the full extent of Wells's involvement with the association prior to 1973—

he served, that is, on none of SEMA's "Specs" committees and on none of its legislative subcommittees. Yet during the early 1970s he did gain a considerable amount of hands-on experience in.the regulatory and legislative realms. For in addition to his duties at Petersen, Wells was also the president and CEO of the National Street Rod Association (NSRA), and in the early 1970s, the NSRA dealt with precisely the same sorts of local- and state-level regulations that bedeviled SEMA at the time.[94] Wells was therefore present at the VESC's reconstructed vehicle discussions in the fall of 1973; there, Prieto and Hogue began to work with him on legislative and regulatory issues.

Prieto and Hogue learned from Wells that the NSRA's strategy for dealing with local variations in the regulatory landscape differed markedly from that of SEMA. Whereas SEMA relied on an informal network of enthusiast and entrepreneur watchdogs to keep it abreast of local developments across the United States, the NSRA had a formal system of state representatives. The duties of these representatives, all of whom were volunteers from within the NSRA's ranks, included not only the task of keeping the association's Los Angeles headquarters informed of what was going on in the rest of the country but also of meeting with local officials, attending public hearings, and leading local members in grassroots campaigns. By 1973 the NSRA had thirty of these state representatives in place, and on balance, Wells was pleased with how the system worked.[95]

By contrast, Prieto and Hogue knew that recent legislative developments had exposed serious flaws in SEMA's more informal system. In Pennsylvania, for example, Stahl and Lipton had served SEMA well as early-warning watchdogs, but once the association learned of the "Pennsylvania Crisis," it was entirely up to SEMA's L.A.–based staff to find a way to deal with it. While this system had worked well enough in other states, in Pennsylvania the problem had dragged on for years.[96] After learning of the NSRA's system, Prieto and Hogue began to wonder whether an official state-level committee or a set of official state-level representatives might have made the task of dealing with Keystone State officials a bit less taxing.

Discussions ensued among the members of SEMA's board of directors, and in the late summer of 1974, they decided to emulate the NSRA's system.[97] By the beginning of 1975, thirty-two formal state committees were up and running, and during the course of 1975 and 1976, SEMA added several more.[98] Coordinating the efforts of its local representatives proved to be unexpectedly difficult, however. In early 1977, SEMA therefore lured Dick Wells from the NSRA and hired him to serve as their executive director, largely in order to draw on his accumulated expertise in the establishment and maintenance of state-level representatives. Under Wells's guidance, the system took firm root.[99]

SEMA never actually *solved* its state-level regulatory problems of the 1960s and 70s, certainly no more than it *solved* its troubles with NHTSA in Washington. But it did succeed in implementing a system with which to manage the uncertainties implicit in the National Traffic and Motor Vehicle Safety Act's decentralized approach to in-use automotive regulation. At the heart of this system was a deceptively simple and level-headed strategy that stood in marked contrast to the heated, knee-jerk rhetoric that all too often filled the popular periodicals of the 1960s and 70s: cooperate, compromise, and comply.

This approach served the speed equipment industry well in its working relationships with NHTSA, the VESC, and countless local agencies concerned with automotive safety and noise during the 1960s and 1970s. But this is only half of the story. For in its defensive maneuvers against the so-called "war on high performance," SEMA dealt not only with suspension geometry, custom wheels, and exhaust decibels but also with tailpipe emissions and photochemical smog. It also dealt with the California Air Resources Board in Sacramento, that is, and with the Environmental Protection Agency in Washington. And in its legal wrangling with both of these pollution-control agencies, SEMA continued to cooperate, comply, and attempt to negotiate working compromises. The difference, however, was that it was largely forced to do so on its own, without the active assistance—and, indeed, often in the face of the active *resistance*—of ordinary rodders.

CHAPTER NINE

"THIS DREADFUL CONSPIRACY," 1966–1984

Long before the passage of the National Traffic and Motor Vehicle Safety Act in 1966, tentative measures to combat urban air pollution through the regulation of automobile emissions had already emerged at the local and federal levels. Pioneered in California, most of these new regulations targeted the products of the OEMs exclusively, at least at first. But in the mid-1960s, some began to seek to curtail or even to altogether ban performance-oriented engine modifications in order to achieve the greater good of cleaner air.

For an industry based on the manufacture and sale of high-performance parts and accessories, this was an alarming development. Through SEMA, the speed equipment industry therefore launched a passionate rhetorical, technological, and legal battle to defend hot rodding—to prevent the government from "sealing the hoods" of America's cars. This effort was broadly similar to SEMA's contemporaneous engagement with matters relating to automotive safety and noise, for through it all, SEMA rarely lost sight of its commitment to cooperation, compliance, and level-headed negotiation. However, in their efforts to prevent the legal prohibition of performance-oriented engine modifications, the leaders of the speed equipment industry found that ordinary rodders were inconsequential at best and an outright liability at worst. Throughout the 1960s, the 70s, and into the early 80s, SEMA therefore dealt with environmental regulations and environmental regulators largely on its own.

The story of how it did so begins in downtown Los Angeles on a hazy summer day when gasoline cost thirty-six cents a gallon, the American-made V8 was the undisputed king of the highway, and a thick blanket of photochemical smog obscured the skyline. It was 1966.

CALIFORNIA ACTION AND FEDERAL INACTION, 1966–1970

That day, Eric Grant and Miles Brubacher of the California Motor Vehicle Pollution Control Board (MVPCB) called together a small group of journalists and hot rod industrialists to explain an addition to the state's motor vehicle code, Section 27156. After a brief introduction, they told their assembled guests that this new law prohibited the advertising, sale, and installation of any "add-on . . . or modified part which adversely affects the emissions performance of any emissions-related component of a vehicle intended for street use in California."[1] Consequently, *any* equipment manufactured to *any* specifications other than those set forth by the OEMs for a specific application were now illegal for use on model-year 1966 and newer automobiles. Much to the relief of their astonished guests, however, Grant and his associate quickly added that, for practical purposes, 27156 simply meant that aftermarket products sold in California would no longer be allowed to exceed the design parameters of certified OEM products. And in an age in which Detroit's parts bins included all of the add-on components needed to assemble a 500-plus horsepower "Super Stock" drag racer, Section 27156 was therefore unlikely to seriously limit the business of manufacturing speed equipment. Convinced that they had nothing to fear, most of those who attended the meeting simply filed what they learned in the back of their minds and returned to their businesses as if nothing had changed.[2]

The disregard with which these aftermarket representatives received the MVPCB's news that day was nothing new. Indeed, speed equipment industry insiders had greeted the enactment of nearly every environmental regulation of the early to mid-1960s in exactly the same manner—and with good reason. From California's pioneering Motor Vehicle Pollution Control Act of 1960 to the federal government's similar act of 1965, none of these regulations had ever applied to hot rodding. What's more, the environmental regulatory framework had evolved so slowly and with such an air of uncertainty over the previous two decades that it seemed unlikely to most that meaningful reforms of any consequence for their industry would surface anytime soon.

Consider the record: photochemical smog first appeared in the greater L.A. area in the mid-1940s, but its cause remained unconfirmed until 1950. That year a biochemical researcher at Cal-Tech, A. J. Haagen-Smit, discovered a critical link between this urban smog and the automobile. Specifically, he found that unburned hydrocarbons, carbon monoxide, and oxides of nitrogen in automobile emissions underwent a chemical reaction in direct sunlight that converted them into the problematic haze. Haagen-Smit's results were controversial, but in 1955 an independent group, the Air Pollution Research Foundation, verified his work to the satisfaction of the scientific community. However, Congress remained skeptical, as did the California legislature, and both simply voted to fund further research.[3]

Five years later, finally convinced that the automobile was indeed the major cause of urban smog, the California legislature passed a landmark bill, the Motor Vehicle Pollution Control Act of 1960. This act created the Motor Vehicle Pollution Control Board and charged it with the task of developing emissions-control standards by the middle of the decade. Right on cue, in 1964, the MVPCB announced the establishment of pollution-control requirements for all new cars sold in California, effective model-year 1966. From that point on, the American automobile market was forever split: cleaner "California models" were produced for the Golden State, and ordinary models for the rest of the country. Meanwhile, Congress passed the first federal Clean Air Act in 1963, a weak measure that established the framework for federal regulations but did very little else. Two years later, however, Congress passed its own version of California's 1960 law, the federal Motor Vehicle Air Pollution Control Act of 1965, which empowered the Department of Health, Education, and Welfare (HEW) to develop emissions-control standards for new vehicles sold in the 49-state market beginning in 1968.[4]

All of these measures were conceived with the OEMs in mind, and none of them mentioned the high-performance aftermarket. For the hot rodding fraternity, therefore, the air-pollution problem of the 1950s and 60s was neither here nor there: Hot Rod, Popular Hot Rodding, and the rest of the enthusiast periodicals never brought it up, and ordinary rodders and speed equipment manufacturers never gave it much thought. Environmental regulations were a problem for the OEMs, that is, not for hot rodding; thus, given the loose interpretation of the MVPCB, even the advent of California's Section 27156 did little to shake their indifference.[5]

During the fall of 1966, however, many industry insiders did begin to panic. The catalyst for this shift came in September, when the passage of the National Traffic and Motor Vehicle Safety Act served as a wake-up call for SEMA and the high-performance industry, as we have seen. Matters grew worse for the rodding fraternity later that month when HEW submitted its recommendations for federal emissions-control standards to Congress. Broadly resembling the measures that California's MVPCB adopted in 1964, HEW's proposals called for the establishment of tailpipe-emissions limits for all new cars sold in the United States, effective model-year 1968. More to the point, HEW also called for the prohibition of parts or accessories that might adversely affect the emissions of these regulated vehicles. Although this particular recommendation failed to make it into the

final congressionally approved package of anti-pollution guidelines that soon followed, the very fact of its inclusion in HEW's proposals sufficed to confirm what
the National Traffic and Motor Vehicle Safety Act already suggested: the long-
unrestricted activities of the high-performance industry had begun to attract the
government's attention on multiple fronts. With this in mind, Robert Herzberg,
the Washington correspondent for *Hot Rod Industry News*, concluded his summary of the legislative developments of the 89th Congress the following month
with an expression of concern. Readers, he warned, should "look out for anti-hot
rod legislation" in the 90th.[6]

The leaders of the high-performance trade took Herzberg's message to heart,
and together they began to plan for their defense. Through SEMA, they spent the
rest of the fall of 1966 addressing each of the legislative fronts in which they perceived a threat. Thus, they secured the services of a Washington lobbyist, opened
a dialog on the requirements of the National Traffic and Motor Vehicle Safety
Act with the federal Department of Transportation and National Highway Traffic Safety Administration (NHTSA), and even initiated talks with the California
MVPCB over Section 27156—the very law about which they had expressed very little
concern only months before. Their efforts first began to pay off the following summer,
when Dr. William Haddon Jr. of the Department of Transportation amended his
interpretation of the National Traffic and Motor Vehicle Safety Act in ways that
rendered it far more manageable for the high-performance industry.[7]

Negotiations with California's MVPCB, on the other hand, went nowhere fast.
In the winter of 1966–67, SEMA representatives met regularly with board officials
to try to persuade them to amend 27156 so that it would no longer constrain aftermarket products to the limits set by OEM designs. SEMA's argument was simple:
if the MVPCB truly had an interest in improving the quality of the air, then its
guidelines should be *performance*-based rather than *design*-based. If, for example, an aftermarket manufacturer could prove that a given high-performance part
whose design specifications happened to fall outside OEM limits was nonetheless
capable of producing tailpipe emissions within the range of the state's requirements, then the MVPCB should allow the item to be sold. But board officials
remained convinced that design-based guidelines were absolutely necessary, and
they had the evidence in hand to prove it.[8]

Federal- and state-supported research conducted in the 1950s and 60s demonstrated time and again that photochemical smog was the result of the interaction
of three compounds found in automobile exhaust: carbon monoxide (CO), unburned hydrocarbons (HC), and oxides of nitrogen (NOx). Theoretically, therefore, if the average automobile could be made to generate less NOx, HC, and

CO, then the severity and the rate of incidence of smoggy days would taper off. In practice this was very difficult to achieve, however, and the MVPCB spent much of its time in the early 1960s developing a set of guidelines that would actually lead to cleaner cars.

Beginning in 1961, the MVPCB required all cars sold in California to be equipped with a simple hydrocarbon-control device known as a positive crankcase ventilation valve (PCV). For decades, most of the internal combustion engines used in passenger cars were equipped with ventilation tubes that allowed the crankcase to "breathe," relieving the counterproductive internal pressures generated under normal operating conditions by the movement of the reciprocating assembly. Unfortunately, this ventilated air was typically saturated with unburned hydrocarbons picked up from the oil pan. PCV-equipped engines, on the other hand, recycled these hydrocarbon-rich crankcase vapors by feeding them back into the incoming fuel-air mixture through a simple one-way valve.[9]

MVPCB officials also sought to control exhaust-borne hydrocarbons, ruling that in 1966 all new cars sold in California would need to be equipped with exhaust-system devices to reduce their hydrocarbon output. Most of the OEMs met this requirement through the use of an afterburner, also known as an air injection system, which used a belt-driven pump to force fresh air into the exhaust manifold(s). There this oxygen-rich air reacted with the superheated unburned hydrocarbons in the motor's exhaust, preventing them from escaping through the tailpipe. MVPCB officials also found that subtle changes in camshaft profiles, ignition timing curves, and carburetor fuel-air ratios also helped to reduce hydrocarbon emissions and, in certain circumstances, carbon monoxide concentrations as well.[10]

Oxides of nitrogen proved to be more difficult to control, and not until the early 1970s would the MVPCB require substantial reductions in exhaust-borne NOx. The problem in this case was that the better an engine performed in terms of HC and CO emissions, the worse it tended to do in terms of NOx. This was especially true of engines with high compression ratios, for the simplest way to reduce NOx was to reduce combustion temperatures through lower compression. All things being equal, however, lower compression was counterproductive in terms of HC and CO. By the mid-1960s, the MVPCB was well aware of this dilemma, but it also knew of a potential solution, exhaust gas recirculation (EGR). Using a one-way valve, EGR allows a small amount of exhaust gas to mix with the incoming fuel-air mixture, resulting in lower combustion temperatures and lower concentrations of NOx. However, not until the very end of the 1960s would the MVPCB's successor, the California Air Resources Board, be comfortable enough with EGR technology to require its use on California-market automobiles.[11]

Nevertheless, by the time SEMA began to try to win concessions from the MVPCB in the winter of 1966–67, board officials felt that they had come a long way in their effort to control emissions, and they were very happy with the cleaner new cars sold in California. Tuned with leaner fuel-air mixtures, reconfigured with conservative ignition curves, and equipped with PCVs and exhaust after-burners, these new vehicles were just what board officials hoped the OEMs could produce. As far as they were concerned, rodders were meddlers whose engine modifications tended to "add to the smog problem."[12] When they met with SEMA's representatives, board officials therefore flatly declared that any changes to the OEM package of ignition timing, carburetion, manifolding, and camshaft profiles that the MVPCB approved for a given automobile would upset the balance of a system designed to work in concert and would therefore continue to be unacceptable. What's more, board officials presented SEMA's representatives with anecdotal testimony from the California Highway Patrol that suggested that in police spot-checks conducted throughout the state since 1961, automobiles with modified powerplants were far more likely than their run-of-the-mill, general-service counterparts to have disconnected, circumvented, or otherwise inoperative PCVs. In short, the board was none too pleased with the rodding fraternity, and it was altogether disinclined to rule in SEMA's favor and relax the guidelines set by Section 27156.[13]

To a large extent, the MVPCB was right, and SEMA knew it. Radical, long-duration racing camshafts did indeed result in higher concentrations of HC and CO at the tailpipe, for example, as did excessive carburetion and ignition advance. The problem, of course, was that in the 1960s many equipment manufacturers, speed shop owners, and ordinary enthusiasts did not differentiate between engine modifications appropriate for improved everyday performance and those that were only appropriate for use in quarter-mile racing. As a result, there were more than a few "overbuilt" cars tooling around in the mid-1960s, cars that were hopped-up for occasional strip use but that actually spent the majority of their time on the streets. *These* were the vehicles that concerned the MVPCB, that the Highway Patrol noticed in its spot checks, and that therefore endangered the future of performance tuning in California. Convinced that they needed to disassociate themselves from these overbuilt rods and instead project an image of the ordinary rodder and the typical parts supplier as reasonable, responsible adults who wanted cleaner air as much as anybody, the leaders of the high-performance industry therefore launched a concerted public-relations campaign in the early months of 1967.

Not coincidentally, this campaign began at precisely the same moment that SEMA also began to try to work with the nation's automotive safety authorities.

SEMA's decision to change its name from the *Speed* Equipment Manufacturers Association to the *Specialty* Equipment Manufacturers Association in March 1967, for example, served not only to distance the association from the scourge of street racing but also to establish that irresponsible enthusiasts who drove hydrocarbon-spewing racecars on the streets were the exception, not the rule. Likewise, SEMA's efforts to promote cooperation and compliance applied not just to NHTSA and the VESC, but also to the MVPCB. Where the emissions-related campaign of 1967 differed from its safety-oriented twin, however, was in its explicit formulation and emphatic promotion of the idea that equipment manufacturers, retailers, and ordinary enthusiasts needed to begin to differentiate between aftermarket parts intended only for the strip and those intended for the street.

In March 1967, for example, Ray Brock, the editor of *Hot Rod Industry News*, urged his readers to exercise some commonsense restraint in their high-performance sales:

It's no secret that many dealers have been guilty of "over-selling" performance equipment. If your customer has a . . . machine which he uses principally on the street, let's face facts, he doesn't need four Weber carburetors, a roller cam with rev-kit, and a fixed-advance magneto. Not only will this customer be generally unhappy with this hard-starting, rough-running, gas-eating combination, but he will also have an engine which spends a good share of its time spitting unburned hydrocarbons into the atmosphere. At this point, the Air Pollution authorities enter the picture and get unhappy.[14]

Eight months later, Brock repeated his appeal, reminding his readers of their responsibility to sell a customer

only that speed equipment which he can really use if the car is to be operated on the street. The California Air Pollution Control Department [sic] has shown us statistically that excesses in carburetion, compression, camshaft timing, and the like can contribute greatly to the unburned hydrocarbon content in the atmosphere. In other words, a six-carburetor log manifold, 12½:1 compression and a hi-rev roller cam will cause the average engine to produce more smog than horsepower when driven on the streets. Don't oversell the exotic speed equipment. The more cooperation we give to the people in charge of limiting smog, the longer we are all going to be in business.[15]

SEMA and *Hot Rod Industry News* also began to push retailers both to educate their customers about the nature and maintenance of antipollution devices and to encourage the retention of these devices on modified engines. The popular magazines also picked up on this theme in 1967, encouraging their readers to think twice before disabling their antipollution equipment.[16]

However, SEMA's leaders knew that even several hundred thousand law-abiding hot rodders, speed shop owners, and equipment manufacturers weren't likely to change the MVPCB's mind simply because they were willing to comply with the law. Thus, they also sought to bolster their organizational clout, quickly adding the air-pollution crisis to their standard list of reasons why *every* retailer and manufacturer in the business ought to join SEMA. This applied not only to those California firms that were beginning to feel the pinch of Section 27156 but also to those back East. For indeed, Congress's adoption of the MVPCB's 1966 new-car standards for cars sold throughout the United States beginning in 1968 appeared to many industry insiders as an indication that their legal troubles in California were about to spread. They feared, that is, that 27156 would soon "go federal."[17]

SEMA therefore ran a series of urgent advertisements in *Hot Rod Industry News* in 1967. One ad stated that "the high performance and custom equipment industry is currently engaged in a fight for its very existence," a fight that "the Specialty Equipment Manufacturers Association is currently leading" but that it is certain to lose without sufficient "ammunition" (money, in the form of membership dues) for the duration of the fight.[18] Manufacturers across the United States responded to SEMA's appeal by sending in their checks and membership applications, but never to the extent that association leaders hoped. In 1968, 1969, and throughout the 1970s, SEMA therefore continued to push for new members and for more overt member activism. As late as 1973, however, Noel Carpenter noted that *Hot Rod Industry News* had some 18,500 monthly subscribers, whereas SEMA itself still had only 600 dues-paying members.[19] Irritated with this count, Lou Baney, who became the association's managing director in 1972, personally began to work the phones. According to Carl Olson, SEMA's technical and legislative coordinator in the mid-1970s, Baney would "[get] on the phone every day, and for an hour he'd call different manufacturers, saying, 'next month, when I pick up *Hot Rod Magazine* and see your ad, if I don't see the little logo that says "proud member of SEMA," I'm gonna come over there and kick your ass! And *then* I'll get serious!'"[20] Baney's tactics worked: by the end of 1975, SEMA had 800 members, and by the time he left his post in 1977, it had 1,500.[21]

Back in the late 1960s, however, federal and state developments slowed to a crawl. By the end of the decade, in fact, only two significant new pieces of antipollution legislation had emerged, both of which actually worked to the industry's long-term advantage. First, the California legislature passed a resolution in 1967 terminating the MVPCB and establishing in its place a stronger agency with a broader mandate, the California Air Resources Board (CARB, or ARB).[22] Though

by no means a friend of the high-performance trade, this new body would ultimately prove to be far more willing than its predecessor to at least consider the
aftermarket's input. Second, Congress voted to amend the Clean Air Act of 1963
with the Air Quality Act of 1967. Although the primary aim of this new law was to
strengthen the federal government's ability to enforce the emissions guidelines set
to take effect the following year, it also declared that with the exception of California, no state or local government would be permitted to establish emissions-control
standards exceeding the stringency of those set by the federal government.[23] With
the passage of this act, therefore, aftermarket businessmen were relieved, for one of
their worst nightmares—having to deal with fifty individual sets of anti-pollution
guidelines, in much the same way that they had to deal with fifty individual sets
of in-use automotive safety specifications—could no longer come true.

In the midst of this regulatory lull, business boomed, and those in the know
were confident that the high-performance trade would continue to prosper in the
1970s and beyond. "Long live performance and the industry which created [the]
excitement of driving we . . . all enjoy today!" exclaimed Ray Brock in his final
Hot Rod editorial of the 1960s. "It's here to stay."[24]

RADICALIZATION: SEMA, THE EPA, AND THE
CALIFORNIA ARB, 1970–1977

Brock was right: the high-performance industry was there to stay. So too,
though, was environmental regulation, whose proponents began to push for
stronger action on the federal level in 1970. That year on Capitol Hill, Senator
Edmund Muskie of Maine proposed a package of amendments to the Clean Air
Act of 1963 designed to fundamentally alter the federal government's conservative approach to air pollution. Before 1970, federal initiatives always "pa[id] heed
to economic and technological feasibility," subordinating the pursuit of cleaner
air to the maintenance of industrial prosperity.[25] Muskie's new proposal called
for a shift in the government's priorities, for as he argued before the Senate in
September of 1970, the responsibility of Congress is not to make "technological
or economic judgments," but rather "to establish what the public interest requires
to protect the health" of individuals.[26]

Enacted later that year as the Clean Air Act of 1970, Muskie's amendments
required the OEMs to reduce their vehicles' current (1970) carbon monoxide and
hydrocarbon output by 90 percent by model-year 1975, technological feasibility
notwithstanding; reductions of oxides of nitrogen were to take effect the following year. To better enforce these requirements, the legislation also established the
Environmental Protection Agency (EPA), modeled after California's ARB. Sig-

nificantly, too, the new act declared illegal any engine modifications that might adversely affect emissions performance, effectively federalizing California's Section 27156. Finally, it also required the OEMs to provide their customers with five-year, 50,000-mile warranties on all emissions-related components.[27]

For the mainstream automobile industry, this new federal legislation was nothing less than a declaration of war—an unreasonable, unwarranted, and manifestly unjust attack on their operations that demanded countermeasures. American OEMs therefore adopted an openly hostile, reactionary stance against the act, which they portrayed as an unnecessary radicalization of the federal air-pollution agenda.[28] In fact, General Motors, Chrysler, Ford, and American Motors were joined in this assessment by none other than A. J. Haagen-Smit, the head of the California ARB. In February of 1971, Haagen-Smit declared in a prepared speech before ARB members in San Francisco that "[the standards] recently proposed by the Environmental Protection Agency for oxidants, hydrocarbons, carbon monoxide and oxides of nitrogen are very restrictive. Whereas the ARB [has] indicated that its air quality standards [are] long-range goals, federal law [now] states that the federal standards must be met within three to five years. It is clear from the ARB staff report on hydrocarbons and oxides of nitrogen that the proposed federal standards for these compounds cannot be met in the short time schedule prescribed and under the present program." The EPA, Haagen-Smit continued, was pursuing an unrealistic schedule for the reduction of automobile pollution, and he wondered aloud whether the ARB would be forced to adopt extreme measures such as "restrictions on land use [and] limit[s] on the growth of cities," in order to meet its new federal obligations.[29]

Ultimately, however, Haagen-Smit and the California ARB responded to the Clean Air Act of 1970, not through urban planning, but rather through the development of even stricter California-only mandates during the 1970s and 80s. Chrysler, Ford, GM, and American Motors, on the other hand, responded to the Muskie amendments by dragging their feet politically and technologically. Politically, the American OEMs publicly proclaimed that the targets of the act were unattainable within the specified timeframe, adding that even if they *could* be met, the cost to consumers would be astronomical. After the 1973 oil crisis, they added that emissions controls and fuel efficiency were fundamentally incompatible too. Technologically, they deployed emissions-control systems that were crude at best, and when customers complained about their temperamental, gas-guzzling, pollution-controlled cars of the early 1970s, the American OEMs maintained that these problems were simply real-world evidence that the mandates of 1970 would not work.[30]

Foreign OEMs reacted differently. In fact, Japanese manufacturers in particular tended to see opportunity in the new mandates, which most of their cars easily met. When Honda announced in early 1973 that its CVCC-equipped line of economy cars met the very same 1975 new-car standards that the American OEMs denounced as technologically unfeasible, it also reported that it had fitted its CVCC system to a small-block Chevrolet V8 and to one of Chevrolet's inline four-cylinder engines, both of which then met the standards with ease. American OEMs remained obstinate, however, and as late as 1977, Ford proclaimed that it was prepared, for 1978, "not to violate the law, but rather for a change in the law, or else to close down." To a limited extent, the American OEMs' stonewalling worked, for the EPA did grant them several extensions to 1975's deadlines. At the same time, however, their product quality and their sales slipped, and their image among the general public suffered.[31]

Like the foreign OEMs, the leaders of the speed equipment industry approached the new legislation in a non-confrontational manner. In the fall of 1970, SEMA's managing director, Eric Grant, delivered a speech in which he urged the industry to cooperate with the new laws. For *whatever* the ultimate technological solution to the air-pollution crisis and the government's targeted mandates might turn out to be, the high-performance aftermarket "will learn to live with it, improve it, and make it individually unique, legally acceptable, and most important, a little bit better than the product was originally."[32] Others— insiders and outsiders alike—were similarly optimistic. The director of NHTSA, Doug Toms, praised the aftermarket's record of compliance with his agency's automotive safety requirements at a gathering of SEMA dignitaries in January 1971, arguing that a similar approach to the new anti-pollution measures would surely earn the attention—and the respect—of the EPA. "If you give your legislators facts, not fantasy; help, not hindrance, and alternate solutions to the ones you oppose," he advised, "you will stand a far better chance of succeeding in your efforts to preserve your business and industry."[33] Later, in an April speech before the Specialty Equipment Manufacturers Association in Dallas, SEMA vice-president Vic Edelbrock Jr. concurred. "We are *not* in an impossible situation," he explained, adding that a level-headed, cooperative approach to environmental regulation in the short term remained, as before, essential to the industry's ability to survive in the long term.[34]

Most within the trade followed their leaders' advice: while Chrysler, Ford, American Motors, and General Motors spokespersons took to the press and to Capitol Hill in a series of confrontational attacks on the new regulations and the EPA, aftermarket gurus quietly retreated to their shops in search of a solution. Or more precisely, they began to labor in search of a way to prove what they and others in the know suspected—namely, that performance-tuned engines could outperform their OEM counterparts not only in the quarter mile but also in the emissions booth.

The notion was not as far-fetched as it might seem. These manufacturers understood, of course, that radical, long-duration camshafts, excessive carburetion, fixed-advance magnetos, and many other popular engine modifications did precisely the opposite, resulting in engines far dirtier than their unmodified counterparts. But milder performance modifications more appropriate for everyday use were an entirely different matter—or so they hoped. For they knew that high-performance engine tuning was above all else a quest for *circumstantial efficiency*. Drag-racing engines, for example, needed to be at their most efficient at the upper end of the powerband, where complete combustion, fuel consumption, and overall engine flexibility were of little consequence and the goal of achieving maximum horsepower per cubic inch through maximum airflow meant everything. This was why racing parts were often inappropriate for street use. OEM engines, on the other hand, needed to be reasonably powerful and flexible, but only insofar as the significant constraints associated with mass production, maintenance, fuel quality, and cost-effectiveness allowed. This was why OEM powerplants responded so well to bolt-on performance accessories. For in the design and manufacture of these add-on parts, the speed equipment industry was not constrained in these critical ways—at least not to the extent that the OEMs were most of the time. Thus, high-performance manifolds could be costly. Exhaust headers could be difficult to install. And until the EPA and the California ARB began to phase out leaded gasoline in the mid-1970s, high-compression pistons could assume the use of high-octane fuels.[35] For this was what the speed equipment industry was all about, at least when it came to street-use parts: the elimination of mass-production compromises. It therefore stood to reason that an engine tweaked for optimal airflow, precise fuel-air mixture delivery, and complete combustion through the use of street-performance aftermarket parts would generate not only more useable power, but also a cleaner exhaust. To prove this, aftermarket technicians simply began to measure the emissions impact of their street-use products. And by the middle of 1971, the preliminary verdict was in: an engine equipped with any of a number of off-the-shelf performance products could—and usually did—produce less pollution.[36]

A closer look at one particular product, the high-performance intake manifold, might help to explain why. A manifold is a simple device with no moving parts that distributes the incoming fuel-air mixture from the carburetor to each of the cylinders. On a typical 1960s V8, the manifold rested in the cradle of the "V," between the cylinder banks, with a single carburetor mounted centrally on its crown. Unfortunately, this positioned the carburetor much closer to the engine's middle intake ports than to its outer ports. OEM manifolds, particularly those of the 1950s and 60s, therefore tended to dump more than enough fuel into the inner cylinders and less than enough into the outer ones. Part of the engine thus ran rich, part of it ran lean, and none of it ran just right. It ran well enough to meet the demands of the average consumer and the OEMs' bean counters, that is, but not well enough for enthusiasts.

In theory, the remedy for this was fairly straightforward: construct a manifold with equal-length intake runners, and you will achieve an even distribution of the fuel-air mixture (and unlock a lot of hidden power). Geometrically, however, this was virtually impossible with single-carburetor manifolds, so the best practical solution was to simulate equal passages by manipulating the intake velocity of each individual runner. By the 1970s, this particular trick had long been the business of a number of speed equipment firms, most notably Southern California's Edelbrock Equipment Company and its longtime rival Offenhauser.

Both of these companies planned to release new lines of single-carburetor intake manifolds in 1971, and due to their concern over the new Clean Air Act, both closely studied their new products' "emissability" during the course of their research and development. That summer at Edelbrock, an otherwise unmodified 396 c.i.d. Camaro was fitted with the company's new TM-2 "Tarantula" manifold and run through a series of federal emissions trials. Compared with the results of a baseline test performed prior to the installation of the Tarantula, the modified engine generated 14.7 percent less hydrocarbons, 27.6 percent less oxides of nitrogen, and 34.5 percent less carbon monoxide on average, throughout the powerband. On the track, the Tarantula-equipped Camaro also picked up approximately 2 mph and more than a tenth of a second in the quarter-mile. At a press conference held in the fall, Edelbrock explained that these remarkable emissions reductions were due to the Tarantula's ability to evenly distribute the incoming air-fuel mixture to each of the cylinders—precisely the same reason cited for the vehicle's performance gains. Specifically, whereas the OEM manifold allowed the air-to-fuel ratio delivered to each of the cylinders to vary from 15.7:1 to 20.4:1, the Tarantula permitted a variation of only 14.9:1 to 16.6:1. This dramatically reduced over-rich and over-lean conditions, making for a cleaner and more powerful V8. Subsequent testing performed in 1972 verified that Edelbrock's TM-1 and TM-2 manifolds were capable of achieving impressive emissions reductions on 350 and 402 c.i.d. engines as well. For its part, Offenhauser performed a similar series of tests on its new "Dual Port" manifold during the summer of 1971, achieving emissions reductions and horsepower gains that lent additional credence to the notion that the quest for improved street performance and the pursuit of cleaner air might well be compatible.[37]

In December 1971, *Hot Rod Industry News* featured a far more comprehensive analysis of the emissions-performance relationship that its editor, Don Prieto, had conducted that summer. Whereas the tests at Edelbrock and Offenhauser focused on the ways in which the addition of a single, isolated piece of equipment would affect emissions, Prieto's goal was to analyze the exhaust of an engine fitted with a typical combination of aftermarket products. He began with an unmodified small-block Chevrolet V8, "selected primarily because it is an engine that most represents the type of equipment sold by the high-performance industry." He then rebuilt the motor using as many off-the-shelf performance products as possible. Compared with the results of baseline tests conducted on the original motor, the modified powerplant generated fewer HC, NOx, and CO emissions across the powerband while also delivering markedly improved performance. Prieto's conclusion? Performance-tuned engines could in fact make for cleaner air.[38]

Others performed similar tests. Early in 1972, for example, a member of the advertising staff of *Hot Rod*, Bob Weggeland, teamed with Ollie Morris of the Offenhauser Equipment Company to rebuild the 327 c.i.d. small-block engine in his wife's 1968 Corvette with one eye toward improved performance and the other toward improved emissions. After performing a standard 7-mode emissions test on the original low-mileage 327, Weggeland and Morris rebuilt the engine, broke it in, and retested it. Like Prieto, Weggeland and Morris used ordinary off-the-shelf speed equipment when rebuilding their mill, but they also sought to more accurately replicate typical performance projects by avoiding "fancy super tuning" tricks. Their results were therefore all the more impressive: 58 additional horses, a quarter-mile E.T. reduction of 1.8 seconds, and nearly 8 more miles per hour at the traps. More to the point, carbon monoxide emissions dropped from 7.25 percent to 5.30 percent at maximum rpm, from 1.80 percent to 0.04 percent at idle, and from 2.25 percent to 1.65 percent overall. Hydrocarbon emissions also dropped, from 2,250 parts-per-million (ppm) to 1,400 ppm at maximum rpm and from 1,050 ppm to 637 ppm, on average, across the powerband. However, idle-speed hydrocarbon emissions actually jumped from 200 ppm prior to the rebuild to 300 ppm after. Nevertheless, in light of the substantial HC and CO reductions that they achieved in every other segment of the 7-mode test and in light of the fact that they did not "super tune" their V8, Weggeland and Morris were pleased with their results.[39]

So too, by the summer of 1973, were the editors of *Popular Hot Rodding*, who launched an emissions-performance rebuild project that January that was similar to those featured elsewhere in 1971 and 1972.[40] Teaming up with several aftermarket manufacturers, they rebuilt a 1967 small-block Chevrolet V8 and successfully improved both its quarter-mile and emissions-booth performance. Their project differed from those of 1971 and 1972, however, in the sorts of comparisons it sought to make. For whereas Prieto, Weggeland, and Morris aimed to demonstrate that a high-performance engine could be made to run cleaner than its OEM counterparts, the editors of *Popular Hot Rodding* sought instead to compare the emissions numbers generated by their performance-tuned engine with those that the California ARB required. Significantly, they chose a 1967 model, originally sold in Arizona, which lacked the complicated emissions-control equipment required on cars sold that year in California. Compared with the ARB's specifications for 1967 cars, their performance-tuned engine achieved a staggering 57.5 percent reduction in hydrocarbon emissions and an 80 percent reduction in carbon monoxide.[41] What's more, compared with the ARB's requirements for 1970 and 1971 models, their modified 1967 car produced 51.4 percent less hydrocarbons and 80 percent less carbon monoxide.[42] Finally, when compared with the state's requirements for 1972 and 1973 models, their project engine ran 38.2 percent cleaner in terms of hydrocarbons and 68 percent cleaner in terms of carbon monoxide.[43] Thus, their 1967 Chevrolet, equipped with a high-performance V8 and no pollution-control devices (except PCV), actually undercut California's limits for OEM cars that were six full model-years newer. Clearly, street-use aftermarket high-performance parts weren't poisoning the air—certainly no more so than their OEM equivalents.

Meanwhile, industry leaders had wasted little time putting this new argument to work. In the fall of 1971, Vic Edelbrock Jr. was elected president of SEMA, and his intimate knowledge of the performance-emissions relationship served as the cornerstone of his strategy for the association's ongoing regulatory negotiations. Before running off to the EPA and the ARB to spread the new performance-emissions gospel, however, Edelbrock spent the first half of 1972 seeing to it that SEMA would be able to defend its claims. That spring, to enable the association to better manage its independent research activities, his administration split the SEMA Technical Committee into three new and functionally independent groups: the Noise, Safety, and Emissions Committees.[44] During the summer of 1972, as the new Emissions Committee feverishly accumulated a mountain of testing data on the emissions-performance relationship from a number of individual firms, SEMA's governing body drafted a policy statement, a formal declaration of the association's position on everything from marketing ethics to emissions controls. The resulting document, released in July, recognized the high-performance aftermarket's "responsibility to provide safe products" for street use as well as products "designed and produced in such a manner that they will not . . . pollute the air."[45] In addition, a second statement released at the same time urged the editors of enthusiast periodicals not to publish any material "that would encourage anyone to degrade or alter any motor vehicle component intended for emission control or occupant safety."[46] Having secured the commitment of SEMA's members to the production of environmentally responsible equipment, Edelbrock confidently dispatched his representatives to the offices of the EPA and the ARB. He would not be disappointed.

Since the passage of the Muskie amendments in 1970, SEMA had frequently nagged the EPA for a clarification of its position on the aftermarket. Following a series of meetings at which SEMA's representatives presented their new data on the emissions-performance relationship, the EPA finally issued a ruling in February 1973. According to its statement, aftermarket parts—including everything from OEM-style replacement sparkplugs to high-performance manifolds—would henceforward be acceptable in the 49-state market as long as the party responsible for their installation had a "reasonable basis" for believing that they would not adversely affect emissions. Significantly, the EPA would require no official certification process or exemption hearings, for as far as the agency was concerned, the willingness of aftermarket manufacturers to stand behind the results of their own tests would suffice to provide end users with the requisite "reasonable basis."[47] This was a favorable ruling for the industry, to say the least.

Far less promising in the long term, though equally welcome in the short, were the results of SEMA's ongoing dialogue with the California ARB. In May of 1973, after reviewing the emissions-performance data SEMA presented that winter, the ARB issued a new "Policy on Replacement Parts" that granted the speed equipment industry a temporary reprieve from the restrictions of Section 27156. More specifically, the board declared that until it could complete its plans for a statewide aftermarket-parts certification program, "any part offered in the market as a replacement for original equipment will be presumed to be in conformity in the absence of specific evidence to the contrary."[48] High-performance carburetors, ignitions, manifolds, camshafts, exhausts, and any of a number of other components could therefore be sold and used in California on pollution-controlled motor vehicles, at least for the time being. Hailing the decision as "a breakthrough . . . that may well lead the way to sensible and fair guidelines for replacing of stock parts without violating emissions laws," SEMA nevertheless cautioned businessmen and enthusiasts alike not to read too much into the interim policy.[49] After all, it would only be a matter of time before the ARB's plans for a permanent certification program were finished, plans whose requirements remained unknown and therefore potentially problematic. Still, even if only for a brief while, the ARB's decision meant that for the first time in nearly seven years, street-use high-performance products for new cars were wholly legal in the Golden State.

Just as EPA and ARB officials released these favorable rulings in 1973, how-ever, a new problem surfaced in Detroit. The OEMs announced that in order to provide their customers with the five-year, 50,000-mile emissions warranty re-quired by the Clean Air Act, they would need to be granted complete control over the maintenance and repair of the entire automobile for the duration of the war-ranty period. Thus, replacement-parts options would be restricted to those offered through the OEMs, and any and all repair work would need to be completed by an official dealership.[50] Not surprisingly, SEMA argued in response that Detroit's proposal would be disastrous for the aftermarket. "Restricting parts replacement to OEM pieces," the association explained in a press release, "restricts trade and eliminates the market for independently produced parts." After listening to both sides of the argument in the spring of 1973, the EPA came up with what it thought was an ideal solution: certification. Prior to sale, aftermarket equipment could be brought to a national testing center, run through the federal emissions trials, and certified as having passed or failed. Passing equipment could then be sold and installed without voiding anyone's emissions warranty.[51]

SEMA was less than enthusiastic about the EPA's idea, however. In a strongly worded response, the association suggested that "certification . . . is an unproduc-tive way of approaching the [problem]," primarily because "simply certifying a part, either publicly or privately, does not thereby impose a duty on auto manufac-turers to accept" its legitimacy. In addition, SEMA objected to the way the EPA proposed to carry out its program, arguing that the testing procedures the agency favored—the 40 CFR 85 test used at the time to certify new cars to the five-year, 50,000-mile criteria—would "work to the disadvantage of smaller parts manufacturers who [would] be unduly burdened by the costs" of such a process. Instead, SEMA proposed to develop what it called "Dynamic Performance Stan-dards," a testing regimen that would simulate (in a much faster, cheaper, and simpler manner) the conditions experienced during the course of a full 40 CFR 85 test. Finally, SEMA also suggested that neither Detroit nor the aftermarket should be held responsible for the emissions performance of modified vehicles. Instead, consumers—average enthusiasts, in this case—ought to bear the burden of proof, if necessary.[52]

Impressed, the EPA decided to consider SEMA's yet-to-be-developed Dynamic Performance Standards, conditioning its agreement on the aftermarket's ability to produce a more systematic evaluation of the emissions-performance relationship. More specifically, the agency wanted further proof that high-performance prod-ucts were indeed reliably benign when installed by ordinary enthusiasts in real-world settings.[53] Fortunately for SEMA, its Emissions Committee had already established a rigorous long-term testing program in January of 1973 in order to more systematically evaluate the emissions performance of a group of commonly modified vehicles (or rather, commonly modified engines) equipped with various combinations of high-performance parts. Dubbed the SEMA Combination Testing Program and largely modeled after Don Prieto's earlier work, this project appeared to be the perfect means through which to obtain the information requested by the EPA and to develop the proposed Dynamic Performance Standards. However, the Combination Testing Program was a complicated project not slated for completion until 1975, and the EPA was pushing for results posthaste.

Stalling for time, SEMA dispatched Edelbrock engineer Jim McFarland to the EPA in June, where he went over the association's data in greater detail with agency officials. A series of similar presentations followed over the next few months, culminating in a September 1974 meeting at which Rick Kozlowski of the EPA warned that his agency was growing weary of the industry's delays. To speed along its Combination Testing Program, SEMA therefore reassigned the project to its Product Evaluation Subcommittee.[54] Chaired by Don Prieto, this group saw to it that the program's first phase was completed on schedule the following spring.

The results of phase one of the Combination Testing Program, presented to the EPA in the summer of 1975 and summarized in a lengthy *Hot Rod* feature that fall, overwhelmingly corroborated SEMA's earlier claims regarding the "emissability" of performance-tuned engines. The objective of phase one was to measure the impact of a combination of aftermarket products on the emissions composition of a group of typical "stock, production automobiles built before [the] national clean air legislation came in." Consequently, only 1968 and 1969 model-year vehicles were used for the study, and in order to ensure that they were representative, it was decided "that the cars obtained should be of common engine/transmission combinations sales-weighted according to the national car population and to the aftermarket industry." Selected, therefore, were 327- and 396-cubic-inch Chevrolets, 318 and 383 Mopars (Chryslers), and 289 and 390 Fords (although the Fords, "due to circumstances beyond the control of the Combination Testing Group," were not included in the formal evaluations). Moreover, the aftermarket equipment chosen for the tests was installed "with no particular attention to supplied instructions and *without* optimizing performance by retuning the car[s] to other-than-stock specifications."[55] Doing so, SEMA officials believed, would enable the industry to demonstrate that even when installed carelessly, their products were entirely benign—a belief that echoed its "Specs" program's commitment to worst-case assumptions.[56] The testing procedures were straightforward, though time-consuming: each of the vehicles first received a complete engine overhaul, followed by a break-in run and a series of baseline tests before being fitted and tested with six different combinations of aftermarket products from Edelbrock, Weiand, Cragar, Appliance, Doug Thorley, and Mallory. Testing was conducted at Edelbrock's 7-mode emissions laboratory, and the results were verified at an independent research center.

"At the end of the program," according to *Hot Rod*'s feature, "there was an overwhelming body of scientific evidence acceptable to the manufacturers and to the [EPA] that aftermarket parts of the type described will reduce output of photochemical smog components significantly even when installed right out of the boxes with no attempt at optimization of engine tuning." In addition, the EPA tentatively concluded that the 7-mode, Dynamic Performance procedure used in the project to sample emissions levels did in fact appear to be an acceptable alternative to the 40 CFR 85 program.[57] Following the release later that year of the results of phase two, which verified that high-performance parts could also reduce the emissions of newer (1974-model) vehicles equipped with OEM pollution-control devices, the agency was convinced.[58] Thereafter, SEMA enjoyed the full support of the EPA on the warranty issue.

However, only Congress could actually compel Detroit to accept the legitimacy of aftermarket products. For according to the letter of the law (the Clean Air Act of 1970), the OEMs could reject the use of add-on parts—and void the emissions warranties of those who used them—regardless of whether they had the EPA's "approval." So during 1976, SEMA and the EPA met regularly to hammer out the details of a congressional proposal.[59] Their plan was simple: the OEMs would be barred from requiring the exclusive use of their own replacement parts during the warranty period, the EPA would establish a voluntary automotive aftermarket product self-certification program using Dynamic Performance Standards, and the end user would be assured, via a product-labeling system, that the parts—high-performance and replacement-duty alike—that he chose to install on his car would neither prompt the termination of his emissions warranty nor cause his vehicle to fail an emissions inspection.[60] As part of the Rogers Bill of Amendments to the Clean Air Act of 1970, their plan cleared both houses of Congress and was signed into law in the fall of 1977.[61] The "warranty flap" was over, and the 49-state market was again secure.[62]

SEMA also submitted the results of both phases of its Combination Testing Program to the California ARB in 1977, hoping that the favorable results of the program—and the EPA's response to it—would convince the board to allow the industry to self-certify its products for use in California as well.[63] The ARB remained wary of SEMA's data, however, and in August 1977 it released its own recommendations for the long-term enforcement of Section 27156. Essentially, the ARB proposed to establish a waiver program through which aftermarket manufacturers could apply, on a part-by-part basis, for individual Executive Order (EO) exemptions to Section 27156. To do so, manufacturers would need to submit to the ARB "a description of the device, drawings, installation instructions, a list of vehicle makes and model-years" for which it was designed, "part numbers associated with the device, and any relevant test data which support the request."[64] SEMA offered no objections to these basic requirements. However, other details of the ARB's proposal—that aftermarket manufacturers use full-scale emissions tests when preparing waiver applications, for example, and that they refrain from advertising or selling contested goods during the certification process—troubled industry leaders. Further wrangling, including the threat of legal action, secured for the aftermarket a slight tempering of some of the more objectionable features of the plan, but the ARB's final ruling embraced most of its original (and strict) EO requirements.[65]

Determined to forestall any additional setbacks, SEMA turned to its trump card, thus far held in reserve: the enthusiast. Average rodders had of course assisted greatly in the industry's early attempts to get a handle on state-level safety regulations in the late 1960s and the very early 70s. In addition, enthusiasts had been kept abreast of the latest environmental developments since the mid-1960s through the editorial and feature coverage of the popular periodicals. For example, *Hot Rod* published numerous attacks on environmental regulations in the late 1960s and 70s in an attempt to motivate enthusiasts to stand up for their hobby, either by writing to their congressmen (which many did) or by joining with their buddies to form local action groups (which fewer did).[66] Working with "Motorists United," an enthusiast group, *Popular Hot Rodding* also ran a number of motivational pieces in the early 1970s.[67] And throughout the 1970s, the National Street Rod Association (NSRA) attempted to persuade performance buffs to do their part to save their hobby, often through anti-regulatory membership drives.[68] However, not until the end of the 1970s was a serious, concerted effort made to form a national enthusiast group specifically to support the counter-regulatory efforts of the high-performance industry. SEMA finally took this step in 1977, creating an "Enthusiast Division" that spring. The plan was to allow ordinary enthusiasts to join SEMA as "Supporting Members" for a yearly fee of $10, which would help to fund the association's state-level programs. In a press release that May, SEMA's executive director, Dick Wells, confidently projected that some 16,000 "concerned car owners" would join the association within six months.[69]

However, this new program failed to generate much interest. This was especially true during those first six months, for before the ARB issued its final ruling on Section 27156 in the winter of 1977–78, the Enthusiast Division did not receive any real publicity in the popular magazines.[70] But even then, most enthusiasts declined to participate, and with good reason: by 1978, SEMA had already warded off most of their regulatory problems on its own. High-performance engine tuning remained legal in the 49-state market, for example, and even in California, lax enforcement meant that many rodders were still able to drive their cherished performance-modified automobiles.[71] In addition, few of the doomsday prophesies that appeared in the popular periodicals during the 1960s and 70s had actually come to pass. Why, then, should the average rodder bother to worry when confronted with contentions that the legislative situation was "getting worse?"[72] Moreover, SEMA's support throughout the 1970s for periodic, performance-based inspection programs throughout the country did little to endear the organization to the average rodder. SEMA's Enthusiast Division therefore flopped, and as before, SEMA was on its own.[73]

Still, if in 1980 the average speed shop owner, performance-parts manufacturer, or SEMA board member had stepped back to survey the events of the 1970s, he would have had plenty of reason to smile. The potentially disastrous performance-aftermarket and emissions-warranty provisions of the Clean Air Act of 1970 had been altogether eliminated. So too had the worst-case outcome in California, for although aftermarket parts now needed to undergo a strict evaluation in order to receive an EO exemption, the prospect of a total ban on engine modifications in that state—very real in the late 1960s—had been all but quashed in the 1970s. Finally, speed equipment sales had continued to break records throughout the 1970s.[74] Little wonder, then, that Hot Rod's editor, Lee Kelley, so confidently exclaimed in 1980 that "hot rodding is going to continue to flourish," for the state of the high-performance trade was better than it had ever been.[75]

1984

Sacramento had yet to play its final hand, however. Back in 1973, the California legislature passed a bill, SB473, that empowered the ARB to develop a statewide emissions inspection program. As originally drafted, SB473 called for the program to begin in the mid-1970s with change-of-ownership inspections, followed by the gradual introduction of annual testing for all cars by the end of the 1970s, but the legislature quickly voted to delay the program so that its change-of-ownership phase would begin on January 1, 1979, and the universal phase on January 1, 1981.[76] The first phase did begin in 1979, but technical difficulties with the testing procedures necessitated an indefinite postponement of the second. Thus, most California rodders were still able—de facto, if not de jure—to modify their cars in whatever manner they wished as the 1970s came to a close.

However, in 1980, under mounting pressure from environmentalists, the EPA, and the ARB, the legislature finally established a statewide biannual motor vehicle inspection program. Set to begin in the spring of 1984, the program, known as "Inspection/Maintenance," or "I/M," required all vehicles registered in California to pass both a tailpipe emissions evaluation and an under-hood visual inspection every other year. Thus, automobiles whose emissions components were modified or removed would fail, regardless of the results of their tailpipe tests, as would those equipped with unauthorized high-performance equipment. Come 1984, therefore, the California hot rod would indeed be an endangered species.[77]

In reality, it wasn't as bad as it seemed, for vehicles built prior to the pollution-control era—before 1966, that is—were exempted from the I/M program. Owners of by-the-book 1930s hot rods, 1940s and 50s street machines, and even a number of early to mid-1960s muscle cars thus had nothing to fear. On the other hand, owners of newer modified vehicles faced the grim reality that in 1984 their cars would have to be de-tuned. Golden State enthusiasts panicked, and *Hot Rod's* mailbags swelled with letters from concerned rodders wondering how this Orwellian nightmare had come to pass and what, if anything, they could do to bring it to an end. Responding to his beleaguered Pacific-Coast readership in the final issue of his editorial tenure at *Hot Rod*, Lee Kelley explained that his biggest disappointment as editor of *Hot Rod* was

> our failure to motivate you, the automotive enthusiast, to organize nationally to save your hobby. This is my last chance, so I'm going to tell [it to] you like it is. Tomorrow is 1984, and you will not be allowed to modify your vehicle . . . Oh, so now you're worried and you want to know what you can do to stop this dreadful conspiracy? Well, I don't have an answer for you now; a year ago *Hot Rod* tried to get you all to join the Enthusiasts Division of the Specialty Equipment Market Association, but so few of you responded . . . that the organization is on the verge of being dissolved. I know of no other national group that will stand up for your rights as an automotive enthusiast, so if you're not willing to do some hard work to organize such a group, maybe you'd better take up knitting, because your street-driven performance cars are going to be impounded just as sure as there's an Environmental Protection Agency. And if you want to know who to blame, just look in the mirror; a little action on that person's part could have changed the whole picture.[78]

In Kelley's eyes, the opportunity for average enthusiasts to act had long since passed, and the bitter reality they now faced was no one's fault but their own.

For the speed equipment industry, the prospect of a mandatory statewide inspection program was no less troubling. For although SEMA had been calling for the establishment of a California inspection system for years, it had done so only because it believed that any such system would be performance-based rather than design-based.[79] California's plans to require an under-hood visual inspection along with a tailpipe evaluation therefore drew an angry response from the industry, whose representatives charged that the ARB was clearly far more interested in persecuting performance enthusiasts than it was in achieving air-quality improvements.[80] It was an argument as old as the ARB itself—older, actually, for SEMA first tried it with the MVPCB in 1966. It was also an argument that proved to be as ineffective in the early 1980s as it was in the late 1960s, for the ARB was not about to change its plans.

In the short term, this meant that aftermarket companies and California rodders would have to start playing dirty. California law, for example, required manufacturers and retailers to state in their advertisements that high-performance components that had not been granted EO exemptions were "not legal for use in California on any pollution controlled motor vehicle."[81] By 1985, however, they were permitted to say instead that non-exempted products were "legal in California only for racing vehicles which may not be used on highways."[82] As far as some equipment manufacturers were concerned, this new wording basically let them off the hook, for technically, what their customers chose to do with these "racing-only" products was out of their hands.[83]

Enthusiasts came up with some interesting schemes of their own. Some removed their high-performance parts before their cars were inspected and then re-installed them once their renewal stickers were firmly affixed to their plates. Others kept complete pollution-controlled engines on hand just to use for their biannual inspections.[84] Many others, often on the advice of the popular magazines, focused their tuning efforts on *internal* engine modifications that were virtually undetectable during a visual inspection.[85] Nevertheless, these were short-term patches, not long-term solutions, and everybody knew it.

In the long run, then, the ARB's visual I/M requirement meant that speed equipment manufacturers needed to begin to obtain EO exemptions. And for their part, consumers needed to begin both to actually heed the manufacturers' warnings and to actively seek out exempted aftermarket products (or exempted model-year cars). By the time the biannual program began in the spring of 1984, therefore, many manufacturers had applied for EO exemptions, and by the end of the 1980s, many had received them.[86] Nevertheless, the EO process was time-consuming and expensive. Consequently, aftermarket firms often sought them only for a select few of their best-selling components, which meant that a number of products that were perfectly legal in the rest of the country remained illegal in the Golden State. Section 27156, that is, still haunted the California scene.

But it did not crush it. After 1984, aftermarket manufacturers learned to live with Section 27156, well aware that their failure to secure a more lenient interpretation of this law during the 1960s, 70s, and 80s was an aberration. For in all but a tiny handful of cases, their defense of hot rodding and speed equipment manufacturing against what they perceived to be unreasonable regulatory requirements and unruly agencies was remarkably successful. With the benefit of hindsight, in fact, the advent of I/M in the State of California actually appears to have been the closing act of the regulatory saga, at least as it pertained to hot rodding. For in the years that have passed since the introduction of I/M, the battle to save the American hot rod industry has lost most of the dire urgency with which it was fought in the late 1960s and early 1970s. This is not to say that I/M somehow marked the definitive end of the industry's encounter with environmental regulations and regulators, certainly no more so than its deals with NHTSA and the VESC in the mid- to late 1970s marked the end of its involvement with automotive safety. Likewise, SEMA has not backed off, and in many ways, it remains as active in the regulatory scene today as it was thirty-five years ago. However, since the early 1980s it has sustained a working resolution to its air-pollution problems through its willingness to continue to cooperate with the powers that be. In other words, because the industry has learned to incorporate the requirements of regulatory agencies into its overall approach to the design and manufacture of high-performance parts, and because government agencies have come to accept the input and concerns of the high-performance industry, environmentalism, safety activism, and automotive enthusiasm have learned to coexist.

REGULATION AND THE HOT ROD INDUSTRY

In an analogous case, Leonard Reich argues that the recreational snowmobiling industry's successful defense against the regulatory pressures that *it* faced in the 1960s, 70s, and 80s required, in equal measure, both the active participation of its industrial organization and the grassroots efforts of ordinary snowmobiling enthusiasts, a combination of forces decidedly absent in the case of the mainstream automobile industry. His overarching implication therefore is that if the OEMs had enjoyed the support of a strong industrial organization and of a rabid band of enthusiasts, then perhaps they would not have fared so poorly in the 1960s and 70s.[87]

Perhaps. But in the case of the speed equipment industry, the second of Reich's prerequisites was lacking. Only very rarely was the speed equipment industry able to rely on the average rodder to come to its aid, and SEMA was almost always on its own. Exactly how was it, then, that this association was nevertheless successful? Contrast the ways in which the high-performance aftermarket and the mainstream automobile industry handled their respective crises. Prior to the passage of the Clean Air Act in 1970, the OEMs more or less complied with federal and state anti-pollution initiatives, enjoying generous new-standard lead times and a civil, if not always friendly relationship with the authorities.[88] After 1970, however, relations between the OEMs and the government soured, as did those among the OEMs themselves. Thus, it was ultimately with an openly hostile and fractured voice that the mainstream automobile industry attempted to deal with the advent of environmental regulations. Consequently, it "lost a lot of battles against the federal and local state."[89]

On the other hand, the high-performance industry maintained its cooperative approach throughout the regulatory era. Not the passage of the Clean Air Act of 1970, the lingering difficulties associated with the "warranty flap," the ARB's certification ruling of 1977, or even the advent of I/M in 1984 persuaded the industry to change it course. Instead, through it all, SEMA continued to meet with the EPA and the ARB, patiently presenting its case and far more often than not successfully arriving at some form of negotiated compromise. In addition, individual speed equipment companies put aside their day-to-day differences and stuck together throughout the period in question, enabling SEMA to speak with a unified voice and facilitating genuine inter-firm cooperation on important projects like the Combination Testing Program. SEMA succeeded, in short, because of the powerful esprit de corps it instilled among its members, because of its cooperative approach, and, above all, because of its willingness to strike a compromise when necessary.

Whether a similar approach involving inter-firm collaboration would have worked for the OEMs is unclear. For as John Rae has explained, the automobile companies "were not permitted to cooperate in the development of emission control techniques," having been advised by the Department of Justice in the 1960s "that any such collaboration would be considered to be in violation of the antitrust laws."[90] One wonders, however, whether this is actually what prevented them from working together. After all, dozens of speed equipment companies managed to work together on emissions-related programs in the 1960s and 70s without winding up before a federal judge. Perhaps the difference lies in the extent to which Edelbrock, for example, cooperated with its rivals Offenhauser, Weiand, and Holley: not once did these aftermarket companies collaborate to *develop* emissions-friendly products. Instead, they pooled their resources to *test* their extant lines for emissions-control compatibility. Furthermore, there was absolutely nothing illegal about the ways aftermarket companies lobbied and negotiated collectively, through SEMA, in Washington and Sacramento. In short, there may well have been plenty of room for legal cooperation among the OEMs in the 1960s and 70s. That they instead chose division over unity and confrontation over compromise stands as the most significant difference between their approach to environmental regulation and that of the high-performance industry.

However, the negotiated compromises that the aftermarket's regulatory approach necessarily entailed did create a major stir among the enthusiasts who turned to the pages of the popular periodicals each month for the latest on the legislative state of the rodding art. By and large, rodders wanted nothing to do with regulatory compromise, and rarely did they miss an opportunity to express their disapproval of the notion. When *Hot Rod* titled its 1981 review of the all-new Z-28 Camaro "The Best One Yet," for example, angry readers quickly fired back, denouncing the magazine for its willingness to rank an "under-powered, over-regulated fraud" above the classic Z-28s of years past. In fact, one particular enthusiast concluded his brief tirade with the admonition that he, like most true rodders, placed a "heavy negative value on obvious mass market designs, compliance with dumb regulations, and poor first impressions."[91] Regardless of the fact that it was a strategy that ultimately worked to the advantage of high-performance businessmen and enthusiasts alike, in other words, a number of rodders held an abysmally low opinion of *Hot Rod's* (and, by extension, SEMA's) cooperative stance. As far as they were concerned, performance enthusiasm should never have been open to compromise.

One wonders, though, whether the uncompromising nature of these 1970s rodders was indeed something new—something, that is, that grew out of their experience with government regulation in the 1960s and 70s—or whether it was actually an integral part of their psyche all along. Superficially, of course, it would appear to be the former, for it is difficult to reconcile the image of a 1950s NHRA member (organizing events with his local police, for example) with that of an angry 1970s NSRA member (railing against all things governmental in the pages of *Hot Rod*) without coming to the conclusion that the average rodder's worldview grew more radical during the regulatory era. But if we try to pinpoint when this shift took place, that it happened at all becomes increasingly uncertain. For 1940s and 50s hot rodders were in many ways just as hardheaded as their 1970s counterparts. They too hated compromise, especially when it came to their hot rods. They too got excited about intake manifolds, multiple carburetors, and loud exhausts. And especially, they too got upset whenever lawmakers attempted to rein them in.[92] In short, they too were automotive and, more broadly, *technological* enthusiasts.

But theirs was an "elemental" technological enthusiasm, an enthusiasm for the technology of the automobile itself rather than for the ends that automobile use might ultimately deliver.[93] Until the mid-1960s, however, the differences between *their* enthusiasm for automotive technology and, say, that of the architects of the Interstate Highway Act would have been difficult to discern. For as Thomas Hughes has shown, the United States as a whole was a nation of technological enthusiasts in a broader, Progressive sense for much of the twentieth century.[94] As a result, the rodder's efforts to improve his automobile's performance meshed quite nicely, conceptually, with the American public's more generalized faith in the transformative power of technological progress. It mattered not a whit that the rodder sought, not to improve the lives of his fellow Americans, but rather to travel across a dry lake bed at a higher rate of speed. Instead, what mattered was that hot rodders were always able to justify their pursuits in terms that the broader American public understood and embraced.[95] During the 1960s, however, mounting concerns over the social, economic, and environmental impact of technological and industrial development began to erode the public's faith in what they once regarded as the wellspring of American greatness: technological advance. According to Hughes, these doubts ultimately led, by the early 1970s, to the end of a century's worth of progress he refers to as the "American Genesis."[96]

This is what had changed. Ordinary rodders, still motivated by their enthusiasm for automotive technology, suddenly found that their pastime was no longer necessarily compatible with the ways many Americans had come to conceive of technology in general and automobility in particular.[97] Unable or unwilling to comprehend this broader shift, rodders were baffled and angered whenever news of proposals to ban the internal combustion engine, to outlaw engine modifications, or to enforce lifestyle changes aimed to lessen our dependence on the automobile reached them.[98] Hence their disdain for the EPA and for NHTSA. And hence, therefore, their radical and altogether uncompromising attitude toward the regulatory process.

Yet we would surely be mistaken if we were to conclude, on the basis of the average rodder's intransigence, that the high-performance aftermarket—and, by extension, hot rodding itself—were somehow saved from regulatory oblivion in spite of the machinations of the ordinary enthusiast. After all, even with SEMA's string of negotiated victories, the speed equipment industry would never have survived had the lingering power of performance enthusiasm suddenly ceased to compel the smitten, young and old, to spend their hard-earned dollars on the latest and greatest in rodding gear. Over-sold, overly zealous enthusiasts aside, in other words, most hot rodders did in fact do their part to save performance tuning, however unwittingly. For without their frequent equipment binges, which enabled the high-performance industry to attend to its legal affairs without the fear that its war chests might someday run dry, such a battle would never have been possible.

CHAPTER TEN

THE BEST OF TIMES, THE WORST OF TIMES, 1970–1990

During the 1970s and 80s, the speed equipment business boomed. Year in and year out, individual high-performance parts manufacturers posted record sales, as did speed shops, wholesale distributors, and retail chains. The industry's *rate* of growth was lower than it had been in the 1960s, but still, nearly everybody's bottom line improved each year. SEMA's ranks grew too, as did the number of booths and industry-insider attendees at its annual show.[1] Meanwhile, SEMA officials enjoyed considerable success in their negotiations with NHTSA, the EPA, and the California ARB, and by the end of the 1970s, most of those in the know were cautiously optimistic that they and their industry had weathered the worst of the regulatory storm.[2] For them, the future was bright.

But ordinary rodders weren't nearly as sanguine, for in several important ways, the 1970s and 80s were difficult times for American performance enthusiasm. This was especially true of the early to mid-1970s, when pollution controls, rising insurance premiums, fuel shortages, lower octane gasoline, low-performance OEM cars, and often-draconian local inspection programs made buying, building, modifying, and driving high-performance vehicles far more difficult than ever before. At the same time, Watergate, the lingering sting of the Vietnam War, and economic "stagflation" did little to brighten the mood. Perhaps Ed Almquist was right, therefore, when he labeled the entire decade "sad," for in the eyes of many enthusiasts, the 1970s were a colossal disappointment, and the 1980s proved to be little better.[3]

Yet throughout the period in question, ordinary rodders continued to buy aftermarket products at a record clip, and they also developed some new performance niches that played vital roles in the ongoing growth and overall prosperity of the speed equipment industry. For as the era of the OEM high-performance car came to a close, American enthusiasts scattered. Some worked hard to transform the low-performance, pollution-controlled compacts, imports, and sporty domestics of the time into genuine high-performance vehicles. But many others wanted nothing at all to do with regulated (in their eyes, *compromised*) performance tuning. Some therefore continued to modify, drive, and race the muscle cars and street machines of the 1950s and 60s. Others turned to full-size vans and pickups, which were saddled with fewer emissions-control and safety equipment than ordinary passenger cars. But there were also those for whom the gloom and doom of 1970s performance tuning (and 1970s America) were simply too much to take in stride. Gripped by nostalgia for a simpler, happier, and more prosperous time, these enthusiasts eagerly turned to "street rods," prewar coupes and roadsters inspired by the classic hot rods of the 1950s but equipped with reliable later-model powertrains.

Compacts, imports, street machines, muscle cars, vans, trucks, and street rods: these niches defined hot rodding in the 1970s and 80s and enabled the speed equipment industry to prosper during an otherwise difficult time for the American automobile. Simply put, these years were at once the best of times and the worst of times. Or more precisely, they were the best of times for the speed equipment industry precisely because they were in many ways the worst of times for new-car automotive enthusiasm in the United States.

THE GATHERING GLOOM

During the 1970s, SEMA lobbied, and in many cases closely consulted, with local, state, and federal officials to ensure that high-performance tuning and other sorts of end-user modifications remained legal. Its efforts paid off handsomely too, especially for the businesses that manufactured and sold aftermarket products. SEMA's regulatory saga unfolded slowly, however, and as the association's representatives haggled with NHTSA, CARB, and the EPA over certification programs, testing procedures, and the precise meaning of stipulations like "replacement-style parts," ordinary enthusiasts suffered. Lacking clear direction from above, local inspection officials in particular were quick to develop their own interpretations of federal and state mandates. As a result, many enthusiasts soon found that their hot rods, customs, and street machines no longer conformed to the law of the land, in spite of what they read in the magazines each month about SEMA's many regulatory triumphs.

For example, Steve Fickler of Atlantic, Iowa, wrote in despair to the editors of *Popular Hot Rodding* in 1972, explaining that his "semi-custom 1963 Corvette was denied license until [he] make[s] the following 'minor' changes:

a. Remove headers and outside pipes.

b. Remove cam, pistons, high-rise manifold, big Holley and slant plug heads.

c. Remove mags and 60 series tires.

d. Remove wheelwell flares and custom spoilers.

e. Install front bumper.

f. Remove air shocks and tube front axle."

Fickler went on to explain that he faced a terrible decision: he could restore his Corvette to factory condition, undoing years of careful work, or he could park it in his garage and never drive it again. The editors sympathized, of course, and they took his letter as an opportunity to remind their readers that unless they made themselves heard, Fickler's case would soon be unexceptional.[4] Similar letters poured in over at *Hot Rod*, and throughout the 1970s the popular magazines published countless inspection-station horror stories highlighting what this state here or that city there was up to.[5]

SEMA was well aware of this pressing issue of local interpretive variation, and association members did their best to identify and deal with problematic interpretations wherever they emerged. Indeed, by the early 1980s they had succeeded in doing so most of the time. But in the meantime, enthusiasts like Fickler paid dearly for the ways local officials chose to understand the National Traffic and Motor Vehicle Safety Act of 1966 and the Clean Air Act of 1970.

Other legislative and regulatory developments of the 1970s and early 1980s further dampened the average rodder's mood. Federal and state officials proposed to slap punishing excise taxes on high-horsepower vehicles, for example, or to limit the number of cars per household, or to gradually phase the internal combustion engine out of existence.[6] Across the country, suburban towns that had grown up around established quarter-mile strips pushed for their elimination via noise-control ordinances; on a broader scale, conservationists and environmentalists called for restrictions on automotive racing as a means of reducing fuel consumption and airborne pollution.[7] These and other similar regulatory efforts—mandatory seatbelt-use laws, the national 55-mph speed limit of 1974, and the EPA's ongoing efforts to define and combat end-user emissions-control "tampering," among others—simply added to the frustration and the deeply felt sense of abandonment that enthusiasts across the country felt toward their elected representatives.[8] Still, their disappointment with developments in Washington, Sacramento, and Des Moines paled in comparison with the way they felt about the goings-on in Detroit.

From the point of view of the average enthusiast, of course, the 1970s had begun on a very high note. Muscle cars were all the rage, horsepower ratings were at an all-time high, and OEM advertisements spoke the rodders' language, boasting of quarter-mile E.T.s and performance accessories. But as early as model-year 1971, all of this began to change. Compression ratios fell, as did advertised (and actual) horsepower ratings. So too did the number of high-performance vehicles the OEMs produced and sold.[9] In time, low-performance four- and six-cylinder engines multiplied, while large-displacement V8s all but vanished. Moreover, compared with the power-pack-equipped cars of the 1950s and the muscle cars of the 1960s, the high-performance V8 options that did survive into the 1970s were unimpressive at best.[10] Meanwhile, California customers had fewer and fewer new-car engine options, and by the end of the 1970s, nominally high-performance vehicles like Camaros were available to Golden-State consumers only with ARB-pleasing but performance-robbing automatic transmissions.[11]

Hot Rod in particular tried to keep its readers upbeat about all of this by reminding them that rodding as they knew and loved it began with four-cylinder engines prior to World War II, and also by undertaking several V6 projects and reporting favorably on the OEM performance cars that remained available.[12] Few enthusiasts were sold, though. For they understood that pollution controls and smaller engines were here to stay, but they also knew that extensively modifying these new cars of the 1970s might at best result in vehicles that could match, but not exceed, the off-the-showroom capabilities of those produced a decade earlier.[13]

As we will soon explore in greater detail, some enthusiasts responded to this dismal regulatory and new-car atmosphere of the 1970s by accepting the challenge of working on these newer cars, while others chose either to move into less-well-regulated automotive niches or to hold on to their older street machines. But regardless of their response, everyone with an interest in high-performance motoring had to learn to deal with the so-called "fuel quality crisis" of the 1970s and 80s. Simply put, the quality—octane ratings—of the pump gasoline available to American consumers began to deteriorate during the early 1970s and steadily worsened right up through the 1980s. And this, far more than any of the other unsettling developments of the period in question—including the oil-supply crises of 1973 and 1979—cast a pall over the activities of ordinary rodders.

This fuel quality crisis began in 1970, when ARB members first discussed the possibility of phasing tetraethyl-lead-based additives out of California's gasoline by the end of the decade. By the fall of 1971, the EPA was contemplating a similar initiative.[14] Both agencies understood that tetraethyl lead was a dangerous toxin, and both therefore believed that its elimination was desirable. Significantly, though, both were also motivated by the knowledge that in order to meet their respective emissions guidelines for 1975, most OEMs planned to equip their vehicles with catalytic converters, sensitive emissions-control devices that required the use of unleaded fuel.[15]

For the ARB, the solution was clear: leaded fuel would have to go; and during the course of the 1970s, it slowly phased the additive out of circulation in the Golden State. For the EPA, however, the matter was a bit more complicated. Lacking a clear mandate from Congress to require the elimination of tetraethyl lead, the agency instead focused on the task of making sure that by the end of 1974 there would be an adequate supply of unleaded fuel throughout the 49-state market for the catalyst-equipped cars to come. Further federal action on the subject would not come until 1984, when Congress voted at long last to empower the EPA to eliminate leaded fuel from the American market altogether by the end of the decade.[16]

California's phase-out of the 1970s and the federal follow-up of the 1980s mattered a great deal to performance enthusiasts for a very specific reason: unleaded fuels tend to carry octane ratings that are much lower than their leaded counterparts. Consequently, as leaded fuels disappeared from the market during the 1970s and 80s, pump-octane ratings slipped. Whereas premium leaded fuels of 97 to 101 octane were available throughout the United States as late as 1970, for example, their unleaded premium replacements of the 1970s and 80s topped out at only 91 to 93. This was good enough for the newer low-compression engines of the 1970s, but it was entirely unfit for the high-compression, high-performance motors of the 1950s and 60s.[17]

Here's why. Lower-octane fuels burn more quickly than higher-octane grades, and they are also prone to ignite prematurely when exposed to high temperatures or high combustion-chamber pressures. Therefore, in a high-performance, high-compression engine, the use of lower-octane fuels tends to result in what is known as "detonation," an ill-timed explosion of the fuel-air mixture that occurs while the piston is still traveling upward on its compression stroke. This results in severe mechanical stresses that can quickly lead to total engine failure. What's more, an engine with a generous amount of ignition advance will suffer an equally destructive fate, "pre-ignition," when fed fast-burning low-octane fuel. Needless to say, engines with high compression ratios *and* a lot of ignition advance—those of the late 1960s muscle cars, for example—absolutely could not last for long on low-octane fuel. Consequently, in anticipation of the unleaded era to come, the OEMs began to produce engines in the early 1970s with less ignition advance and much lower compression ratios. This is why the horsepower ratings and quarter-mile capabilities of typical OEM offerings declined so sharply in the early 1970s, for the mainstream automobile industry had little choice but to prepare for lower-octane fuels.

Further exacerbating the situation was the fuel-supply crisis of the fall of 1973. Obviously, less gas meant less driving, and this was especially true for those who tooled around in 5-10-mpg muscle cars. As of January 1, 1974, it also meant *slower* driving, for on that day, a Nixon administration order imposed a national 55-mph speed limit in order to conserve fuel.[18] Perhaps most significantly, though, the production of high-octane unleaded fuels used more crude per unit than the production of low-octane unleaded. Therefore, in the face of oil shortages both real and projected, as leaded fuels gradually vanished, so too did the very notion of high-octane gasoline.[19]

Fuel quality and fuel availability thus were problematic for the average rodder of the early to mid-1970s, and fuel quality would remain a sore spot through the 1980s. Indeed, if there was a single issue that consistently appeared in the enthu- siast periodicals' monthly letters sections, especially after the California phase-out ended, this was it. Rodders from every corner of the United States wondered whether low-octane unleaded gasoline would harm their engines' valves (tetra- ethyl lead also acted as a valvetrain lubricant); whether there were any alternative higher-octane fuels that they might be able to use as a substitute, such as aviation gasoline; and, of course, whether there were any reasonable, real-world techno- logical "fixes" that would allow them to keep their compression ratios high and their ignition-advance curves aggressive in the new unleaded era.[20] This is where the speed equipment industry stepped in.

In fact, for many aftermarket manufacturers, both the long-term decline of oc- tane ratings and the sudden exposure of American vulnerability to fuel shortages were actually welcome. To be sure, they realized that the return of low-octane fuels meant that fewer sets of high-compression pistons would be sold in the years to come. Nevertheless, lower octane ratings also generated opportunities to inno- vate—and to turn additional profits. For on the one hand, the cars of the early to mid-1970s performed poorly. With low compression, lean carburetion, and con- servative ignition curves, these cars often struggled to produce 200 horsepower even with more than 400 cubic inches of displacement. Consequently, sales of moderate-compression pistons, larger carburetors, and cleverly engineered in- take manifolds all grew during the 1970s.[21]

On the other hand, the high-compression, high-performance automobiles of the 1960s—many of which remained in service in the 1970s—still required high- octane fuel. Thus, another challenge for the speed equipment industry was to de- velop specific solutions to enable these cars' owners to continue to operate them on the pump fuels available in the mid-1970s without inducing detonation *and* without requiring the installation, say, of power-robbing low-compression pistons. To accomplish this, several firms began to work to perfect a type of anti-detonant that had first appeared in the 1940s: water injection. This concept involved intro- ducing small amounts of water vapor into the fuel-air mixture, cooling the inner surfaces of the combustion chambers and reducing the risk of detonation. Back in the 1940s and 50s, firms that sold water injection systems did fairly well, but as pump octane ratings climbed in the 1950s and 60s, the need for them had all but vanished.[22] Crower, Edelbrock, Holley, and several others therefore picked up where these older firms left off, further developing the water-vapor concept to deal with the reemergence of low-grade fuels in the 1970s and 80s.[23]

Aftermarket products that enabled equipment firms to "cash in" on the fuel supply crises also appeared. As early as February of 1974, *Hot Rod* began to suggest to its readers that performance tuning could make their cars not only faster but also more efficient. *Popular Hot Rodding* followed suit in March, and in the fall, *Hot Rod Industry News* ran a lengthy series designed to encourage retailers to more aggressively market the "economy-performance package."[24] Here, as with the emissions-performance link, the key was moderation. Certain types of street-use aftermarket products *could*, in other words, produce both moderate power and slight fuel-efficiency gains. This is because at least in theory, engines tuned for complete combustion should be able to squeeze more power out of every drop of fuel. For some manufacturers, the only challenge was to prove to skeptical customers that their street-use parts were capable of doing this, and advertisements and feature articles replete with testing data and percentage-gain claims therefore began to appear regularly in the popular periodicals.[25] Other firms needed to develop altogether new products specifically designed to improve low-end power while also reducing fuel consumption. This was particularly true of high-performance camshaft companies, many of which introduced all-new "economy" bumpsticks in the 1970s.[26] Either way, the implication for the speed equipment companies of the period was that fuel shortages needn't be disastrous—at least not for them.

But for ordinary American rodders, the fuel-supply and fuel-quality crises of the 1970s and '80s were indeed disastrous. They understood that there were ways to deal with both of these problems, whether through the purchase of a water-injection system, the use of octane-boosting fuel additives, the installation of a mileage-enhancing camshaft, or any of a number of other means the aftermarket made available. However, they also understood that none of these expensive add-ons actually improved the performance of their cars in any meaningful way. Mileage-enhancing camshafts and intake manifolds, for example, did tend to increase low-end torque, at least marginally, but they always sacrificed a bit of top-end horsepower in order to do so. Products designed to combat detonation, on the other hand, added virtually nothing: they simply enabled enthusiasts to *maintain* the performance of their high-compression engines on the lower-quality fuels of the period. And for the die-hard rodder who grew up modifying the power packs of the 1950s and the muscle cars of the 1960s, this was a miserable state of affairs. For the essence of rodding had always been performance *enhancement*, not performance *maintenance*.

▼

The fuel-related difficulties of the 1970s and 80s were troublesome, but as we have seen, they were but part of a much broader set of problems plaguing performance enthusiasts at the time. For long before the first OPEC shock and long before their precious leaded fuels began to disappear, rodders recognized that things were no longer going their way. The question, of course, was what to do about it. For as *Hot Rod's* Terry Cook explained in early 1973: "The overriding tone of the mail [the magazine receives] seems to be pessimism, gloom, the depressed and defeated attitude of the street enthusiast who is forced to cope with insurance rate-jacking and the emission/safety regulations we all face today. We know that these things can be discouraging. Most of today's cars are lackluster in comparison to the crop of five years ago. So what should we all do, take up spelunking? I say *buffalo chips!* . . . there is plenty not only to be happy about, but to do."[27]

Cook's solution? *Find a way.* Or rather, find a niche, a high-performance segment in which one's enthusiasm could continue to thrive in spite of the depressing broader context of the time. And this is precisely what American rodders did: they found novel and creative ways to carry on.

"PLAYING THEIR GAME"

For a number of enthusiasts, the challenges of the 1970s were far too daunting, and the new cars of the period were far too dull to even consider. Many therefore looked to the cars of the 1960s, 50s, 40s, and even the 1930s and 1920s. These, in their minds, were the only acceptable sources of raw material for a rodding project. However, others understood that in the past, diligent work on newer-model cars was what always worked to advance the state of the rodding art. Back in the 1930s, the widespread shift from inline four-cylinders to the newer, flathead V8 brought higher lakes speeds, better on-road performance, and booming sales to then-fledgling speed equipment firms. In the early to mid-1950s, too, a willingness to experiment with the new overhead-valve V8s of the period once again delivered higher speeds, improved road-going capabilities, and a fatter bottom line to aftermarket companies. The same was true of the 1960s as well, when the rodding fraternity found ways to improve the performance of even the fastest of the "factory hot rods." Thus, for some of the enthusiasts of the 1970s and 80s, the task at hand was clear: work to find ways to improve the performance and appearance of the newer, pollution-controlled vehicles of the time.

Doing so wasn't always easy, and the results were rarely awe-inspiring. Particularly during the latter half of the 1970s, in fact, when emissions-control regulations for new cars tightened further, speed equipment manufacturers and rodding journalists alike often felt compelled to reassure themselves and their customers that the endeavor was actually worth their while. *Hot Rod*'s editors admonished those who were tempted to give up and turn their backs on the newer cars of the period in favor of those of the past, reminding them in 1978, for example, that "if the suppliers of products which enable us to modify our cars can no longer produce and sell those products for later-model cars, which represents a substantial portion of their sales volume, it's anyone's guess as to whether or not they'll be able to stay in business."[28] The following year Jim McFarland concurred, albeit in a far more positive and upbeat manner, when he explained to everyone within the rodding fraternity that "SEMA manufacturers work with what O.E.M. provides. If it's a small-block Chevrolet V8, then we work with small-block Chevrolet V8s. If it's a turbo-supercharged four-cylinder Pinto, then we work with that. Or a V6. Or an in-line six-cylinder. Or an X-body. Or a turbine. We don't originate the plan. We *play* the game, using *their* toys, not ours."[29]

During the 1970s and well into the 1980s, "playing their game" typically meant one of two things. For some, it involved attempts to transform newer low-compression, low-performance V8 cars into genuine high-performance machines. This was especially true during the early 1970s, before the sharp decline of pump-octane ratings made the installation of hot ignitions and high-compression pistons virtually impossible. Nevertheless, *throughout* the period in question, those who lived in jurisdictions with open-minded inspection officials were often able to make good use of street-performance intake manifolds, carburetors, headers, and other smog-legal accessories. Indeed, even in more restrictive areas—California after the beginning of I/M in 1984, for example—*hidden* performance accessories like hotter camshafts, larger pistons, stroker cranks, and re-mapped ignitions remained viable options as long as the car in question was still able to pass the tailpipe-sampling test.[30] Still, those who sought to recreate the automotive thrills of the 1960s with the cars (and regulations) of the 1970s and early 1980s were often disappointed.

For many others, "playing their game" meant embracing the much smaller V6 and four-cylinder compacts of the period: Vegas, Pintos, Omnis, Chevettes, and the like. Fortunately for those with these types of cars, hot cams, better-flowing intake manifolds, multiple carburetors, exhaust headers, and turbochargers worked as well on these smaller engines as they did on large-displacement V8s. However, because these compacts were newer-model cars, they were subject to the same emissions-control requirements as their larger stablemates, and this meant that those who modified them had to work within a tight set of constraints to ensure that they remained street-legal. In theory, therefore, those who worked on these so-called "mini cars" during the 1970s and 80s were largely in the same boat as those who modified the much larger V8-powered sedans and coupes of the era: smog-legal aftermarket modifications were in abundance, but there was a limit to what these parts could deliver.[31]

That said, owning and modifying a compact car did have its advantages. They were far more economical to operate than were the larger-engined cars of the time, even when modified to the hilt and driven at the limit. Also, because they were smaller, they were also lighter, which meant that especially when modified, they were rarely at a serious power-to-weight disadvantage against the two-ton, low-compression V8 models of the period. Because their engines were smaller, their weight distribution tended to be more favorable, even when equipped with front-wheel drive; and this, in turn, made them nimbler and better-handling. Finally, smaller-engined cars were often the first to receive sophisticated engine technologies such as overhead camshafts, multiple valves, and all-aluminum construction. Rodders who modified these cars thus were often working on the cutting edge.[32]

Still, many rodders weren't sold. Equipment manufacturers and magazine editors alike therefore worked quite hard to convince their customers that turning wrenches on a ninety-horsepower four cylinder was indeed a respectable pursuit. Some did so by reminding rodders that the 1960s were over, that "the days of the V8 are numbered," and that "four and six-cylinder engines are here to stay."[33] Others rightly pointed out that hot rodding began in the 1920s and 30s with four-cylinder engines; four-cylinder tuning in the 1970s and 80s was therefore a legitimate undertaking.[34] Finally, most of the periodicals also sought to sway their skeptical readers by sponsoring four- and six-cylinder projects, including everything from turbo installations to complete performance rebuilds.[35]

In the end, "mini car" parts sold well enough to generate a lot of buzz among manufacturers, especially during the mid-1970s, but domestic compacts in particular never grew to account for more than a tiny fraction of the industry's overall sales.[36] In fact, even those who promoted compacts as rods-in-waiting often seemed to be doing so tongue-in-cheek, or at least with a heartfelt sense of regret. For example, Hot Rod's disappointment with the new four cylinders from Pontiac and American Motors in the spring of 1977 was almost palpable: "We know that today's economic realities have put the kabosh [sic] on our 455-inch monster motors," a staff writer explained, "but when you've been raised on muscle, AMC's 121-inch, 300-pound, 8.0:1 in-line [engine] . . . seems to cry for the rodder's touch."[37] His preference, that is, would have been to write about the possibilities inherent in those "monster motors" of yesterday rather than those of the anemic mills of the time. Countless died-in-the-wool hot rodders felt exactly the same way; hence the negligible share of the speed equipment market ultimately won by these compact "mini cars."

There was one important exception to this small-car malaise, however: *imported* compacts. Japanese brands like Honda, Toyota, and Datsun (Nissan) had worked for years to establish a significant presence in the American market with their tiny four-cylinder, front-wheel-drive sedans and hatchbacks, and by the early 1970s, their efforts had begun to pay off. Soon their sales eclipsed those of Volkswagen, long the leading import in the United States.[38] Before long their well-engineered, well-built, efficient, and cheap cars began to fall into the hands of performance enthusiasts, and established firms like Edelbrock and Offenhauser quickly added high-performance parts for these cars to their catalogs.[39]

However, Japanese import tuning would remain a small and virtually insignificant niche until the early 1990s. Instead, the overwhelming majority of enthusiast and manufacturer involvement in imported-compact tuning in the 1970s and 80s centered on the brand that the Japanese were gradually overtaking: Volkswagen. High-performance parts and accessories for Volkswagen's rear-engine, air-cooled Beetle first appeared in the early 1960s, and by the end of the muscle car era, "hot VWs" were a popular and legitimate part of the American rodding fraternity.[40] During the early 1970s, the hot setup for a Beetle enthusiast was what was known as the "California Look," or "Cal Look." Lowered, de-chromed, and equipped with a highly modified engine, the Cal-Look Beetle was all about blistering straight-line speed, solid handling, and a clean and purposeful appearance. The fastest of these cars were capable of quarter-mile E.T.s in the low 12s, about half the time it took an unmodified VW to cover the same distance.[41] Looking back on thirty years of small-car tuning in the United States, in fact, *European Car's* Kevin Clemens remarked in 1999 that "the origins of import car tuning, or at least the point of critical mass," lies with the Cal-Look Bug.[42]

By and large, Clemens is right. However, air-cooled Volkswagen sales plummeted in the mid-1970s, and by the end of 1977, the hard-top Beetle was no longer sold in America. Beetle tuning continued to thrive as a high-performance niche, but during the late 1970s and the 1980s, the so-called "second generation" Volkswagens rose to prominence among import enthusiasts. These new cars, the Rabbit and the Scirocco, were nothing like the Beetle and the Karmann-Ghia they replaced. They were still small, of course, and they were also relatively inexpensive, but their layout differed considerably: like their Japanese competition, the Rabbit and its sibling were front-wheel-drive cars with front-mounted, water-cooled, inline four-cylinder engines.

Volkswagen began to import the Rabbit ("Golf," in Europe) in 1975,[43] and second-generation tuning began in earnest a couple of years later when VW equipped the car's single-overhead-cam, eight-valve engine with electronic fuel injection. In 1983 the company also began to import its Golf-based GTI.[44] A high-performance model with more horsepower and a stiffer suspension, the GTI has since become the car of choice for second-generation VW enthusiasts. During the 1970s and early 1980s, however, there were only the low-performance Rabbit and its sportier cousin, the Scirocco. Still, aftermarket companies and enterprising enthusiasts were quick to unearth ways to improve their performance: turbos were a popular option, as were stroker cranks, larger pistons, reground cams, and performance exhausts.[45]

However, like all other new cars sold in the United States at the time, Rabbits and Sciroccos were subject to the mandates of the EPA and the California ARB. As a result, the modifications performed on them had to be smog-legal, particularly for those who lived in jurisdictions with strict interpretations of the Clean Air Act or Section 27156. In time, second-generation aftermarket specialists began to offer EPA-approved and even CARB-exempted engine accessories. But especially during the late 1970s and the early 1980s, VW enthusiasts instead devoted much—not all, but much—of their attention to suspension and handling modifications unrelated to the performance of the engine itself: lowering kits, larger-diameter brake discs, suspension stress- and sway-bars, fiberglass body panels, and lightweight alloy wheels, among others.[46]

Some of the companies that catered to these second-generation VW enthusiasts were latter-day incarnations of firms that had once specialized in air-cooled parts. Oettinger, for example, was a German firm that produced a variety of products for the Rabbit, including a sixteen-valve cylinder-head conversion that did for the VW engine what the conceptually similar sixteen-valve Roof, Rajo, and Frontenac kits did for the Model T mill sixty years before. Oettinger's heritage? It was the same firm that once produced the popular Okrasa engine kit for the Beetle.[47] Similarly, Darrell Vittone's Techtonics Tuning, founded in 1975, had its roots in the same Riverside, California, VW dealership that once housed EMPI.[48] Others that produced Rabbit parts, such as Drake Engineering, were old-school American hot rod companies. Once known as Meyer and Drake, the company that purchased the Offenhauser racing-engine facility in the 1940s, Drake Engineering produced a turbo kit, a reground cam, and other performance parts for the water-cooled VW in the late 1970s and 80s.[49]

Altogether new firms were common as well. German Motor Parts of North Carolina, for example, opened its doors in the mid-1970s as a part-time parts-importing business, and contemporary descriptions of its origins faintly echo those of the Southern California hot rod firms of the 1930s and 40s. As *Dune Buggies and Hot VWs* explained in February of 1980, "German Motor Parts . . . was formed a few years ago by German Jo Klitsch, who came over here some years ago . . . With his knowledge of the German market, it was not long before friends who owned Rabbits started to ask him to get parts from Germany on some of his regular visits to his homeland. So it was that Jo found his hobby turning into a full time business with several employees and a large warehouse."[50] New Dimensions of Santa Clara, California, had similar origins, as did Neuspeed, Automotive Performance Systems, Autotech, and many others. That is, new second-generation VW companies often began as part-time German-parts importers founded by enterprising enthusiasts with import-tuner friends.[51] In time, most began to produce their own parts and accessories for water-cooled Volkswagens; eventually, during the 1990s, many of them began to offer parts for Hondas, Nissans, and Toyotas too. Together with those who worked on the domestic compacts and the low-performance V6s and V8s of the period, the early import tuners who shopped at German Motor Parts and New Dimensions helped ensure that at least in part, hot rodding and speed equipment manufacturing remained forward-looking endeavors.

So too did "vanners" and "street truck" enthusiasts. Not unlike the custom-car crowd of the 1950s, the vanners of the 1970s were far more interested in interior and exterior modifications for improved comfort and style than they were in those that delivered all-out speed. Brand-new full-size domestic vans were their raw material of choice, and the modifications they undertook ran the gamut from simple and inexpensive to lavish and costly.[52] Most stuck closer to the former, fitting their vans with alloy wheels, plexiglass porthole windows, and denim, velour, or tweed-trimmed interiors. Many also added elaborate murals to the flanks and rear doors of their vans. A few went much further though, installing everything from big-rig-style smokestacks to wet bars, fireplaces, sofas, and even full-size beds. What was supposed to have gone on inside these decked-out vans is probably best left unmentioned, but what certainly warrants emphasis here is that these hefty vehicles were seldom built to outrun anything.

Nevertheless, vans and vanners were highly visible elements of the hot rodding community of the 1970s, and modified vans frequently graced the covers of the popular magazines. News of regional van meets was common as well, as were how-to articles, feature spreads, and even full-scale, magazine-sponsored project vans.[53] Speed equipment manufacturers took notice of the trend as well, for vanners bought a lot of custom wheels, suspension sway-bars, and mileage-enhancing manifolds, camshafts, and ignitions.[54] In short, vans were popular, prominently featured, and, for those to whom it mattered, profitable too.

But not everyone was thrilled about this, and throughout the 1970s, letters to the editors of the rodding periodicals complaining about the attention they devoted to these low-performance beasts were common. Some suggested that the magazines were driving away loyal performance-obsessed readers.[55] Others complimented the editors whenever an issue with "less garbage on vans" appeared.[56] A few did venture to defend the vanners and the magazines' van coverage, explaining, for example, that all of the "recent articles on vans" deserved the editorial space they received precisely because "it appears that this is 'where it's at' today."[57] The debate raged on for several years, but in the end the vanners proved to have been nothing more than a flash in the pan—and an embarrassing one at that, for in 1984 *Hot Rod*'s Leonard Emanuelson actually apologized to his readers for the ways in which his magazine lost its way in the 1970s. In the future, he promised, "we will never forget that we are HOT ROD Magazine, as happened during the van era."[58]

As the age of the vanner came to a close, however, that of the street truck dawned. During the 1970s, most street-truck enthusiasts worked on new-model, full-size domestic pickups, fitting their V8 engines with high-performance parts and their exteriors with chrome wheels, running lights, sidesteps, and the like. During the 1980s, some also modified compact imported trucks, though those who did tended to be far more interested in flash and flair than in speed: lavish paint jobs, detailed graphics, and lowered or even hydraulically adjustable suspensions trumped performance-oriented engine and handling modifications.[59] But whether full-size and powerful or compact and customized, street trucks were not universally loved, and their place within the rodding world was as hotly contested as was that of the van. For indeed, among those who wanted nothing to do with compacts, imports, vans, and the rest of the uninspiring and over-regulated vehicles of the time, these street trucks simply served to confirm that hot rodding's best days had long since come and gone.

UNBRIDLED NOSTALGIA

For die-hard, tire-smoking gearheads, the vehicles for sale on new-car lots in the 1970s and 80s were utterly disappointing. Vans were slow, low-compression V8s were an embarrassment, compacts and imports were short four cylinders, and trucks were neither here nor there. On the other hand, used-car lots and classified ads teemed with high-performance V8 models from the 1950s and 60s. Free of the EGR valves, catalytic converters, dished pistons, and exhaust afterburners that strangled the performance of most newer vehicles, these older cars were much faster than anything built in, say, 1975. They were also easier to modify legally, for there was no need to worry about things like emissions-equipment compatibility, visual-inspection requirements, or EO numbers. In the simplest of terms, there-fore, the "street machine" enthusiast of the 1970s and 80s firmly believed that the cars of the 1950s and 60s were the best the OEMs had ever built—and, by exten-sion, that these cars were the best-available raw material for high-performance projects.

Hence the widely agreed-upon definition of a street machine: a road-going vehicle with a highly-modified engine that was originally manufactured between 1949 and 1972. Neither of these dates is arbitrary, for 1949 marked the beginning of the postwar, full-bodied, overhead-valve era; and 1972, the debut of the last of the muscle cars.[60] Those who built and drove street machines were therefore those who wished that the postwar horsepower race—and, in particular, its muscle car phase—had never come to an end. This is not surprising, given that most were twenty-something baby boomers who came of age in the heart of the muscle car era. Raised on large-displacement, high-compression V8s, they lamented the emergence of automotive regulations and heaped ridicule on the EPA, the ARB, and NHTSA; they also mocked the OEMs for caving in to government mandates and abandoning the high-performance market.[61] Sales of headers, manifolds, and other traditional V8 items to these street machine enthusiasts were brisk, easily out-pacing every other high-performance niche by considerable margins in the 1970s.[62]

Street-machine enthusiasts were clearly a nostalgic bunch, for they wished above all else that the 1970s had turned out more like the 1960s. For many others, though, the bygone era worthy of nostalgic longing wasn't that of the factory hot rods built by Chevrolet and Ford, but rather that of the genuine hot rods built by ordinary enthusiasts. Theirs, that is, was a longing for the mid- to late 1940s and especially the early to mid-1950s, a time when prewar coupes and roadsters were abundant, when new cars were of interest only when their wrecked hulls yielded low-mileage overhead-valve V8s, and when hot rods couldn't be bought but had to be built. And in their longing for the return of the spirit of this earlier time, they weren't alone.

Writing about the fortunes of the custom-car enthusiast, John DeWitt explains that the so-called Kustom Kulture declined from prominence during the late 1960s and all but vanished during the early 1970s. Then, in the late 1970s and early 80s, it reemerged in a powerful flurry of postmodern expression that was linked to a generalized nostalgia among the American public for the simpler pleasures of the 1950s. Pat Ganahl, in his masterful epic on the American custom car, essentially agrees, as does Timothy Remus in his shorter but still excellent book.[63] However, traditional-style rods and customs did *not* reemerge together, as DeWitt suggests, in a *single* wave of nostalgia that swept across the enthusiast community during the late 1970s and early 1980s.[64] Instead, in much the same way that the original custom car itself grew out of the by-then-well-established rodding culture of the early 1950s, the reemergence of the Kustom Kulture in the late 1970s and early 1980s actually grew out of an earlier nostalgic rebirth of traditional-style rodding that occurred in the late 1960s and early 1970s.

At least at first, this earlier renaissance had very little to do with the *cultural* ac-coutrements of the 1950s—the leather jackets, drive-in restaurants, slicked-back hair, and Doo Wop records that the later reemergence of the Kustom Kulture featured.[65] Instead, what emerged during the late 1960s and early 1970s was a de-sire to recreate the *technological* state of the rodding art as it was in the late 1940s and early 1950s. It was a desire, that is, to relive the automotive pleasures of the era of the traditional hot rod.

Back in the late 1940s, dual-use hot rods were the norm, particularly among California enthusiasts. During the early 1950s, however, as hot rodding spread throughout the United States and became an increasingly competitive endeavor, rodders built more specialized machines. Some converted their prewar roadsters into dedicated lakesters and dragsters, while others turned theirs into polished, comfortable, street-driven "cool rods." Within another five to ten years, though, most of the performance-modified vehicles driven on the streets in the United States were based on *postwar* coupes and convertibles, and by the mid-1960s, street-use hot rods based on *prewar* roadsters were rare. Little deuce coupes and A-V8s did appear frequently in magazines, car shows, and popular music, but out on the streets, the 1960s belonged to the muscle car.[66]

Yet it was precisely then, when the so-called factory hot rods were at their peak, that widespread interest in prewar cars began to reemerge.[67] And as it did, it quickly picked up where the cool rods of the early 1950s had left off. Popularly known as "street rodding," this rediscovered niche thus involved the construction of nicely finished prewar cars for the street. Small at first, street rodding expanded dramatically during the late 1960s. By 1969, feature articles on the phenomenon had begun to appear in the popular magazines, and how-to manuals were avail-able on bookstore shelves; by 1970, the activity had its own national organization and had begun to sweep across the rodding world. Soon street rod meets, features, and how-to projects were among the most widely covered aspects of the American rodding scene.

Street rodders were a diverse group, but their nostalgia for 1940s- and early-1950s-style hot rods generally stemmed either from a desire to relive their own past or from a less precise and far less personal desire to return, conceptually at least, to the fabled glory days of the American hot rod. For some, that is, it was all about recreating the beloved and long-lost roadsters of their youth, while for others it was all about participating in a bygone era they had missed.[68] But regardless of its source of inspiration, their enthusiasm for the 1940s and 50s was often ridiculed by rodders who followed other niches.

By the middle of the 1970s, for example, street-machine enthusiasts in particular had branded the street-rod crowd as nothing more than a bunch of graybeards, backward-looking old farts who refused to accept that the 1950s were over and that 1960s muscle cars were better-engineered and faster than prewar-bodied rods ever were. Firing back, street rodders blamed the muscle car for ruining the build-it-yourself creativity of the 1940s and 50s; they also reminded their rivals that they too were stuck in the past, for the 1960s had long since come and gone as well.[69] Tempting though it is to dismiss these claims and counterclaims as nothing more than the petty squabbling of rival factions, there is at least some vital truth to both sides' claims. Street rodders were indeed *older*, for example, but they were by no means *old*: in 1976, their average age was 32, slightly older than the typical street-machine buff but certainly not over the hill.[70] Likewise, *both* groups were indeed stuck in the past, at least inasmuch as neither embraced the new cars of the 1970s and 80s. However, whereas the attitudes and activities of street-machine enthusiasts can be read as pure reaction, those of the street-rod crowd cannot. To be sure, they longed for the 1950s, and they worked on cars manufactured in the 1920s and 30s. But as they did, most actually sought to combine the *look* of the 1950s with the *technology* of the 1960s, 70s, and 80s (fig. 18). Or to put it another way, the street rods that they built were decidedly postmodern.[71]

Consider the typical 1970s street rod. Based on an early-1930s Ford (coupe or roadster), it was powered by a modified late-model Chevrolet V8, almost certainly a 327 or 350 small block. Its transmission was a 1950s or 60s Ford or Chevrolet automatic with a modified torque converter, and its front and rear suspension came from a 1950s or 60s sedan. Disc brakes were likely as well, though drums from larger cars and trucks were equally common. Assembled with meticulous attention to every detail, from its Cragar rims and Hedman headers to its polished paint, gleaming chrome, and finely stitched interior, the typical street rod was built to look good. But it was also driven regularly, and thus air conditioning, high-end audio systems, power seats, and other luxuries were common on them too. For the goal, above all else, was to own a 1950s-style rod that could be driven comfortably, safely, and reliably on 1970s highways.[72]

Toward the end of the 1970s, however, the aforementioned far more generalized nostalgia for the 1950s swelled, and soon a number of enthusiasts began to reconsider what the "street rod" ought to be. Perhaps inevitably, some concluded that the deliberate blending of the old with the new was a conceptual blunder: surely, they reasoned, 1950s-style rodding ought to attempt above all else to be true to the American hot rod as it actually was in the early 1950s—flathead

Figure 18. A nicely finished street rod, circa 1991: 1940s and 50s style, powered by 1970–90s technology. (Reproduced courtesy of the photographer, Robert C. Post).

engines, primered finishes, crude brakes, and all. Thus was born the "repro rod," and in time, "repro customs" appeared as well, as did nostalgic reproductions of even earlier cars (fig. 19). Nevertheless, those who flocked to 1950s-themed car shows and cruise nights during the early to mid-1980s were far more likely to have done so in an old-meets-new street rod than in an out-and-out period-correct reproduction.[73]

Blending the old with the new was never straightforward, of course: 1957 Chevrolet rear-ends are designed to bolt into 1957 Chevrolets, not 1932 Fords;· likewise, 1968 Dodge steering assemblies, 1941 Ford brakes, and 1965 Hydramatics were never meant to work in 1929 Model As. Thus, considerable ingenuity was required to create a street rod — that and the will and the means to fabricate everything from wiring looms, transmission adapters, and pedal assemblies to driveshafts, wheel spacers, and engine mounts. This was particularly true during the late 1960s and the very early 1970s, for at that time there were no companies within the American speed equipment industry that catered specifically to this emerging phenomenon. Whereas countless firms made manifolds, camshafts, ignition systems, custom wheels, and other performance-related accessories that *could* be used on a street-rod project, that is, no one built the more specialized products early street rodders needed, like interior and exterior trim, custom

Figure 19. A latter-day reproduction of a 1920s Model T "racer." This particular example is equipped with a period-correct Rajo overhead-valve conversion. (Photographed at a Hershey, Pennsylvania, show in 1980. Reproduced courtesy of the photographer, Robert Deull.)

chassis parts, and most of the specialized brackets and adapters necessary to cobble together a prewar-bodied car with mechanical underpinnings sourced, in many cases, from a half-dozen or more makes and models. There were some fiberglass reproduction body panels available for prewar cars by the mid-1960s though, and a handful of firms produced complete fiberglass bodies as well (genuine 1932 bodies were by then quite scarce). Thus, in the late 1960s, it was possible to purchase a brand-new fiberglass 1932 Ford roadster body to use as the basis for a street rod project. Beyond the body, though, the rodder was on his own.[74]

Then, during the early 1970s, the speed equipment industry caught up. Soon brand-new chassis were available, as were other frame- and suspension-related products, engine adapters, and brake components. In fact, by the end of the decade it was possible to build an entire street rod from scratch, using only reproduction parts.[75] Some large and well-established firms even offered complete do-it-yourself street rod kits.[76] But by and large, the firms that manufactured street-rod parts during the 1970s and 80s were tiny mail-order companies that offered a diverse range of products that were either produced to order or manufactured in small batches. In fact, the made-to-order and small-lot model was the norm among *all* of the companies that manufactured street-rod parts, even the larger ones. This is because there was no established best-practice approach to street rodding. Many enthusiasts based their projects on genuine or reproduction 1932 Ford roadsters, but there were many other prewar bodies (reproduction and genuine) that were nearly as popular. More to the point, the number of engine-transmission-suspension combinations fitted to these bodies was nearly infinite. Thus, it might have made sense for a given manufacturer to know *how* to produce a set of engine mounts that would allow a late-model 427, say, to be mounted in a reproduction 1929 Model A chassis fitted with a genuine 1927 Model T body, suspension components from a 1965 Corvair, and a 1969 model-year automatic transmission. But it probably would have been unwise for that company to manu- facture more than a handful of sets at a time. And since diversity was the norm among street rodders (and flexibility the norm among the companies that catered to them), retail speed shops were often hesitant to carry these highly specialized parts. Consequently, this particular niche largely remained a catalog and mail-order business throughout the period in question.[77]

But how that business grew! And as it did, the popular magazines' street-rod coverage expanded. Feature articles and how-to technical pieces were soon com- mon, as were write-ups about regional and national street-rod meets. Many of these meets featured elements borrowed from early-1950s hot rod club gatherings, including "reliability runs" and voluntary safety inspections. One important dif- ference, however, was that whereas a 1950s club meet generally involved a bunch of young men and teenage boys gathered around a handful of cars, a 1970s street- rod meet was a family affair that was more akin to a holiday picnic writ large than to an assembly of testosterone-overloaded teens. Also, whereas the clubs of the 1950s were generally affiliated with the racing-oriented NHRA, most of those who flocked to the Midwest each year for the Street Rod Nationals were affiliated with an altogether different group, the National Street Rod Association.[78]

Founded informally by a small group of ex-drag racers in Memphis, Tennes- see, toward the end of the 1960s, the NSRA was formally established in 1970 as a privately owned, California-based corporation.[79] According to Dick Wells, the association's first president (and, under the terms of its incorporation, its majority owner), the NSRA was at first a shoestring operation that he managed in his spare time. But all this changed after Ray Brock, Wells's partner in a new street-rod- oriented publishing venture called *Rod Action*, visited Wells's home one night to discuss their magazine. As Wells later recalled,

> I said, "Well, why don't we go into my office?" So we went into this spare bedroom where there's this typewriter and an ironing board set up at counter-height that I

used to stuff envelopes. And he said, "Well, what's this?" And I said, "Well, this is the headquarters office of the National Street Rod Association."

Brock was appalled, and the next day he arranged for the NSRA's headquarters to move into a vacant office in *Rod Action*'s suite in Burbank.[80] From there, Wells oversaw the NSRA's affairs until the late 1970s, organizing its regional and national meets and managing its growth. He also pioneered a decentralized watchdog system to monitor and address local safety, noise, emissions, insurance, and registration regulations—a highly successful system from which SEMA eventually learned a great deal.[81]

Perhaps most importantly, however, the NSRA of the 1970s managed to maintain a coherent definition of what exactly counted as a "street rod" and what did not. For as the 1970s wore on and a more generalized nostalgia for the 1950s emerged, there were some within the NSRA's fold who began to push for the inclusion of early-1950s customs and street machines as genuine street rods. Against this current, the NSRA held firm in its insistence that "street rods" were 1948-and-earlier-bodied vehicles, period. To pacify its internal dissenters, the association launched a street-machine division, and it also organized a number of street-machine-only events during the 1970s.[82] But the NSRA understood that its 1948 divide was about a great deal more than simply separating the prewar street rod from the postwar street machine. And for that matter, it was also about a great deal more than the need to separate the graybeards from the youngsters. Instead, the 1948 divide was based on a fundamental difference in approach. For whereas street-machine enthusiasm was predicated on the notion that the cars of the late 1950s and 1960s were the best there ever were, street rodding was instead based on a desire to recreate the look and the feel of the past by way of the technology of the present.

Still, both the street rod and the street machine were exercises in nostalgia, automotive movements that must be understood for what they were: significant elements of a much broader reaction against the underwhelming new cars of the 1970s and 80s. During the early to mid-1980s, however, the fundamental reasons for this generalized reaction began to disappear as the horsepower ratings, quarter-mile capabilities, and overall driveability of the automobiles sold in America improved dramatically, even as exhaust emissions and overall fuel consumption continued to decline. And the key to this budding new-car renaissance was a technology that was decidedly unfamiliar to performance-car enthusiasts: the microprocessor.

PLUG-AND-PLAY HOT RODDING

What ailed the new cars of the 1970s was the inability of the OEMs to deal with the advent of emissions controls, low-octane gasoline, and tightening efficiency expectations in a simple, reliable, and cost-effective manner. Reducing exhaust emissions, for example, necessitated the adoption of power-robbing devices like air pumps, EGR valves, and vacuum-operated throttle-valve positioners. Dealing with low-octane fuels, on the other hand, required drastic reductions in ignition advance and static compression ratios, both of which reduced the efficiency of the combustion process. There were some exceptions, of course. Honda's CVCC-equipped compacts were efficient, reliable, clean, and inexpensive. But they were by no means fast, nor were they sporty.[83] As we have seen, aftermarket water-injection systems, performance intake manifolds, and freer-flowing exhaust headers *could* improve the performance and in many cases the fuel efficiency of a 1970s vehicle without elevating exhaust-emissions levels. However, accessories of this sort were expensive—and perhaps more to the point, the performance returns they delivered paled in comparison with those that similar parts for unregulated 1950s and 60s vehicles produced. Hence the reaction of the 1970s: the bumbling, stumbling, molasses-slow new cars of the time simply weren't worth the effort.

In the late 1970s and early 1980s, though, the situation began to improve as the OEMs applied "little black boxes" to their vehicles—modules sourced from the rapidly advancing field of microelectronics. It began with relatively simple devices in the 1970s that replaced vacuum- or mechanically operated mixture controls on OEM carburetors as well as similar ignition-system controls. By monitoring and precisely adjusting the fuel-air mixture or the amount of ignition advance in light of variables such as engine temperature, these devices rendered the cars to which they were fitted more reliable, more efficient, more powerful, and cleaner at the tailpipe. During the early 1980s, the OEMs began to augment these computerized ignitions and carburetors with what are known as "knock sensors." Mounted in the engine block, these sensors detect the onset of preignition or detonation (typically lumped together in the enthusiast lexicon as "knock"), signaling the ignition-control module to back off on the advance and, in some cases, signaling the carburetor control to enrich the mixture too. Higher compression ratios and more aggressive ignition curves thus began to reappear in the early 1980s even as pump-octane ratings continued to fall, for knock sensors served as an effective safety net that prevented the inevitable pings and knocks from going too far.[84]

However, the computerization of the American-market automobile really only came into its own in the mid- to late 1980s, when foreign and domestic manufacturers alike began to replace the venerable carburetor with sophisticated, microprocessor-controlled fuel injection systems. Fuel injection itself was hardly new: Chevrolet's Rochester system debuted back in 1957, for example, and by the early 1970s similar mechanical systems had been in use on Mercedes-Benz, Porsche, and other high-end makes for years. Aftermarket mechanical fuel injection systems—for example, the Hilborn—were by the 1980s old news too. *Electronic* systems were much less familiar though. Bosch's multi-port D-Jetronic system first appeared in the American market on the 1967 Volkswagen Squareback, and its successor, L-Jetronic, made its debut on 1975 model-year Beetles. Soon, BMW and a number of other European firms used L-Jetronic as well.[85] Nevertheless, these systems were by no means common in the 1960s and 70s, and American and Japanese manufacturers shunned them entirely. But as emissions-control requirements and fuel-efficiency expectations continued to tighten in the early 1980s, nearly every OEM that sold cars in the United States began to turn to them. For by measuring and processing dozens of engine-related variables hundreds of times a second, these systems could adjust the amount of fuel delivered with a level of precision unmatched by any "black-boxed" carburetor.

At first domestic manufacturers in particular tried to keep their electronic fuel-injection systems simple. For example, GM's TBI "throttle body" setup, which was introduced in 1981, used a single injector mounted in a central throttle body. Its computer contained a simple electronic "map," or set of calculations, that used variables such as engine temperature and engine speed to determine precisely how much fuel the injector should spray into the incoming airstream. Soon GM started working on a multi-port injection system as well.[86] Similar to Bosch's older D- and L-Jetronic setups, GM's multi-port system used a set of injectors (one per engine cylinder), which allowed for an even greater degree of precision in the computer's delivery of metered fuel. Throttle-body and multi-port systems coexisted throughout the 1980s as Ford, Chrysler, Nissan, Honda, and Toyota switched from carburetion to fuel injection, but by the 1990s, multi-port systems were the norm.

The advent of computerized ignitions and carburetors and, eventually, of multi-port fuel injection systems in the late 1970s and 80s meant above all else that OEM technology was finally catching up with the legislative mandates of the 1970s. For at long last, the mainstream industry was able to offer new cars that ran smoothly, reliably, efficiently, and cleanly on low-octane, unleaded fuel. And this, in turn, meant that by the early 1980s domestic manufacturers in particular were able to tentatively turn back to the profitable high-performance segment of the new-car market. Soon more powerful (and faster) V8 Camaros and Mustangs appeared, although a number of enthusiasts and rodding journalists remained skeptical of this fledgling high-performance renaissance. After all, these new and fuel-injected cars were significantly better than those of the mid-1970s, but they still paled in comparison with those of the 1960s. On balance, though, most were genuinely enthusiastic about the apparent return of OEM muscle.[87]

But there was one considerable catch. Literally and figuratively, automotive electronic control units (ECUs) were black boxes. Engine-sensor inputs flowed into them at one end, and electric pulses to strategically trigger individual fuel injectors flowed from the other. What exactly went on within these ECUs was entirely mysterious to nearly everyone within the hot rodding fraternity. So too, therefore, was the answer to the age-old question every died-in-the-wool hot rodder asked of every OEM powerplant: *how can I tweak this thing to maximize its power?* Rodders and technical editors alike wondered aloud whether ECUs would be able to deal with high-lift cams, larger pistons, freer-flowing exhausts, and the like. They also wondered whether it might be possible to actually modify the operating parameters of those little black boxes too.

At least at first, the outlook wasn't promising on either front. Prior to the early 1980s, the only contact any rodder in the United States was likely to have ever had with any sort of electronic fuel-injection system was with the Bosch D-Jetronic and L-Jetronic systems fitted to late-1960s and 1970s air-cooled Volkswagens. Neither of these systems fared well once a high-lift camshaft or a set of larger pistons were installed, and many of the enthusiasts who owned these cars found it simpler, cheaper, and far less aggravating to simply replace the entire system with a pair of performance-oriented carburetors. Technically, this was illegal in all fifty states because it necessarily entailed the removal of essential emissions-control devices that were integrated into the Bosch system; but realistically, it was the only way to modify these engines for improved performance.[88] The same was true of the later domestic systems of the early 1980s, for their ECUs contained fixed maps that could only handle engine-sensor inputs that fell within a predetermined and preprogrammed range.[89]

By the mid-1980s, however, several aftermarket options had emerged. First, carburetor and manifold specialists like Edelbrock and Holley began to offer performance-oriented fuel-system components that integrated seamlessly with the black-boxed carburetors of the late 1970s and the simple fuel-injection systems of the early 1980s. Second, a number of newer startups offered complete replacement ECUs that were calibrated to handle larger displacement, high-performance engines. Finally, a handful of companies also began to sell plug-in chips known as "P-ROMs."[90]

Of these options, P-ROMs were by far the most promising, at least in theory.

They were also the simplest, for most of the automotive ECUs in use by the mid-1980s were calibrated using read-only-memory (ROM) chips. These tiny chips plugged into their respective ECUs in much the same way that 1980s video-game cartridges plugged into their consoles. Thus, replacing an OEM chip with a high-performance P-ROM was only slightly more involved, conceptually, than slipping in a different video game.[91] Among the first to produce these plug-and-play P-ROMs was a new firm known as Hypertech, whose approach to their manufacture was as follows: remove the OEM chip, plug it into a computer interface, download and decipher its coding, recalibrate it, burn the recalibrated coding onto a batch of new P-ROMs, and sell them to customers who wanted instant horsepower.[92]

Hypertech's chips were successful during the 1980s, as were those of several other firms, for they did indeed improve the performance of the vehicles into which they were installed. However, chip-swapping had its limits on at least two levels. First and foremost, in the words of *Hot Rod*'s Rick Voegelin, "There isn't a program in the world that will make a stock Chevette run 9-second elapsed times."[93] In other words, a P-ROM swap could *improve* a car's performance, in much the same way that a set of headers could, but it could never entirely transform a car's potential on its own. This, of course, was a well-established performance-tuning dictum. Less well-understood, however, was the second limiting factor related to P-ROM chips, and that was the simple fact that aftermarket P-ROMs could *recalibrate* but could not entirely *reprogram* factory ECUs. That is, a high-performance chip could call for more fuel at certain points on the ECU's preprogrammed map, but it could not alter the overall contours of the map itself. Thus, even aftermarket P-ROM–equipped cars were limited by the larger, predetermined parameters of their OEM ECUs, and not until the early to mid-1990s would more versatile ECUs begin to appear that could be substantially altered via aftermarket chips.[94] Still, the P-ROM was a widely heralded development within performance-tuning circles, for it signaled that even the black-boxed car could be made to do the rodder's bidding.

Not everyone was thrilled with the advent of the high-tech hot rod, though. The California ARB and the federal EPA, for example, both were concerned that plug-and-play chip tuning could all too easily evolve into a widespread method of emissions-compliance evasion. CARB in particular worried about the fact that these plug-in chips, most of which did not have official EO exemptions from the provisions of Section 27156, were virtually impossible to detect during a visual emissions inspection. The OEMs were concerned about the phenomenon as well, for they feared that chip-tuned cars would develop engine and emissions-control maladies that *they* would have to fix under warranty. By the end of the

1980s, therefore, regulatory interventions as well as voluntary measures on the part of the OEMs meant that nominally tamper-proof, soldered-in chips were fast becoming the norm.[95]

Nevertheless, with a recalibrated ECU but a couple of hours with the soldering iron away, enthusiasts were not at all deterred, and neither were those who produced the chips. For they had glimpsed the high-tech future of high-performance tuning, and they were not about to turn back.

❖

The 1970s and 80s were difficult for American performance enthusiasts, particularly for those who remembered a time when the notion of "automotive progress" referred, not to emissions reductions and the advent of ever-smaller, ever-more-efficient cars, but rather to the quest for more horsepower, more torque, and ever-lower quarter-mile E.T.s. However, with the emergence of sophisticated electronic controls in the mid- to late 1980s—which the OEMs developed first and foremost as a way to meet the tightening emissions-control and fuel-efficiency mandates of the Nixon, Ford, Carter, and Reagan years—the new-car malaise ushered in by the Clean Air Act of 1970 began at long last to give way to guarded optimism among hot rodders. Nostalgic street rods and street machines remained popular, and they remain so to this day. But by the end of the 1980s, performance enthusiasts no longer *needed* to look to the past. Instead, with their screwdrivers, rolls of solder, and P-ROM chips in hand, they were able once again to look back to the future.

CONCLUSION

In 1991, aftermarket sales, which had been on the rise for decades, actually fell by 4 percent. It was only a temporary dip, however, and sales picked up again in 1992 and have risen dramatically each year since. Thus, particularly with the benefit of fifteen-plus additional years' worth of hindsight, 1991 seems to have been at most an aberration, a brief lull associated with the economic recession that followed on the heels of the Gulf War.[1] Closer examination of the data, however, reveals that though this postwar recession was indeed the *immediate* trigger for the aftermarket losses suffered that year, deeper structural problems within the industry actually meant that it was a setback that was at least a couple of years, if not decades, in coming.

Consider the trends of the 1970s and 80s. Through 1975, much of the industry's annual growth was associated with the muscle car boom of the 1960s and, to a lesser extent, the lingering afterglow of the first OEM horsepower race of the 1950s. In fact, add-on street and strip components for domestic high-output, V8-powered street machines accounted for more than 60 percent of the performance industry's aggregate sales right up through the Ford administration. New-model muscle cars disappeared in the early 1970s, however, and by 1975 there were very few cars left on showroom floors with more than 200 horsepower under their long and pin-striped hoods. Enthusiasm for new American cars dwindled accordingly, as did late-model speed equipment sales.[2] Nevertheless, *total* high-performance sales continued to rise each year as equipment manufacturers developed new niches that more than made up for their losses in the street-machine market, including imports, domestic compacts, vans, street trucks, and street rods. Combined with lingering demand for street-machine products, not to mention traditional drag-racing gear, these new market fragments drove the industry's ongoing growth, enabling it to continue to expand during the course of an otherwise difficult period.[3]

Lurking just below the surface was a fundamental structural weakness, however, one that would ultimately contribute to the industrywide slump of 1991. Again, the recession of the early 1990s was the immediate cause of that dip, but recessions had never seriously affected the industry before—certainly not in the 1970s, at any rate, when the high-performance aftermarket boomed while much of the rest of the American economy suffered intermittently from stagflation, energy crises, and manufacturing-job losses. The difference in the late 1980s and the 1990s was that the specialty equipment industry of the time was particularly soft in a critical niche that had not been weak in a time of economic recession since the very early 1930s: new-car products.

Toward the end of the 1980s, that is, the street rod niche remained strong, as did the nostalgic custom market, the street truck market, and the performance aftermarket for 1950s, 60s, and early 70s street machines. However, with the exception of street trucks, each of these niches was naturally limited by the supply of the older-model cars on which it was based, and in the long run, each was therefore destined to level off eventually. This is precisely what occurred in the mid-1980s: muscle cars and other street machines rose in value—and slowly crept out of the reach of ordinary enthusiasts—as they became scarcer and thus collectible. What's more, vans turned out to have been a passing fad, as did Detroit's compacts, which quickly fell out of favor among enthusiasts as soon as the oil-supply fears of the 1970s began to fade from memory. On the other hand, European imports continued to grow in popularity during the 1980s, though with the exception of water-cooled Volkswagens, European tuning was for the most part a higher-dollar, lower-volume enterprise centered on Mercedes-Benz, BMW, Porsche, and other expensive makes. In short, what was missing in the late 1980s was a strong, reliable, entry-level, new-car niche for speed equipment.

In fact, the entry-level, new-car niche had begun to dry up years earlier, when the production of affordable OEM performance cars effectively came to a halt. For when Chrysler, GM, and Ford abandoned the high-performance market in the early 1970s, die-hard enthusiasts turned either to the classic street machines of the 1950s, 60s, and early 70s or to the street rods of the 1930s and 40s for their projects. Newer-model Camaros, Mustangs, and Firebirds remained popular among enthusiasts right up through the 1980s, of course, but most newer models did not fare as well. Enthusiasm for the new Ford Escorts, Chevrolet Citations, and Chrysler K-Cars of the early to mid-1980s never really materialized at all, for example—certainly not to the extent that it had in the 1960s for entry-level Corvairs, Tempests, and Falcons. Why this was the case depends on whom you choose to believe. Perhaps it was because of government regulations and annual emissions inspection programs, which made it difficult for many hot rodders to legally extract acceptable performance gains out of these newer, smaller-engined vehicles. Another possible explanation has less to do with regulations per se than with the perplexing computerized controls that the OEMs began to use to meet them in the late 1970s and early 80s. However, a handful of aftermarket companies did begin to produce high-performance P-ROM chips that enabled enthusiasts to work with these newer, black-boxed cars. Yet they still preferred their old-school street machines and street rods to the available high-tech Cavaliers and Shadows. But why?

All things being equal, the simplest explanation tends to be correct, and this is almost certainly the case here. Simply put, most of the new cars of the 1970s and 80s weren't particularly popular among hot rodders precisely because they weren't the sorts of cars that enthusiasts *could* get excited about. Journalists often tried to drum up enthusiasm for Detroit's new and smaller-engined cars, informing readers, for example, that "the V6 may well be the small-block Chevy of the 1980s."[4] But this was nothing more than wishful thinking: as it happened, the small-block V8 remained by far the most popular engine for performance enthusiasts of the 1980s, ultimately rivaled neither by new V6s nor by turbocharged and fuel-injected four cylinders. More generally, V8-powered cars remained the dominant choice among ordinary rodders, even though it meant they often had to stick with much older models. Most rodders looked at their 130-hp V6 or 80-hp inline-4 new-car options, did the math, and figured that even a well-used, V8-powered 1955 Chevrolet—or 1970 Duster, or 1967 Camaro, or 1965 Mustang— promised a lot more bang for the buck.

Also, few were pleased with the *appearance* of the new cars of the time, particularly those of the early to mid-1980s. When *Hot Rod* reported in 1982 that the all-new, Carroll Shelby–built Dodge Omni GLHS was faster around the track at Willow Springs than a 1965 Shelby GT350 Mustang, for example, the enthusiasts who elected to respond were nonplussed (to put it kindly). "It appears that Mr. Shelby has been far too busy under the hood to notice what [the Omni GLHS] really looks like," wrote Dean Schwartz of Jordan, Utah. "That little black box is really UGLY with a capital UGH!"[5] Others concurred: *perhaps* the newer cars of the time *could* be made to be faster than the much-ballyhooed classics of years gone by, but the classics were still more appealing because they were easier on the eyes. Besides, they were simpler, more reliable, and easier to work on.[6] But when the supply of these older, simpler, and better-looking cars began to dry up and their prices began to rise toward the end of the 1980s, there wasn't anywhere for the low-buck, ordinary enthusiast to turn. With affordable new cars uninspiring and older models increasingly out of reach, they were left with few reasonable options—few, that is, that were attractive enough to compel them to make high-dollar aftermarket purchases in a time of economic recession.

Hence the 1991 dip: the specialty equipment industry was vulnerable to the economic recession following the Gulf War precisely because automotive enthusiasm as a whole was in the midst of a widespread lull in the late 1980s and very early 1990s comparable to those it had faced in 1927–30 and in 1953–57. For in the late 1920s, when Ford no longer manufactured Model Ts, no one in the high-performance field knew for certain what exactly would or could replace the Tin Lizzy as the preferred basis of their speedsters and racers. In time, of course, the Model A would fill this role, but not until the very early 1930s. Likewise, when Ford phased its flathead V8 out of production in 1953, no one knew which of the new domestic engines would eventually take its place among enthusiasts. Certainty, in this instance, did not come until the late 1950s, when Chevrolet's low-cost, small-block V8 rose to prominence within the rodding world. What took place in the late 1980s and the very early 1990s was remarkably similar: the absence of high-output, domestic V8-powered new automobiles once again left the enthusiast community without clear-cut, affordable new-car options.

In this case, the next big thing, "import tuning," did not begin to emerge as a mass phenomenon until the early 1990s. And for several years thereafter, it was not clear to anyone involved that it would ultimately prove to be anything more than just another passing fad—like vans, for example, or the domestic compacts of the 1970s. But by the middle of the 1990s, its importance was undeniable.

FRONT WHEEL DRIVE

Import tuning essentially involves the construction of a customized, performance-oriented vehicle out of any of a number of otherwise unassuming four-cylinder, front-wheel-drive compacts—Volkswagen Golfs, Honda Civics, Scion xBs, and the like (fig. 20). The modifications involved vary wildly, encompassing everything from the addition of oversized rear spoilers and other fiberglass body accents to lowered springs, lightweight wheels, stainless-steel exhaust systems, high-performance camshafts, ECU chips, and even add-on turbos and superchargers. In certain parts of the country, compact Chevrolets, Dodges, and Fords also receive this tuner treatment, though lowered Neons and turbocharged Cavaliers grow increasingly scarce the farther one gets from UAW strongholds like southern Michigan and northern Ohio. Still, many prefer the more inclusive "sport compact tuning," particularly those among the old-school American hot rod firms that have diversified into this niche in recent years. Among participants, however, a simpler (if less precise) terminology prevails: "tuning" for the niche as a whole, and "tuners," curiously enough, for those who design and manufacture the necessary add-on parts, for those who build and drive this type of modified car, *and* for the cars themselves.

Historically, however, "import tuning" is perhaps the most accurate way to describe this phenomenon because this particular niche, and each of its many twenty-first-century derivatives, descends directly from the cultural and techno-logical milieu of the European car enthusiast of the 1980s. As we have seen, Rabbit, GTI, and Scirocco tuning began in earnest in the late 1970s, and by the mid-1980s it had all the requisite trappings of a distinct car-enthusiast niche—popular periodicals and specialty aftermarket firms, most notably. Significantly, it also had an established *aesthetic*. Or rather, more accurately, most of those who participated in this niche shared a general approach to the process of automotive customization, an approach that the attitudes and actions of today's sport compact and import tuner crowds echo forcefully.

In brief, this approach embodies a commitment to the notion that a better-handling car is almost always preferable to one that is only capable of straight-line speed. Consequently, 1980s water-cooled VW enthusiasts placed a great deal of emphasis on lowered and stiffer springs, strut-tower braces, alloy wheels, and lightweight body and interior materials (though heavier-than-stock stereo components often found their way into their cars as well). Engine modifications, such as they were in the mid-1980s, generally consisted of attempts to maximize the output of tiny, 1.6 and 1.8 liter engines through the use of radical camshafts, balanced and lightened internal components, forced induction, freer-flowing exhaust systems, and sky-high engine rpms. The traditional rodder's mantra that there is no substitute for the cubic inches and low-end torque of a big V8, that is, was not highly valued among these early import tuners.

In fact, the top-dog engine among these enthusiasts during the late 1980s and the very early 90s was Volkswagen's double-overhead-cam, sixteen-valve four cylinder. Available in second-generation GTIs and Sciroccos, this engine pro-duced 134 horsepower from the factory at its evolutionary peak in 1993, and it responded well to simple (if sometimes pricey) aftermarket modifications such as freer-flowing intake and exhaust systems, high-performance camshafts, and ECU chips.[7] A less exotic single-overhead-cam, eight-valve engine was Volkswagen's cheaper alternative, and it too could be modified productively (and often less ex-pensively). In time, VW dropped the sixteen-valve mill from its American-market range, however, replacing it in the third-generation 1994 GTI with a compact six-cylinder engine known as the VR6.

The VR6 first appeared in the United States in the Scirocco's replacement,

Figure 20. A front-wheel-drive burnout. The early-model Volkswagen Rabbit pictured here has been retrofitted with a late-model Volkswagen / Audi 1.8T engine. (Photo by Jared Holstein, featured in Dan Barnes, "Canadian Rabbit Stew," *European Car,* April 2004, 66. Reproduced courtesy of the editor of *European Car,* Les Bidrawn.)

the Corrado, in 1992. In GTI form, it developed 172 horsepower from its unusual "inline-V6" format: staggered in banks of three that were separated by only fifteen degrees, the VR6's cylinders fit snugly under a single twelve-valve head, which meant that the entire engine was only marginally longer and wider than a typical inline four. Thus, the virtue of the VR6 was that it easily fit in the tiny engine bay of the Golf-based GTI.[8] Unfortunately, its vice was excess weight, and the third-generation GTI was at the time by far the heaviest, most lavishly-equipped, and most-expensive of VW's long-running line of "hot hatches." Enthusiasts and aftermarket manufacturers alike embraced the VR6, however, and it wasn't long before some began to shoehorn this unique mill into older (and lighter) Sciroccos, first- and second-generation GTIs, and even Rabbit pickups.[9] By the time VW's fourth-generation GTI debuted in 1999, the VR6 was an established and beloved mill for which countless performance accessories were available. But the GTI itself had lost a great deal of its luster, for the fourth-generation model was even costlier and heavier than its predecessors, and it was equipped with virtually every luxury option available from VW. Thus, no longer was it the affordable, lightweight, and nimble-handling car it once was.[10]

Not coincidentally, the import-tuner mindset pioneered by the water-cooled Volkswagen enthusiasts of the 1980s began to spread to other imported marques in the early to mid-1990s, just as the original hot-hatch GTI began to gain cylinders and weight. More specifically, many gearheads turned to the ubiquitous and ostensibly economy-oriented Honda Civic. Small, light, inexpensive, and blessed with a sophisticated four-wheel independent suspension, the front-wheel-drive Civic quickly became the overwhelming favorite of the growing import-tuner crowd. (Honda buffs also embraced the Prelude, a sportier model that was to the Civic what the Scirocco and the Corrado were to the GTI.) Though Honda offered a number of four-cylinder engines in its Civic line during the course of the mid- to late 1990s, those equipped with the firm's VTEC system were by far the most prized.

VTEC was a variable camshaft-timing system that sought to eliminate the compromises inherent in engine-camshaft selection. At low rpms, a VTEC engine delivered excellent economy and everyday driveability, but once the needle of the tachometer swept past a predetermined point—5000 rpm, say—the VTEC system suddenly shifted the effects of the camshafts' action to improve top-end power.[11] Driving a VTEC Honda was therefore somewhat akin to driving an early turbocharged car: very little excitement for the first several thousand rpms, followed by a dramatic surge in power for the last few. Thus, any Honda with a VTEC engine packed a screaming high-end punch that proved to be nothing less than an intoxicant for power-hungry, front-wheel-drive enthusiasts. By the late 1990s, lowered Hondas with trunk-lid wings, alloy wheels, high-performance ECU chips, and booming mufflers were common, as were similarly modified four-cylinder Mitsubishis, Toyotas, Nissans, and other Japanese makes. Soon specialist magazines geared specifically toward the interests of the Asian-car enthusiast appeared, and the number of aftermarket firms that produced high-performance accessories for Hondas, Toyotas, and other Japanese brands multiplied (fig. 21). So popular were these Asian brands, in fact, that "import tuning" quickly came to be virtually synonymous with "*Japanese* import tuning."[12]

Today, nearing the end the first decade of the twenty-first century, Japanese import tuning remains a vibrant and vital part of the American hot rodding fraternity. More broadly, sport-compact tuning as a whole, including its Asian, European, and domestic compact niches, has proven to be a major source of aftermarket sales growth.[13] It is a phenomenon that has already spawned its own motion pictures (*The Fast and the Furious* and its sequels, *2 Fast 2 Furious* and *The Fast and the Furious: Tokyo Drift*) and television programs (e.g., SPEEDtv's *Street Tuner Challenge*). Crucially, too, its devotees are virtually obsessed with cutting-edge technology, from carbon-fiber body panels to direct, solder-free ECU flash-tuning.[14] Finally, because it is based on affordable and abundant new-car models, it is a segment that appears to have a bright and prosperous future. In short, sport-compact tuning has proven to be the aftermarket's "next big thing."

This is not to say, of course, that it is the *only* niche that matters to the high-performance industry of today. Far from it, for drag racing, street machines, street rods, street trucks, repro rods, and a number of other niches all contribute to the

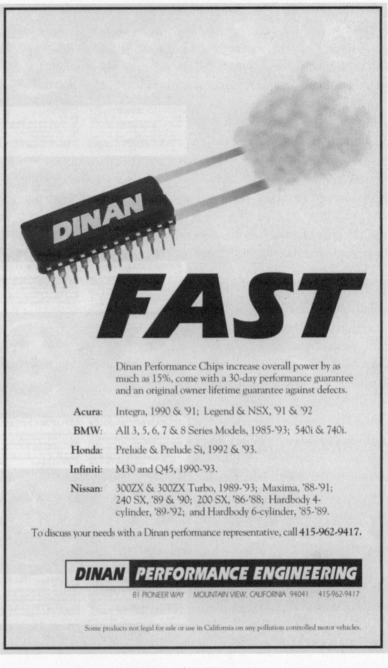

Figure 21. A 1994 advertisement for Dinan P-ROMs. Notice the prevalence of Japanese applications listed in the text of this advertisement, which was published, however ironically, in a magazine for European-car enthusiasts: by the mid-1990s, "import tuning" was increasingly synonymous with "Japanese import tuning." (Dinan Performance Engineering advertisement, *European Car*, January 1994, 29. Reproduced courtesy of the editor of *European Car*, Les Bidrawn.)

industry's bottom line. Likewise, to say that import tuning is for the most part an Asian-car phenomenon focused predominantly on front-wheel drive compacts is by no means meant to imply that there isn't considerable variety at play within its ranks. For indeed, not only does there remain a great deal of interest in compact European models, but there is also a great deal of aftermarket activity surrounding more expensive makes such as Acura, Lexus, BMW, Porsche, Mercedes-Benz, and even Ferrari and Lamborghini. But among bread-and-butter, teen and twenty-something enthusiasts, sport-compact tuning generally (and Japanese import tuning, in particular) is without a doubt *it*.

Not everyone is thrilled about this, though. Die-hard American V8 fans still react with horror to the fact that so many young enthusiasts choose under-powered, high-strung, tin-can imports over genuine domestic muscle. On the other hand, import tuners love to laugh at those who still insist on driving terminally under-steering, nose-heavy, V8-powered dinosaurs. At the same time, European-car enthusiasts lambaste the Asian-car buffs for choosing unrefined Japanese economy over sophisticated German engineering. Meanwhile, the Honda crowd fires back by poking sport at those who drive fourth-generation VR6 GTIs, which stray too far from the lighter-is-always-better philosophy of import tuning and dangerously close to the no-substitute-for-cubic-inches-and-extra-cylinders mentality of the V8 boys.

But is it really fair and reasonable to conclude on the basis of these fissures that fundamental differences must in fact exist between these groups of enthusiasts? In other words, is a lowered and tuned front-wheel-drive Civic really all that different from a raked and blown SS Chevelle or a channeled and fenderless A-V8? In a word, no. Civics, Sentras, Neons, and Golfs are to the 1990s and 2000s what 1932 Ford roadsters were to the 1930s and 1940s, what overhead-valve V8s were to the 1950s, and what muscle cars were to the 1960s. They are affordable. They are abundant. They respond well to aftermarket modifications. And perhaps most tellingly, they are enormously popular among younger enthusiasts. In short, they are today's hot rods.

Nowhere—save, perhaps, at the magazine racks at Barnes & Noble—is this any more evident than it is at the annual SEMA Show in Las Vegas. There after-market displays featuring wildly painted Hondas and VWs with booming exhaust systems, polished wheels, and enormous spoilers are interspersed with displays focused on performance accessories for street machines, street rods, trucks, and old-school dragsters. And in spite of the derogatory things that the V8 crowd might say about import tuners (and vice versa), on the floor of the SEMA Show it's all the same. For there it's all about the business: the business, that is, that ties these warring niches together; the business that dates back to the earliest days of mass-produced automobility; the business that helped to make high-performance motoring a reality for those with ordinary cars and ordinary means. For the business of speed equipment manufacturing has never been about a certain brand of car, a certain type of engine, or a certain class of enthusiasts. Instead, it has always been about the ability to *adapt*. Recall the words of Jim McFarland, written in 1979: "We don't originate the plan. We *play* the game, using *their* toys, not ours."[15]

The toys have changed considerably over the years, but the rules of the game have not. In fact, they have remained unchanged for more than ninety years. For if there's one thing the Model T enthusiasts of the 1910s, the dry-lakes gearheads of the 1930s, the quarter-mile racers of the 1950s, the muscle-car buffs of the 1960s, the street rodders of the 1970s, and the import devotees of today have shared in common, it has been their need for commercially available products to enable them to transform their mass-produced and otherwise mundane cars into personalized, high-performance machines.

Therein lies a crucial point, one that's all too easy to overlook today, particularly with all of the attention that television programs like *American Hot Rod*, *Street Tuner Challenge*, and *Overhaulin'* regularly receive. These shows typically follow the complete transformation of an ordinary car into a customized masterpiece at the hands of highly paid professionals working in controlled shop environments. Average enthusiasts play no role in the customization process featured in these programs, and that is precisely the point emphasized by those who have written about them. According to Kathleen Franz, for example, these shows help to "demonstrate that, by and large, restyling and tinkering with automobiles now takes place in a shop under the directions of experts rather than in the home garage."[16]

There are indeed a number of hot rod shops today where well-heeled enthusiasts regularly drop $100,000 or more on professionally built, show-quality street rods. Likewise, there are many import-tuner shops where amounts well in excess of $10,000 regularly change hands for comprehensive, high-performance engine and drivetrain work. However, far more common than either of these are the mail-order houses and retail parts counters where ordinary enthusiasts spend far lesser amounts on individual components that they then install themselves (or with the help of a friend or two) in their suburban garages. This is no less true today than it was in the 1950s, the 1970s, and the early 1990s. For at its core, hot rodding is and has always been a hands-on affair—a do-it-yourself pastime the sheer contemporary magnitude and near-100-year longevity of which suggest in no uncertain terms that automotive end-user agency is alive and well.[17] It will remain so, too, as long as gearheads young and old continue to imagine what their ordinary cars *could* be. Television programs may well help them to do this, but once they decide to take the next step, most of them won't call *American Hot Rod's* Boyd Coddington to schedule a top-dollar appointment. Instead, they'll dial the advertisers in the latest issue of *Hot Rod*, *European Car*, or *Import Tuner*. And when their parts arrive in the mail, they'll head to their garages, roll up their sleeves, and get to work.

INTRODUCTION

1. This biographical sketch draws together material from Dean Batchelor, *The American Hot Rod* (Osceola, Wis.: MBI Publishing, 1995), 179–80; Ed Almquist, *Hot Rod Pioneers: The Creators of the Fastest Sport on Wheels* (Warrendale, Pa.: SAE, 2000), 90–91; and Steve Hendrickson, "Introduction," in *Hot Rod Magazine: The First 12 Issues* (Osceola, Wis.: MBI Publishing, 1998), 4–5.

2. Almquist claims that they ordered 10,000 copies of their debut issue, but in their October 1948 issue, Petersen and Lindsay claim to have published only 5,000. See "Editor's Column," *Hot Rod Magazine*, October 1948, 5. Hereafter, *Hot Rod Magazine*, also known as *Hot Rod*, is cited as *HRM*.

3. Their January 1948 issue was ready in December 1947, which is when the hot rod exposition opened.

4. "Editor's Column," *HRM*, January 1949, 7.

5. In September 1949, Petersen and Lindsay launched *Motor Trend*, a general-interest automotive publication designed to complement *Hot Rod*'s more specialized coverage, and the following May they added a third, *Cycle* (Batchelor, *American Hot Rod*, 180–81). Although Petersen bought him out in 1950, Lindsay continued to co-publish *Hot Rod* through April of 1952.

6. *Forbes*, 13 September 1982, 153.

7. *Forbes*, 14 October 1996, 286. In this edition, *Forbes* estimated that the sale of the Petersen Publishing empire generated $450 million; subsequent *Forbes* 400 lists corrected the figure to $500 million.

8. Dean Batchelor vividly recalls meeting Petersen for the first time at the October 19, 1947, Southern California Timing Association (SCTA) meet at El Mirage Dry Lake, where Petersen cornered Batchelor's friend and racing partner Alex Xydias in an attempt to sell an advertisement for Alex's fledgling speed shop (Batchelor, *American Hot Rod*, 179).

9. *Hot Rod* first exceeded fifty pages of coverage in April 1951, and its circulation reached 500,000 copies in September 1952.

10. "World's Most Complete Hot Rod Coverage" was the magazine's first slogan; today it remains the largest automotive enthusiast publication on the market.

11. Batchelor, *American Hot Rod*, 181.

12. "Editor's Column," *HRM*, January 1948, 3.

13. Technically, another enthusiast magazine called *Speed Age* beat *Hot Rod* to market by several months, but its coverage was more diffuse: *Speed Age* was not devoted specifically to hot rodding, as was *Hot Rod*.

14. Enthusiasts spent $34 billion in 2006 (Angus MacKenzie, "SEMA Moments," *Motor Trend*, January 2007, 10), up from $26 billion five years earlier (Specialty Equipment Market Association [SEMA], "2002 Automotive Specialty Equipment Industry Update," 2 [SEMA Research Center, SEMA Headquarters, Diamond Bar, California—hereafter, SEMA-RC]).

15. Among the best of these popular titles, apart from Almquist's *Hot Rod Pioneers* and Batchelor's *American Hot Rod*, are Don Montgomery, *Hot Rods in the Forties: A Blast from the Past* (Fallbrook, Calif.: D. Montgomery, 1987); Montgomery, *Hot Rods As They Were: Another Blast from the Past* (Fallbrook: D. Montgomery, 1989); Art Bagnall, *Roy Richter: Striving for Excellence* (Los Alamitos, Calif.: Art Bagnall Publishing, 1990); Tom Medley and LeRoi Smith, *Tex Smith's Hot Rod History, Volume One: The Beginnings* (Osceola, Wis.: Motorbooks International, 1990); Montgomery, *Hot Rod Memories: Relived Again* (Fallbrook: D. Montgomery, 1991); Montgomery, *Supercharged Gas Coupes: Remembering the Sixties* (Fallbrook: D. Montgomery, 1993); Medley and Smith, *Tex Smith's Hot Rod History, Volume Two: The Glory Years* (North Branch, Minn.: CarTech, 1994); Montgomery, *Authentic Hot Rods: The Real "Good Old Days"* (Fallbrook: D. Montgomery, 1994); Montgomery, *Those Wild Fuel Altereds: Drag Racing in the Sixties* (Fallbrook: D. Montgomery, 1997); Peter Vincent, *Hot Rod: An American Original* (St. Paul: MBI Publishing Company, 2001); Jerry Burton, *Zora Arkus-Duntov: The Legend Behind Corvette* (Cambridge, Mass.: Bentley Publishers, 2002); Ron Roberson, *Middletown Pacemakers: The Story of an Ohio Hot Rod Club* (Chicago: Arcadia, 2002); Robert Genat and Don Cox, *The Birth of Hot Rodding: The Story of the Dry Lakes Era* (St. Paul, Minn.: Motorbooks International, 2003); Mark Christensen, *So-Cal Speed Shop: The Fast Tale of the California Racers Who Made Hot Rod History* (St. Paul: Motorbooks, 2005); and Tom Madigan, *Edelbrock: Made in USA* (San Diego: Tehabi Books, 2005).

16. Strictly speaking, H. F. Moorhouse's *Driving Ambitions: A Social Analysis of the American Hot Rod Enthusiasm* (New York: Saint Martin's Press, 1991), Robert C. Post's *High Performance: The Culture and Technology of Drag Racing* (Baltimore: Johns Hopkins University Press, 1994), and John DeWitt's *Cool Cars, High Art: The Rise of Kustom Kulture* (Jackson: University Press of Mississippi, 2002) are the only academic books that deal with hot rodding. Brenda Jo Bright's anthropological work on the lowrider culture of the American Southwest—a culture loosely related to that of the hot rodder—pushes the total to six if we count her M.A. and Ph.D. theses ("Style and Identity: Houston Low Riders" [M.A. thesis, Rice University, 1986] and "Mexican American Low Riders: An Anthropological Approach to Popular Culture" [Ph.D. diss., Rice University, 1994]) as well as her edited volume on automotive art (*Customized: Art Inspired by Hot Rods, Low Riders, and American Car Culture* [New York: H. N. Abrams, 2000]). In addition, Moorhouse published two short essays about hot rodding in the 1980s: "Racing for a Sign: Defining the 'Hot Rod,' 1945–1960," *Journal of Popular Culture* 20 (1986): 83–96, and "The 'Work' Ethic and 'Leisure' Activity: The Hot Rod in Post-war America," in *The Historical Meanings of Work*, ed. Patrick Joyce (New York: Oxford University Press, 1987), 237–57. More recently, Jessie Embry published a wonderful account of Bonneville racing that touches briefly on

hot rodding: "The Last Amateur Sport: Automobile Racing on the Bonneville Salt Flats," *Americana: The Journal of Popular Culture* 2 (2003), available online at www.american popularculture.com/journal.articles/fall_2003/embry.htm. Beyond these contributions, references to hot rodding within the academic literature are fleeting at best. Lizabeth Cohen mentions it briefly in *A Consumer's Republic: The Politics of Mass Consumption in Postwar America* (New York: Knopf, 2003), 309; Kathleen Franz, in an epilogue to her marvelous book, largely dismisses the importance of hot rodding as a latter-day form of end-user "tinkering" (*Tinkering: Consumers Reinvent the Early Automobile* [Philadelphia: University of Pennsylvania Press, 2005], 161–66); Carroll Pursell includes a brief section about hot rods in his recent *Technology in Postwar America: A History* (New York: Columbia University Press, 2007), 104–5; and David Edgerton gives hot rodders passing mention in *The Shock of the Old: Technology and Global History Since 1900* (New York: Oxford University Press, 2007), 97–98.

17. Specifically, "Big Three" refers to Chrysler, Ford, and General Motors; "Big Four" also appears in the sources from time to time (number four was American Motors).

18. The academic literature on mass automobility is varied and vast. Notable surveys include John B. Rae, *American Automobile Manufacturers: The First Forty Years* (New York: The Chilton Company, 1959); Rae, *The American Automobile: A Brief History* (Chicago: University of Chicago Press, 1965); James J. Flink, *The Car Culture* (Cambridge, Mass.: MIT Press, 1975); Rae, *The American Automobile Industry* (Boston, Mass.: G. K. Hall and Co., 1984); Flink, *The Automobile Age* (Cambridge: MIT Press, 1988); and Rudi Volti, *Cars and Culture: The Life Story of a Technology* (Westport, Conn.: Greenwood Press, 2004). On the manufacture of the American automobile, see David A. Hounshell, *From the American System to Mass Production, 1800–1932: The Development of Manufacturing Technology in the United States* (Baltimore: Johns Hopkins University Press, 1984), chaps. 6 and 7, and Thomas J. Misa, *A Nation of Steel: The Making of Modern America, 1865–1925* (Baltimore: Johns Hopkins University Press, 1995), chap. 6. On labor management within the automobile industry, see Stephen Meyer III, *The Five Dollar Day: Labor Management and Social Control in the Ford Motor Company, 1908–1921* (Albany: State University of New York Press, 1981), and Nelson Lichtenstein, *The Most Dangerous Man in Detroit: Walter Reuther and the Fate of American Labor* (New York: Basic Books, 1995). On the regulation of the automobile, see James E. Krier and Edmund Ursin, *Pollution and Policy: A Case Essay on California and Federal Experience with Motor Vehicle Air Pollution, 1940–1975* (Berkeley: University of California Press, 1977); Douglas H. Ginsburg and William J. Abernathy, eds., *Government, Technology, and the Future of the Automobile* (New York: McGraw-Hill, 1980); Jeffrey Fawcett, *The Political Economy of Smog in Southern California* (New York: Garland, 1990); Gary C. Bryner, *Blue Skies, Green Politics: The Clean Air Act of 1990 and Its Implementation* (Washington, D.C.: Congressional Quarterly Press, 1993); and Rudi Volti, "Reducing Automobile Emissions in Southern California: The Dance of Public Policies and Technological Fixes," in *Inventing for the Environment*, ed. Arthur Molella and Joyce Bedi (Cambridge: MIT Press, 2003), 277–88. On automobility and the growth of suburbia, see Kenneth T. Jackson, *Crabgrass Frontier: The Suburbanization of the United States* (New York: Oxford University Press, 1985). On the urban racial geography fostered by automobility, see Howard L. Preston, *Automobile Age Atlanta: The Making of a Southern Metropolis, 1900–1935* (Athens: University of Georgia

Press, 1979), and Ronald Bayor, *Race and the Shaping of Modern Atlanta* (Chapel Hill: University of North Carolina Press, 1995). On the early urban market for automobiles, see Clay McShane, *Down the Asphalt Path: The Automobile and the American City* (New York: Columbia University Press, 1994). On the American highway system, see Bruce Seely, *Building the American Highway System: Engineers as Policy Makers* (Philadelphia: Temple University Press, 1987), and Tom Lewis, *Divided Highways: Building the Interstate Highways, Transforming American Life* (New York: Viking Penguin, 1997). On consumers, end users, and the repair industry, see Ronald Kline and Trevor Pinch, "Users as Agents of Technological Change: The Social Construction of the Automobile in the Rural United States," *Technology and Culture* 37 (1996): 763–95; Franz, *Tinkering*; Kevin L. Borg, "The 'Chauffeur Problem' in the Early Auto Era: Structuration Theory and the Users of Technology," *Technology and Culture* 40 (1999): 797–832; Stephen L. McIntyre, "The Failure of Fordism: Reform of the Automobile Repair Industry, 1913–1940," *Technology and Culture* 41 (2000): 269–99; and Kevin L. Borg, *Auto Mechanics: Technology and Expertise in Twentieth-Century America* (Baltimore: Johns Hopkins University Press, 2007).

19. "Original equipment manufacturers" is a common shorthand for mainstream automotive companies like Ford, Honda, and Chrysler.

20. Moorhouse, "Racing for a Sign"; "The 'Work' Ethic and 'Leisure' Activity"; and *Driving Ambitions*.

21. Post, *High Performance*.

22. DeWitt, *Cool Cars, High Art*.

23. Thomas Misa's study of the American steel industry (*A Nation of Steel*) is a model user-producer analysis, as is Cyrus Mody, "Corporations, Universities, and Instrumental Communities: Commercializing Probe Microscopy, 1981–1996," *Technology and Culture* 47 (2006): 56–80.

24. See especially Thomas P. Hughes, *American Genesis: A Century of Innovation and Technological Enthusiasm, 1870–1970* (New York: Viking Press, 1971); Joseph J. Corn, *Winged Gospel: America's Romance with Aviation, 1900–1950* (New York: Oxford University Press, 1983); and Lee Worth Bailey, *The Enchantments of Technology* (Champaign: University of Illinois Press, 2005), chap. 4.

25. Here I follow the lead of Eugene Ferguson and Robert C. Post. See Ferguson's *Engineering and the Mind's Eye* (Cambridge: MIT Press, 1992); "Enthusiasm and Objectivity in Technological Development" (unpublished manuscript, 1970, held in the files of Merritt Roe Smith, MIT); and "Elegant Inventions: The Artistic Component of Technology," *Technology and Culture* 19 (1978): 450–60. See also Post, *High Performance*, especially chapter 14.

26. "Less overtly rational" motivations among motoring enthusiasts are not limited to hot rodders. Consider, for example, the motorcycling pursuits of the legendary Burt Munro, chronicled in George Begg, *Indian: Legend of Speed* (Christchurch, N.Z.: Begg & Allen Ltd., 2002).

27. See, e.g., John E. Sawyer, "The Social Basis of the American System of Manufacturing," *Journal of Economic History* 14 (1954): 361–79; Hugh G. Aitken, *Taylorism at Watertown Arsenal: Scientific Management in Action* (Princeton, N.J.: Princeton University Press, 1960); Robert S. Woodbury, "The Legend of Eli Whitney and Interchangeable Parts," *Technology and Culture* 1 (1960): 235–53; Nathan Rosenberg, "Technological Change in

the Machine Tool Industry," *Journal of Economic History* 23 (1963): 414–43; Daniel Nelson, *Managers and Workers: The Origins of the New Factory System in the United States, 1880–1920* (Madison: University of Wisconsin Press, 1975); Merritt Roe Smith, *Harper's Ferry Armory and the New Technology: The Challenge of Change* (Ithaca, N.Y.: Cornell University Press, 1977); Alfred D. Chandler Jr., *The Visible Hand: The Managerial Revolution in American Business* (Cambridge, Mass.: Harvard Belknap, 1977); and Hounshell, *American System to Mass Production.*

28. See Michael J. Piore and Charles F. Sabel, *The Second Industrial Divide: Possibilities for Prosperity* (New York: Basic Books, 1984); Charles F. Sabel and Jonathan Zeitlin, "Historical Alternatives to Mass Production: Politics, Markets, and Technology in Nineteenth-Century Industrialization," *Past and Present* 108 (1985): 133–76; John K. Brown, *The Baldwin Locomotive Works, 1831–1915: A Study in American Industrial Practice* (Baltimore: Johns Hopkins University Press, 1995); and Philip Scranton, *Endless Novelty: Specialty Production and American Industrialization, 1865–1925* (Princeton, N.J.: Princeton University Press, 1997).

29. Kline and Pinch, "Users as Agents of Technological Change," and Franz, *Tinkering.*

30. A brief glossary of performance-enthusiast jargon also appears as an appendix to this book.

31. Post, *High Performance,* 126.

32. See, for example, Don Montgomery's discussion of the matter in *Hot Rods in the Forties,* 7.

33. Batchelor, *American Hot Rod,* 8 (emphasis added).

34. By the 1990s import and sport-compact enthusiasts had begun to refer to themselves *and* their cars as "tuners." For a brief discussion of this terminological development, see my conclusion.

35. Post, High Performance, x.

CHAPTER 1: FASTER FLIVVERS, 1915–1927

1. In 1911, annual domestic automobile production totaled approximately 210,000 units; by 1917, the figure stood at nearly 1,940,000. New registrations for calendar year 1917 totaled 1,396,324—up from approximately 330,000 in 1912; per capita ownership therefore stood at twenty people per car in 1917, up from forty in 1915 and close to eighty in 1913 ("4,941,276 Cars and Trucks in the United States," *Automotive Industries—The Automobile* [hereafter, *AI*], 14 March 1918, 534, 538). The per capita figure for 1913 is based on the total number of registered vehicles in the United States that year, 1,253,875 ("4,941,276 Cars and Trucks," 538) and the census figure for 1913 of 97,225,000 (www.census.gov).

2. Production and sales figures for Ford are from Hounshell, *American System to Mass Production,* 224, and the aggregate figures for the entire American automobile industry during the 1910s—as well as the registration data for the same period—are from "4,941,276 Cars and Trucks," 538.

3. On the breakthrough at Ford from an organizational perspective, see Rae, *American Automobile Manufacturers;* from a manufacturing perspective, see Hounshell, *American System to Mass Production;* from a labor-relations perspective, see Meyer, *Five Dollar Day;* and for a broader narrative, see Flink, *Automobile Age.*

4. "Equipped for the Track," *The Fordowner*, September 1914, 22. *The Fordowner* was an independent publication of the Hallock Publishing Company of Cleveland.

5. In May 1915, *The Fordowner* ran a lengthy article condemning lawless driving by owners of ordinary and modified Model Ts, encouraging them to organize racing clubs as a safe, off-street alternative ("Ford Racing," *The Fordowner*, May 1915, 23–25). The following month, the same periodical both welcomed the advent of modified road-going Fords and continued to encourage its readership to form racing clubs ("Ford Racing and Racers," *The Fordowner*, June 1915, 30).

6. "Ford Racing and Racers," 30. Model T differentials are clearly visible from the rear.

7. To commemorate the production of its record-breaking 15,007,034th Beetle, Volkswagen staged an on-track press event in 1972 during which a pack of Beetles and Superbeetles blew past a nicely restored Model T on the straight. See Terry Shuler, ed., *The Origin and Evolution of the VW Beetle* (Princeton, N.J.: Princeton Publishing, Inc., 1995), 122.

8. Kline and Pinch, "Users as Agents of Technological Change," and Franz, *Tinkering*.

9. Apart from automotive publications of the period, an excellent source of technical information on the Model T is Hounshell, *American System to Mass Production*, especially chapter 7.

10. In mid-1922, *Ford Owner and Dealer* reported that the accessory business was a $75-million annual trade ("The Truth of the Ford Accessory Market," *Ford Owner and Dealer* [hereafter, *FOaD*], June 1922, 30–31); by the end of that year, the figure had swelled to $90 million ("The Growth of the Accessory Market," *FOaD*, January 1923, 48). For more on this general-improvement aftermarket, see Franz, *Tinkering*; Kline and Pinch, "Users as Agents of Technological Change"; and James L. Kenealy, *Model "T" Ford Authentic Accessories, 1909–1927* (Seattle, Wash.: Kenealy, 1976). Following a brief lull in 1927 and 1928, this general-improvement aftermarket continued to prosper well into the Model A and V8 eras of the 1920s, 30s, and 40s; see Murray Fahnestock, *Those Wonderful Unauthorized Accessories for Model A Ford* (Arcadia, Calif.: Post Motor Books, 1971), and Dan Smith, *Accessory Mascots: The Automotive Accents of Yesteryear, 1910–1940* (San Diego: D. Smith, 1989).

11. Detroit Radiator and Specialty Company advertisement, *The Fordowner*, August 1915, 75.

12. Among others, these included the Akron Motor & Manufacturing Company, also known as Hal (Akron, Ohio); the Auto Remodeling Company (Chicago); the Beaver Manufacturing Company (Milwaukee); Berg (Chicago); Centri (Oakland, Calif.); the Chevrolet Brothers Manufacturing Company (Indianapolis); the Cooper Manufacturing Company (Marshalltown, Iowa); Craig-Hunt (Indianapolis); D. R. Noonan (Paris, Ill.); the Detroit Radiator Company (Michigan); the Dunn Counterbalance Company (Clarinda, Iowa); Eastern Auto (Los Angeles); the Fordspeed Company (New York); the Green Engineering Company (Dayton, Ohio); the Maibohm Wagon Company (Racine, Wis.); the Mais Company (Indianapolis); the McCadden Machine Works (St. Cloud, Minn.); the Miller Carburetor Company (Los Angeles and Chicago); the Milwaukee Forge & Machine Company (Wisconsin); Morton & Brett (Indianapolis); the PACO Manufacturing Company (Galesburg, Ill.); Rajo (Racine, Wis.); Riley (Los Angeles); the Roof Auto

Specialty Company (Anderson, Ind.); the Ruckstell Sales and Manufacturing Company (Berkeley, Calif.); the Schutte Body Company (Lancaster, Pa.); the Turnbull Company (Wilmington, Ohio); the Walker M. Levett Company (New York); the Waukesha Motor Company (Waukesha, Wis.); the Williams Foundry and Machine Company (Akron, Ohio); the Winfield Carburetor Company (Los Angeles); and the Zenith Automotive Manufacturing Company (St. Louis). Local and national retailers that sold speed equipment but did not manufacture it were also common.

13. New York led the nation with more than 404,000 registered cars in 1918. Rounding out the list, in descending order, were Illinois, Ohio, Pennsylvania, Iowa, California, Texas, Michigan, Minnesota, and Indiana. See "4,941,276 Cars and Trucks," 535 (map). On the Midwestern concentration of the American automobile industry, see Flink, *Automobile Age*, 24.

14. Williams Foundry and Machine Company advertisement, *Ford Dealer and Service Field* [hereafter, *FDaSF*], July 1926, 18; Beaver Manufacturing Company advertisement, *Ford Dealer and Owner* [hereafter, *FDaO*], May 1925, 156; Flink, *Automobile Age*, 67 (on Chevrolet's collapse and absorption by General Motors); Rajo advertisement, *FOaD*, November 1920, 113; Ed Almquist, "Ed Winfield: The Reclusive Genius," in *Hot Rod Pioneers*, 4–5; Terry Cook, "Ed Winfield: The Father of Hot Rodding," *HRM*, January 1973, 107; Kem Robertson, "Robert Roof, Man Extraordinaire," *The Alternate*, 15 April and 15 May 2003; Almquist, "Robert M. Roof—A Granddaddy of Speed Equipment," in *Hot Rod Pioneers*, 10–11; "Indianapolis Speedway Adopts the 3-Liter Limit for Future Races," *AI*, 5 June 1919, 1201–7 and 1245 (on the racing careers of Arthur and Louis Chevrolet as well as of their brother Gaston); Mark L. Dees, *The Miller Dynasty: A Technical History of the Work of Harry A. Miller, His Associates, and His Successors* (New York: Barnes Publishing, Inc., 1981); Bagnall, *Roy Richter*, chap. 1; Batchelor, *American Hot Rod*, chaps. 2, 3, and 12; and below, chapter 2.

15. The Laurel Motors Corporation, which succeeded the Roof Auto Specialty Company in 1917, was one high-performance company that did offer ignition systems for the Model T from the very beginning (see, e.g., the fine print of its advertisement in the April 1918 issue of *The Fordowner*, p. 79). For the most part, though, general-improvement ignition systems were preferred even on high-performance engines. See Mechanician, "Putting Speed in Speedster Type of Ford," *The Fordowner*, January 1917, 67–68, and Murray Fahnestock, "Modern Ignition Systems," *FOaD*, January 1924, 72–76. Over the years, *The Fordowner* slowly evolved from an owner- to a trade-oriented publication, and its title changed several times. *Motor Age* evolved in a similar fashion, although its title never changed during the period in question.

16. Dunn Counterbalance Company advertisement, *The Fordowner*, January 1918, 105. Roof's company manufactured five-bearing cranks for the Model T in the mid-1920s; see Robert M. Roof, "Power and Speed: The Overhead Cam Shaft and Five Bearing Crank Shaft for Greater Engine Efficiency," *FOaD*, April 1924, 138, 140, and 142.

17. Walker M. Levett Company advertisement, *The Fordowner*, August 1915, 75; Green Engineering advertisement, *The Fordowner*, January 1918, 85; and "Reader's Clearing House," *Motor Age*, 31 July 1919, 43.

18. Miller advertisement, *The Fordowner*, February 1918, 4–5; Chandler advertisement, *The Fordowner*, November 1916, 53; and Stewart-Warner advertisement, *FOaD*, March 1922, 97.

19. See William Morrow, "The Ford Speedster: Some Notes From One Who Is Experienced in Rebuilt Jobs," *FOaD*, October 1924, 57. Morrow advocated the use of Fordson valves to improve engine breathing, as did *The Fordowner*'s Murray Fahnestock (see "Power and Speed, Article I: Using Fordson Valves in the Ford Cylinder Block," *FDaO*, June 1925, 66, 68, and 70). Fordspeed of New York manufactured larger valves for the Model T (Fordspeed advertisement, *FOaD*, April 1922, 158), as did several others.

20. In 1918, Laurel began to advertise high-performance camshafts as part of its growing line of speed equipment (Laurel Motors Corporation advertisement, *The Fordowner*, April 1918, 79); by the mid-1920s many others had begun to do the same.

21. As early as 1915, for example, how-to articles began to advocate the use of exhaust "cut-outs," or muffler bypass valves, in order to allow the driver of a modified car to enjoy the benefits of an unmuffled, freer-flowing exhaust on demand ("More Speed," *The Fordowner*, January 1915, 15–18). By 1917, however, the editors of *The Fordowner* advised discretion in the use of cut-outs (Mechanician, "Putting Speed in Speedster Type of Ford," January 1917, 67–68); and in 1920, the magazine's successor ran a detailed though inconclusive analysis of the issue by an automotive engineer, G. I. Mitchell ("Does Muffler Impair Engine Efficiency?" *FOaD*, June 1920, 167–68).

22. Overhead-valve conversions were first available from the Roof Auto Specialty Company, which introduced its sixteen-overhead-valve pushrod conversion for the Ford toward the end of 1917, shortly before it merged with Laurel (Robertson, "Robert Roof," 2; Roof Auto Specialty Company advertisement, *The Fordowner*, January 1918, 53; Laurel Motors Company advertisement, *The Fordowner*, February 1918, 3; and "Sixteen Valve Cylinder Heads," *The Fordowner*, January 1918, 44, 46, and 48); Craig-Hunt introduced its conversion at almost exactly the same time as Roof / Laurel. By the end of the decade, Joe Jaegersberger had introduced his "Rajo" line of cylinder heads, and Arthur and Louis Chevrolet lagged only slightly in bringing out their "Frontenac" heads (see "Valve-in-Head for Fords," *Motor Age*, 11 September 1919, 51, and "A New Cylinder Head," *FOaD*, November 1920, 96). Many others followed in the 1920s.

23. Increasing the Model T's compression ratio for a slight horsepower gain involved little more than milling a few thousandths of an inch from the bottom of the head, thereby reducing the size of the combustion chambers and raising the compression ratio. So poor was the Model T's flathead, however, that doing so without modifying the part in other critical ways increased the engine's propensity to self-destruct through detonation. See Fahnestock, "Shapes of Cylinder Heads," *FDaSF*, September 1926, 68, 70, 72, 74, 76, and 78.

24. Turnbull advertisement, *The Fordowner*, January 1918, 6. Of the other available flatheads, Waukesha's was perhaps the most interesting. Its design stemmed from the work of a British engineer named Harry R. Ricardo, who designed tank engines during WWI. In the course of this work, he developed an improved flathead combustion chamber shape. With the aid of the Waukesha Company in the United States, Ricardo patented his concept in the early 1920s (he filed in 1919 and was granted patent number 1,474,003 in 1923). Waukesha then proceeded to manufacture a line of heads for the Model T based on Ricardo's ideas. See Fahnestock, "Shapes of Cylinder Heads," esp. p. 70, and "Ricardo & Company, Engineers, Ltd.," Trade Catalog Collection, Benson Ford Research Center, The Henry Ford, Dearborn, Michigan.

25. Laurel advertisement, *The Fordowner*, April 1918, 79; Maibohm Wagon Company advertisement, *Motor Age*, 23 January 1919, 367; Morton & Brett advertisement, *FOaD*, November 1920, 187; and Fordspeed advertisement, *FOaD*, April 1922, 158.

26. A. P. Hess, "How to Undersling a Ford," *The Fordowner*, March 1915, 23; Fahnestock, "Ford Speedster," *The Fordowner*, June 1916, esp. pp. 30 and 34; and Fahnestock, "Lowering the Ford Chassis: Underslinging the Ford Speedster, Sport Model or Special Car," *The Fordowner*, October 1919, 43–48, 50, 52, and 54.

27. Road-going Mercers, Packards, Cadillacs, and Stutzes, for example, were capable of approximately 80 mph in the late 1910s. See "Reader's Clearing House," *Motor Age*, 20 March 1919, 35.

28. Batchelor, *American Hot Rod*, 37–41.

29. Detroit Radiator and Specialty Company advertisement, *The Fordowner*, August 1915, 75; Walker M. Levett Company advertisement, *The Fordowner*, August 1915, 75; Roof Auto Specialty Company advertisement, *The Fordowner*, January 1918, 53 (quote); Green Engineering Company advertisements, *The Fordowner*, January 1918, 85, and *Motor Age*, 23 January, 1919, 255; Dunn Counterbalance Company advertisement, *The Fordowner*, January 1918, 105; Turnbull Company advertisement, *The Fordowner*, January 1918, 6; McCadden Machine Works advertisement, *The Fordowner*, January 1917, 92; Miller advertisement, *The Fordowner*, February 1918, 4–5; and PACO advertisement, *Motor Age*, January 23, 1919, 161.

30. Craig-Hunt advertisement, *The Fordowner*, February 1918, 108, and "Making the Ford Fleet-Footed," *Motor Age*, 5 June 1919, 40–41.

31. "Equipped for the Track," *The Fordowner*, September 1914, 22.

32. See, for example, any of the Laurel Motor Corporation's advertisements in the 1910s and 1920s, but especially those from the April 1918 issue of *The Fordowner* (p. 79) and the November 1920, July 1921, and May 1922 issues of *Ford Owner and Dealer* (pp. 127, 131, and 23, respectively). See also Rajo advertisements in *FOaD*, November 1920, 113; *FOaD*, July 1921, 117; and *FOaD*, November 1922, 27; and Winfield Carburetor Company advertisements in *FOaD*, September 1924, 13, and *FOaD*, February 1925, 30.

33. A fourth type of source with the potential to be useful in this regard is the feature article—write-ups and photographs, that is, of outstanding examples of the state of the art. Because there were no dedicated enthusiast magazines in the 1910s and 1920s, however, very few such features appeared during those years. Notable exceptions include "Remarkable Ford Racer," *The Fordowner*, April 1915, 30 and 32; "Ford Completely Rebuilt," *Motor Age*, 3 July 1919, 28–29 and 31; and "Snappy Ford Speedster," *FOaD*, December 1920, 82. Words like "racer" and "speedster" were ambiguous in the era of the Model T: sometimes they described track cars, but usually they referred to hopped-up road-going cars.

34. General overviews began with "Ford Racing and Racers," *The Fordowner*, June 1915, 29–30, and continued to appear occasionally through 1918. See, e.g., Fahnestock, "Ford Speedster," *The Fordowner*, May 1916, 28, 30, 32, 34, and 36; Mechanician, "Putting Speed in Speedster Type of Ford," *The Fordowner*, January 1917, 67–68; and E. B. Williams, "Making the Ford Car Fast," *The Fordowner*, April 1917, 48, 50, 52, and 54.

35. The nine-page article on underslinging is Fahnestock, "Lowering the Ford Chassis: Underslinging the Ford Speedster, Sport Model, or Special Car," *The Fordowner*, October 1919, 43–48, 50, 52. On engine lubrication, see Fahnestock, "Power and Speed, Article 3:

Speedster Lubrication Systems," *FOaD*, September 1925, 65–68, 70, and 72; on camshafts, see "Power and Speed, Article 2: Increasing the Lift of the Valves—and High Speed Cam Shafts," *FDaO*, July 1925, 66, 68, and 70; on crankshafts, see Fahnestock, "Power and Speed, Article 5: Drilled and Counterbalanced Crank Shafts," *FDaO*, November 1925, 66, 68, 70, and 72; and on gears and transmissions, see Fahnestock, "Power and Speed: Gear Ratios and Transmissions," *FDaSF*, July 1927, 39–40, 42, and 44.

36. "Ford Speedster," 34 and 36. The same is true of the pictures in "More Speed," *The Fordowner*, January 1915, 15–18.

37. In the June 1916 piece, Fahnestock dwelt at length on the importance of ensuring that an underslung chassis can support the weight of one or two passengers; in the same article, he also wrote at length about chassis balancing for use on rough roads ("Ford Speedster," 34 and 36). See also "Lowering the Ford Chassis," esp. p. 43, and "Secrets of Speed: Semi-Speedster With Lower Chassis," *FOaD*, May 1923, 65–68.

38. "More Speed," *The Fordowner*, January 1915, 15.

39. "Power and Speed, Article 1: Using Fordson Valves in the Ford Cylinder Block," *FDaO*, June 1925, 66.

40. Murray Fahnestock, "Power and Speed: Lowering the Speedster Chassis," *FDaSF*, June 1926, 62.

41. Some claim that Roof's inspiration grew out of "a chance encounter with Henry Ford at a Michigan speedway" in 1916 (Almquist, "Robert M. Roof," 10), but Roof claimed that the idea stemmed more from his firsthand observations of Model T racers at Midwestern events in 1916, particularly at a major race that year at the Chicago Motor Speedway (Robert M. Roof, "'Jazzing Up' Model A Ford Engines," *FDaSF*, March 1932, 16).

42. Extensive details on Roof's background, training, and early business ventures appear in Kem Robertson's recent article, "Robert Roof," as well as in Almquist, "Robert M. Roof."

43. Ernest Henry designed a sophisticated double-overhead camshaft, sixteen-valve racing engine for Peugeot in 1911. In 1912 and 1913, Peugeots fitted with Henry's engine won the French Grand Prix and the Indianapolis 500, respectively (Ray Thursby, "French Engineering: Innovations, Complexities, Oddities, and Successes," *European Car*, May 1997, esp. 96–97). Roof's advertisements from the 1910s often boasted that his heads were of the "Peugeot Type" (Roof Auto Specialty Company advertisement, *The Fordowner*, January 1918, 53, and Laurel Motors Corporation advertisement, *Motor Age*, 2 January 1919, 87).

44. Kem Robertson ("Robert Roof," 2) speculates that this new company was founded in late 1915 or early 1916, but because Roof's work on the design of the new head itself only began in the summer of 1916, the venture must have originated in late 1916 or early 1917.

45. Roof, "'Jazzing Up' Model A Ford Engines," 16.

46. Roof Auto Specialty Company advertisement, *The Fordowner*, January 1918, 53 (emphasis in original).

47. Robertson, "Robert Roof," 6. Laurel's first advertisement appeared in the February 1918 issue of *The Fordowner*.

48. Laurel Motors Corporation advertisement, *The Fordowner*, February 1918, 3. See also Robertson, "Robert Roof," 7.

49. Laurel Motors Corporation advertisement, *The Fordowner*, April 1918, 79. The speedster bodies were of a design left over from Laurel's days as an aspiring OEM (Robertson, "Robert Roof," 6).

50. "Laurel Motors Corporation Dealer's Discount Sheet," Romaine Trade Catalog Collection, Box 6, Laurel folder, University of California, Santa Barbara (hereafter, UCSB-RT). Although this brochure is undated, the extensive inventory of parts and components it lists includes several items introduced in 1923 but none of those that came out in 1924; the brochure therefore dates from 1923. Miller and Zenith are listed by name in the brochure; Dunn, Green, and McCadden are not, but the accompanying pictures and descriptions match those of the products these firms made in the same period. That Laurel produced its own underslinging brackets is apparent from a 1919 article on the subject by Fahnestock ("Lowering the Ford Chassis").

51. Robertson, "Robert Roof," 6, and "Meet the Author—*Again*," a short biographical sketch that accompanied a 1936 article written by Roof ("Using the Ford V-8 for Dirt Track Racing," *FDaSF*, September 1936, 28).

52. "Regarding Speedster Lubrication," *FOaD*, June 1923, 65–69; Roof, "Power and Speed: The Overhead Cam Shaft and Five Bearing Crank Shaft for Greater Engine Efficiency," *FOaD*, April 1924, 138, 140, and 142; "Laurel Motors Corporation Dealer's Discount Sheet," UCSB-RT; "Power and Speed, Article 2: Increasing the Lift of the Valves—and High Speed Cam Shafts," *FDaO*, July 1925; Fahnestock, "Super Chargers Interest Ford Speedster Builders," *FDaSF*, July 1926, 58; and "Robert Roof," 2 (circumstantial evidence, according to Robertson, suggests that D. R. Noonan of Paris, Illinois, was Laurel's initial cam supplier). On Laurel's heads for the Dodge, see Laurel Motors Corporation advertisement, *FOaD*, July 1921, 131.

53. In 1922 and 1923, the company advertised the Roof 8 equipment at $65 and $75, while the B and BB sixteen-valve units went for $125 and $150, respectively ("Laurel Motors Corporation Dealer's Discount Sheet," UCSB-RT, and Laurel Motors Corporation advertisement, *FOaD*, May 1922, 143). A further iteration of the Roof 8 line was available as well. Dubbed the "Liberty Eight," it was a super-power racing head based on the speedster and racing version of the Roof 8 equipment but fitted with larger valves and special carburetors ("Secrets of Speed: Overhead Valves for Added Power," *FOaD*, January 1924, 65–71).

54. Laurel Motors Corporation advertisement, *FOaD*, March 1923, 111. The Type C sold for a whopping $225 in 1923, making it Laurel's most expensive product ("Laurel Motors Corporation Dealer's Discount Sheet," UCSB-RT).

55. Laurel Motors Corporation advertisement, *FOaD*, April 1924, 24. Roof was awarded U.S. Patent Number 1,509,611 for the double-overhead-camshaft version of the Type C in September 1924 (he filed in September 1923), and he received Number 1,561,666 in November 1925 for the Victory Eight (for which he filed in September 1924).

56. Laurel Motors Corporation advertisement, *FDaSF*, March 1926, 75. This head replaced the touring car and truck versions of the Roof 8. Later that year the "Model 40-S" made its debut, replacing the racing and speedster versions of the Roof 8 line (on the 40-S, see "Power and Speed: Overhead Valve Cylinder Heads," *FDaSF*, December 1926, esp. 65–66, and Laurel Motors Corporation advertisement, *FDaSF*, July 1927, 10). Laurel also added a line of overhead valve conversions for Overland cars in 1926 (Laurel Motors Corporation advertisement, *FDaSF*, May 1926, 129).

57. "Secrets of Speed: Overhead Valves," *FOaD*, January 1924, 67–68, and "Power and Speed: Overhead Valve Cylinder Heads," 65–66. Laurel had fifty employees in 1925 (Robertson, "Robert Roof," 9); not until the end of the 1960s would most aftermarket companies employ so many. See below, chapters 4 and 6.

58. Subcontracted forgings were standard practice within the speed equipment industry and would remain so for another sixty years. See below, chapters 4 and 6.

59. Robertson, "Robert Roof," 9.

60. Ibid.

61. Craig-Hunt's conversion was intended for racing, although in due course the company began to advertise its on-road capabilities as well ("Sixteen Valve Cylinder Heads," 48; Craig-Hunt, Inc. advertisement, *The Fordowner*, February 1918, 108; and "Making the Ford Fleet-Footed," 40).

62. In the spring of 1920, the Craig-Hunt Motors Company incorporated in Indianapolis with an initial capitalization of $1 million, absorbing Craig-Hunt, Inc. The new firm's stated purpose was to bring a small, affordable passenger car to market while maintaining the production, marketing, and sales of speed equipment for the Model T ("Craig-Hunt to Make New Low-Priced Car," *AI*, 15 April 1920, 937, and Craig-Hunt Motors Company advertisement, *Motor Age*, 6 May 1920, 94–95). Ultimately, the company proved unable to do both, and toward the end of 1920 it folded altogether (Robertson, "Robert Roof," 6).

63. "Valve-in-Head for Fords"; Trindl Sales Corporation advertisement, *Motor Age*, 9 October 1919, 24; and "The Rajo Valve-in-Head Motor," *FOaD*, October 1920, 132.

64. "Rajo" derives from a combination of the first two letters of its place of origin (Racine) with the first two letters of its founder's first name (Joe). See Batchelor, *American Hot Rod*, 35.

65. "Valve in Head for Fords"; "The Rajo Valve-in-Head Motor"; and Rajo advertisement, *FOaD*, October 1920, 113. Truck applications for cylinder head conversions were common: since the same engine powered all of Ford's small trucks as well as its cars, commercial trucks also stood to benefit from overhead valve technology.

66. On Jaegersberger's background, see Griff Borgeson, "Accessory Trial: Rajo Returns," *Motor Trend*, March 1952, 33 and 37.

67. Rajo advertisement, *FOaD*, November 1920, 113; "Built for Speed," *FOaD*, October 1921, 100; B. J. Paulson, "Sitting on Top of the World," *FOaD*, October 1922, 118; and Rajo advertisement, *FOaD*, November 1922, 27. "Pikes Peak" was (and is) a yearly automobile hill-climbing competition held in Colorado.

68. Rajo advertisement, *FOaD*, November 1920, 113.

69. Rajo advertisement, *FOaD*, November 1922, 27.

70. Rajo advertisement, *FOaD*, October 1920, 113.

71. "Secrets of Speed: Overhead Valves," 65.

72. Ibid., 66.

73. Rajo advertisement, *FOaD*, January 1924, 23. When discussing the Rajo "Model A," "Model B," and "Model C" cylinder heads as well as Laurel's "Model B," I use quotation marks to prevent confusion with Ford's Model A and Model B automobiles.

74. Rajo advertisement, *FDaSF*, May 1926, 9; "Rajo Has New Model," *FDaSF*, July 1926, 140; and "Power and Speed: Overhead Valve Cylinder Heads," 66–67.

75. For some details, see Rajo advertisement, *FOaD*, July 1921, 7; "The Rajo Valve-in-Head Motor"; and "Secrets of Speed: Overhead Valves," 65–67. Production figures indicate that Rajo outsold Laurel by a narrow margin during the 1920s: 4,000 Rajo heads of all types were produced by the end of the Model T era; whereas Laurel sold its 3,000th unit just before its absorption by Zenith (Robertson, "Robert Roof," 6, and Laurel advertisement, *FDaSF*, March 1926, 75).

76. On the C-35's manufacture, see "Power and Speed: Overhead Valve Cylinder Heads," 66–67. On the "quantity production" claim, see Rajo advertisement, *FDaSF*, May 1926, 9, and "Rajo Has New Model," *FDaSF*, July 1926, 140.

77. After World War II an accessory outfit known as "Rajo Motors" surfaced briefly in Chicago (see, e.g., Rajo Motors advertisement, *Popular Mechanics*, July 1949, 246), but between 1928 and 1949, there is no evidence that Jaegersberger manufactured any Rajo-brand equipment.

78. Used speed equipment was common on 1920s and 1930s California hot rods, and some early speed shops—Bell Auto of Bell, California, for one—actually got their start by salvaging high-performance parts from wrecked cars (see Batchelor, *American Hot Rod*, 169; Almquist, "Roy Richter: Bell Auto Parts," in *Hot Rod Pioneers*, 24–25; Bagnall, *Roy Richter*, chap. 1; and below, chapter 2). As for the use of Rajo equipment on the dry lakes, a program from an event held at Muroc on May 16, 1937, indicates that of the ninety-plus entries that day, eight ran modified Model Ts, six of which were fitted with Rajo equipment. One year later, at a now-legendary SCTA Muroc event on May 15, 1938, only ten of the 225-plus entries ran Model Ts, but seven of these were Rajo-equipped. These programs are reproduced in William Carroll, *Muroc, May 15, 1938: When the Hot Rods Ran* (San Marcos, Calif.: Auto Book Press, 1991), 7–8 and 18–21.

79. Ramar Automotive Company advertisement, *FDaSF*, August 1931, 87, and "The Ramar Valve-in-Head," *FDaSF*, October 1931, 77. Ramar operated out of Racine, Wisconsin, as well, and although it is likely that Ramar was simply a reorganization of the old Rajo concern, I have found no direct evidence to support this.

80. Borgeson, "Accessory Trial: Rajo Returns," 33 and 37, and Shucks advertisement, *Speed Age*, July 1952, 4.

81. "Racing Drivers and Their Past Performances," *Motor Age*, 29 May 1919, 21–22; Flink, *Automobile Age*, 67; and Rae, *American Automobile*, 65.

82. Of the thirty-three cars that started the 1919 race, only ten finished, so Gaston came in last place ("Indianapolis Speedway Adopts the 3-Liter Limit for Future Races").

83. "A New Cylinder Head," *FOaD*, November 1921, 96.

84. Ibid.

85. As with the Laurel and Rajo conversions, I have placed these Frontenac model designations in quotes to avoid undue confusion with Ford's vehicles.

86. The smaller the combustion chamber, the higher the compression ratio: because of the inverse physical relationship between pressure and volume, by reducing the volume of the chamber in which the piston compresses the fuel-air mixture, the ratio of compression necessarily rises.

87. "Secrets of Speed: Overhead Valves," 65–71. According to Batchelor (*American Hot Rod*, 36), this eight-overhead-valve line was initially designed by C. W. Van Ranst, an automotive engineer associated with the Chevrolet brothers. Period evidence, however, credits the brothers with the design ("A New Cylinder Head," *FOaD*, November 1921, 96, and "Secrets of Speed: Overhead Valves," 69). The truth likely lies somewhere in between: perhaps Van Ranst made the initial design and the Chevrolet brothers refined it for practical use, or perhaps Arthur and Louis hired Van Ranst to help them turn their prototypes into viable, production-ready designs.

88. On the stillborn Frontenac car, see "A. A. Ryan to Back Frontenac Car Designed

by Chevrolet," *Motor Age*, 5 January 1922, 5; Frontenac Motor Company advertisement, *Motor Age*, 5 January 1922, 96–97; "Indianapolis Site for New Frontenac Car Considered," *Motor Age*, 23 March 1922; and "Stutz Company Gets Opinions of Dealers on Assembled Model," *Motor Age*, 15 June 1922, 31.

89. Batchelor, *American Hot Rod*, 36, and Chevrolet Brothers advertisement, *FDaSF*, March 1928, 46.

90. For more on the technical specifications of these 1922 racers, see "New 'Fronty' Racer Has Ford Features," *AI*, 18 May 1922, 1096. For an example from the advertising campaign built on the success of these 1922 cars, see Chevrolet Brothers advertisement, *FOaD*, June 1923, 194.

91. Arthur Chevrolet, "Building a 'Fronty-Ford' Race Car," *FOaD*, September 1923, 65–72 (quote on p. 65). In 1923, Arthur and Louis also fielded four cars at Indianapolis in collaboration with Herbert Scheel. Dubbed the "Scheel-Frontenacs," these racers did not use Frontenac overhead-valve equipment but rather a rotary-valve mechanism (similar to the "sleeve valves" of Willys-Knight cars) that combined the action of the camshaft and the valves. See "Scheel-Frontenac Entries Are Powered with Rotary Valve Engines," *AI*, 24 May 1923, 1120–23; see also "Rotary Valve Engines," *Motor Age*, 2 January 1919, 35.

92. Arthur Chevrolet, "Building a *Ford* Dirt-Track Racer," *FDaSF*, August 1926, 70, 72, 74, 76, 78, 80, and 122–23.

93. Robertson, "Robert Roof," 6, and "Secrets of Speed: Overhead Valves," 69–70.

94. On the Frontenac double-overhead-camshaft conversion for the Model A, see Batchelor, *American Hot Rod*, 55. Arthur and Louis also formed an airplane company in the 1920s (the Chevrolet Aircraft Corporation), which they sold to the Glenn L. Martin Aircraft Corporation in 1930, just as their automotive business was beginning to decline ("Chevrolet Joins Martin," *Automotive Industries*, 18 January 1930, 105).

95. Three of the ten Model T entries at the May 16, 1937, and May 15, 1938, Muroc dry lake meets were Frontenac-equipped (Carroll, *Muroc*, 7–8 and 18–21).

96. Rajo did, however, produce a handful of other products for the Model T (Rajo advertisement, *Motor Age*, 2 February 1922, 291).

97. See Scranton, *Endless Novelty*, and chapter 3 below.

98. Jean-Pierre Bardou et al., *The Automobile Revolution: The Impact of an Industry* (Chapel Hill: University of North Carolina Press, 1982), 85.

CHAPTER 2: WESTWARD HO, 1928–1942

1. On the wrenching transition from the Model T to the Model A at Ford, see Flink, *Automobile Age*, 230; Bardou et al., *Automobile Revolution*, 98; and Hounshell, *American System to Mass Production*, 278–83.

2. Flink, *Automobile Age*, 230.

3. Franz, *Tinkering*, esp. chap. 5, and Kline and Pinch, "Users as Agents of Technological Change."

4. On the mechanical advantages of the Model A Ford, see Fahnestock, "Another Thrill: High Compression Heads for *Increased Power* and Speed and Better Fuel *Economy*," *FDaSF*, February 1931, 42, 44, and 46, and Batchelor, *American Hot Rod*, 50.

5. For more on the finer points of Model A, B, and C technology, see Batchelor, *American Hot Rod*, 52–53.

6. Gone were Beaver, Berg, Centri, Cooper, Detroit Radiator, Dunn, Fordspeed, Levett, McCadden, Milwaukee Forge, Noonan, PACO, Rajo, Turnbull, Waukesha, the Williams Foundry, and several others—not to mention Craig-Hunt and Laurel, both of which had slipped away earlier. By 1930, the Milwaukee Engineering and Tool Company (Wisconsin), also known as Mallory; the R&R Manufacturing Company (Anderson, Ind.); and the Simmons Manufacturing Company (Cleveland, Ohio) had introduced Model A parts. Within another year, equipment was also available from the Auto Engineering and Machine Company, also known as Ambler (Philadelphia); Martin & Stoner (Chicago); the Ramar Automotive Company (Racine, Wis.); the Reus Brothers Company, also known as Rallum (Baltimore); and the Trojan Auto Products Company (San Francisco). By the mid-1930s, Dreyer (Indianapolis); the Forster Brothers (Chicago); Gerber (rural Iowa); and Alexander, Bertrand, Gemsa, Harman, Lyons, McDowell, Moller, Morales, and Sparks (all of Los Angeles) as well as several others all produced equipment for the Model A. Together with a few survivors from the Model T years—Akron Motor, also known as Hal (Ohio); the Chevrolet Brothers and Morton & Brett (both of Indianapolis); Eastern Auto, Miller, Riley, and Winfield (all of Los Angeles); Green (Dayton, Ohio); Ruckstell (Berkeley, Calif.); and Zenith (Milwaukee), among a handful of others—these new companies comprised the speed equipment industry of the Model A era.

7. On the dry lakes racing of the late 1920s and early 30s, see Batchelor, *American Hot Rod*, chap. 1; Genat and Cox, *Birth of Hot Rodding*, chap. 1; Montgomery, *Hot Rod Memories*, chap. 1; Carroll, *Muroc*, 6; and Moorhouse, *Driving Ambitions*, 26–29.

8. See Batchelor, *American Hot Rod*, chaps. 1 and 3.

9. At the May 16, 1937, Muroc time trials, for example, 64 of the four-cylinder entries used California-sourced cylinder heads, and only 9 used heads from elsewhere. At the May 15, 1938, Muroc meet sponsored by the Southern California Timing Association, 117 of the four-cylinder cars used California equipment, while only 11 did not. See Carroll, *Muroc*, 7–8 and 18–21.

10. Montgomery, *Hot Rods as They Were*, 23.

11. Morton & Brett's Model A cylinder head conversion, for example, was called "The Indianapolis" and was inspired by the Midwestern oval-track scene (Fahnestock, "Secrets of Speed: Hopping Up the Model A—Some Suggestions from Morton & Brett, Builders of Speed Equipment," *FDaSF*, February 1930, 26–28). See also Roof, "'Jazzing Up' Model A Ford Engines," 16–18; Almquist, "Carl 'Pop' Green: Gasoline Alley's Green Giant," in *Hot Rod Pioneers*, 8; and Batchelor, *American Hot Rod*, 55.

12. Harry Miller teamed up with George Schofield to produce a line of cylinder heads for the Model A just before the Great Depression; Miller-Schofield went into receivership, but Miller himself continued to design and build world-class race cars. Zenith went into receivership in February 1932 ("Court Names Receiver," *Automotive Industries*, 13 February 1932, 240), although it reorganized and went on to produce OEM carburetors and other general-improvement and replacement items for a number of years. Many firms only advertised every few months during the 1930s, and the period also featured far fewer new-product announcements than did the Model T era.

13. Terry Cook, "Ed Winfield: The Father of Hot Rodding," *HRM*, January 1973, 107, and Almquist, "Ed Winfield: The Reclusive Genius," in *Hot Rod Pioneers*, 4–5.

14. Cook, "Ed Winfield," and Almquist, "Ed Winfield," 4.

15. Ibid.

16. Cook claims that the Winfield carburetor was initially intended for street use but then found its way onto the tracks as well; however, period evidence suggests that the opposite was the case. See Cook, "Ed Winfield"; see also Hammel-Gerke Company advertisement, *FOaD*, September 1924, 13, and "The Winfield Carburetor," *FOaD*, December 1924, 144.

17. In 1926 Winfield advertised as "Winfield Laboratories" (*FDaSF*, June 1926, 148), but in 1927 he went back to the tried-and-true "Winfield Carburetor Company."

18. During the early 1920s, Winfield established a solid reputation within racing circles, thanks to the on-track performance of his original carburetors. Ovals—and, by the middle of the decade, Indianapolis—thus became a proving ground of sorts for his designs ("Winfield Announces New Model," *FDaSF*, September 1927, 87).

19. Winfield Carburetor Company advertisement, *FDaSF*, July 1930, 1, and Winfield Carburetor Company advertisement, *FDaSF*, June 1930, 41.

20. In any popular history of 1930s hot rodding you'll find Winfield Model-S or Model-SR carburetors in nearly every picture. The SR debuted in 1933 ("New Winfield Carburetor," *FDaSF*, July 1933, 30).

21. Almquist, "Ed Winfield," 5.

22. On the Winfield flathead, see Winfield Carburetor Company advertisement, *FDaSF*, March 1931, 93; "Higher Speeds for Fords," *FDaSF*, April 1931, 86; and Tom Medley, *Tex Smith's Hot Rod History, Volume Two: The Glory Years* (North Branch, Minn.: CarTech, 1994), 98. On the specifications of Ford's standard flathead, see Fahnestock, "Another Thrill: High-Compression Heads for Increased Power and Better Fuel Economy," *FDaSF*, February 1931, 42, 44, and 46.

23. "Higher Speeds for Fords," *FDaSF*, April 1931, 86. Winfield also produced a very limited number of overhead-valve conversions for the Model A during the early 1930s; exactly five of these expensive, hand-assembled heads were produced, all of them custom jobs intended for use on specific oval-track cars that he sponsored (see Stan Ochs's technical discussion of the matter in Albert Drake, *Hot Rodder! From Lakes to Street: An Oral History* [Portland, Ore.: Flat Out Press, 1993], 172).

24. At the May 18, 1938, SCTA Muroc meet, for example, Winfield flatheads outnumbered all other makes of high-performance Model A flatheads *combined* by a factor of more than five to one (forty-two Winfields to a combined eight others). Moreover, Winfield flatheads outnumbered all makes of overhead-valve conversions at the same event, besting its nearest overhead competitor by ten (thirty-two entries at the event ran Cragar overhead-valve conversions). See Carroll, *Muroc*, 18–21.

25. Cook, "Ed Winfield."

26. Almquist, "Ed Winfield," 5.

27. Medley, *Tex Smith's Hot Rod History, Volume Two*, 59–60.

28. Almquist, "Ed Winfield," 5. Jackson went on to become one of the first manufacturers of high-performance ignitions for the flathead V8 engine in the late 1930s. For his part, Meyer produced some of the first intake manifolds and high-compression cylinder heads

for the flathead V8, also in the late 1930s, and Ed "Isky" Iskenderian went on to become a major high-performance camshaft manufacturer after World War II.

29. Almquist, "Ed Winfield," 5. Winfield and Riley went a long way back, and they even shared advertising space in the early 1920s, when both were under contract with Hammel-Gerke (see for example Hammel-Gerke Company advertisement, *FOaD*, September 1924, 13). The Riley overhead-valve head that grew out of Winfield and Riley's collaboration came with Winfield carburetion (Riley advertisement, *FDaSF*, August 1930, 8).

30. See Medley, *Tex Smith's Hot Rod History*, Volume Two, 60, and below, chapters 4–6.

31. On the genesis of the Miller-Schofield Company, which Miller actually began to plan in 1927, see "Miller to Produce Airplane Engines," *Automotive Industries*, 24 September 1927, 459; "Miller to Produce Airplane Engines," *Automotive Industries*, 13 July 1929, 63; and Bagnall, *Roy Richter*, 8. A typical Miller custom-built racing car of the 1920s would have cost in the neighborhood of $10,000—astronomical for the time but still cheaper than a Deusenburg or a Peugeot ("Miller Plans New Stock Car and New Engine," *AI*, 19 July 1923, 143, and Almquist, "Harry Miller: In Pursuit of Mechanical Perfection," in *Hot Rod Pioneers*, 14). On Offenhauser's history at Miller, see William J. Tandy, "Speed: Six Dollars a Pound," *Popular Mechanics*, August 1936, esp. 195.

32. Bagnall, *Roy Richter*, 8.

33. Schofield Incorporated of America advertisement, *FDaSF*, February 1930, 32–33.

34. Ibid.; see also Schofield Incorporated of America advertisement, *FDaSF*, March 1930, 47, and "More Speed for the Ford," *FDaSF*, March 1930, 70.

35. Montgomery, *Authentic Hot Rods*, 23. A lengthened downpipe was required in order to retain the OEM exhaust manifold, however, since the ports on the Valve-in-Head were several inches higher on the motor than the stock ports (due to their location in the cylinder head rather than the block).

36. Bagnall, *Roy Richter*, 8.

37. Schofield Incorporated of America advertisement, *FDaSF*, February 1930, 32–33, and Bagnall, *Roy Richter*, 8.

38. Bagnall, *Roy Richter*, 8; Schofield Incorporated of America advertisement, *FDaSF*, February 1930, 32–33; and Medley, *Tex Smith's Hot Rod History*, Volume Two, 101.

39. Schofield Incorporated of America advertisement, *FDaSF*, May 1930, 65. Bear in mind that editorial and advertising content for monthly periodicals tends to be generated anywhere from a couple of weeks to a month or more before the actual date of publication. Thus, an advertisement for the May 1930 issue of *Ford Dealer and Service Field* would have been drafted and submitted no later than the second half of April, and probably earlier.

40. Schofield Incorporated of America advertisement, *FDaSF*, June 1930, 79.

41. Harry A. Miller advertisement, *FDaSF*, June 1930, 13.

42. John Johnston, "Random Speedster Letters: Schofield-Hess Head," *Vintage Ford Speed Secrets Magazine*, April 2003, 42–43.

43. Almquist, "Bud Winfield: Co-Designer of the Legendary Novi," in *Hot Rod Pioneers*, 6, and Almquist, "Robert DeBisschop: A Key Player in the Offy's Struggle for Survival," in *Hot Rod Pioneers*, 273.

44. Bagnall, *Roy Richter*, 8, and Cragar Corporation advertisement, *FDaSF*, April 1931, 37. Miller had retained sole possession of a portion of his racing-engine facility when he formed the Miller-Schofield company in 1928.

45. Montgomery, *Authentic Hot Rods*, 21, and Harry A. Miller advertisement, *FDaSF*, June 1930, 13.

46. Harry A. Miller advertisement, *FDaSF*, August 1930, 77.

47. Montgomery, *Authentic Hot Rods*, 21, and Carroll, *Muroc*, 9–10 and 18–21.

48. On Miller's collaboration with Preston Tucker (who went on to produce the ill-fated Tucker passenger car), see Almquist, "Harry Miller," 15, and "First Ford V-8 Powered Race Car Ready for Test," *Automotive Industries*, 18 May 1935, 656. On Miller's later racing-engine career, see Fahnestock, "Speedway Madness!" *FDaSF*, July 1935, 16–17, and "Four-Cylinder Race Cars Used for Ease of Assembly," *Popular Mechanics*, September 1937, 355.

49. "Cragar" was derived from "Crane Gartz": (Cra)ne + (Gar)tz = "Cragar."

50. See Batchelor, *American Hot Rod*, 169; Bagnall, *Roy Richter*, 8; and Cragar Corporation advertisement, *FDaSF*, April 1931, 37 (quote).

51. Cragar Corporation advertisement, *FDaSF*, May 1931, 4–5.

52. On Bell Auto, see Batchelor, *American Hot Rod*, 14 and 169–170, and Bagnall, *Roy Richter*, 8.

53. Bagnall, *Roy Richter*, 8–9, and Drake, *Hot Rodder*, 57–58.

54. Batchelor, *American Hot Rod*, 169, and Bagnall, *Roy Richter*, 8–9.

55. Batchelor, *American Hot Rod*, 173, and Montgomery, *Authentic Hot Rods*, 23.

56. Montgomery, *Authentic Hot Rods*, 23. At about the same time, Wight also introduced a third model, the "Cragar Junior," for midget-car racing (Medley, *Tex Smith's Hot Rod History, Volume Two*, 97).

57. At the May 16, 1937 Muroc event, for example, nineteen of the ninety entries used Cragar heads; nineteen others used Winfield flatheads. At the May 15, 1938 event, thirty-two of the entries used Cragar overhead-valve equipment, ten less than those who used Winfield flatheads but eleven greater than Cragar's closest overhead-valve competitor, Riley. See Carroll, *Muroc*, 7–8 and 18–21.

58. On Richter, see "Bell Auto . . . World's First Speed Shop," *Drag Digest*, 11 November 1966, 27; Almquist, "Roy Richter: Bell Auto Parts," 24–25; and Bagnall, *Roy Richter*.

59. Robertson, "Robert Roof," 9, and above, chapter 1.

60. Roof and Fahnestock, "Secrets of Speed: Building a Dirt-Track Racer," *FDaSF*, April 1930, 34, 36, and 38; this piece dates R&R's dual carburetor system to August 1928. On the company's exhaust manifolds, see R&R Manufacturing advertisement, *FDaSF*, July 1929, 76.

61. Roof and Fahnestock, "Building a Dirt-Track Racer."

62. "New R. & R. Speed Equipment," *FDaSF*, July 1931, 110.

63. Roof and Fahnestock, "Building a Dirt-Track Racer," 36.

64. On R&R's custom work, see Fahnestock, "When the Car Owner Demands Speed!" *FDaSF*, January 1931, esp. p. 36; Roof, " 'Jazzing Up' Model A Ford Engines"; Roof, "Secrets of Speed: Using the Ford V-8 for Dirt Track Racing," *FDaSF*, September 1936, 28, 30, and 52; Thomas Howe and Robert M. Roof, "Building a V-8 Midget," *FDaSF*, January 1938, 18 and 34; "Speed and More Speed," *Ford Field*, June 1941, 18–19; and Howe and Roof, "Speeding the 'Sixty' for Midget Racing," *Ford Field*, December 1941, 16 and 27.

65. No R&R equipment ever appeared at Muroc, for example, although gear from some of its Midwestern competitors—including Morton & Brett, the Chevrolet Brothers, and Rajo—did (Carroll, *Muroc*, 7–8, 9–10, and 18–21).

66. On four-cylinder and V8 Ford pricing, see "Comparative Price Chart," *FDaSF*, April 1932, 42. For more on the V8's technical specifications, see Batchelor, *American Hot Rod*, 87.

67. Chevy, Willys, Pontiac, DeSoto, and Essex all sold six-cylinder cars in the same price bracket as V8 Fords ("Comparative Price Chart").

68. Fahnestock, "The Engineering Principles of 4 & 8-Cylinder Motors: Proven Correct by Racing Results," *FDaSF*, April 1932, 30–31.

69. On the incremental growth of V8 horsepower ratings during the early 1930s, see Batchelor, *American Hot Rod*, 87. On the gradual acceptance of V8s among lakes racers, see Montgomery, *Hot Rods as They Were*, 25.

70. Carroll, *Muroc*, 6, and "Comparative Price Chart."

71. On R&R's oval-track V8 gear, see Roof, "Secrets of Speed: Using the Ford V-8 for Dirt Track Racing," 28, 30, and 52. On Green, see Almquist, "Carl 'Pop' Green," 8–9. Period articles on V8 tuning for oval-track applications included R. F. Havlin, "More Speed from the V-8: Some Ideas for Those Who Want Locomotion Faster Than Fast," *FDaSF*, January 1936, 19; Thomas Howe, "Tuning for Top Speed," *FDaSF*, November 1937, 25–26 and 48; and Howe and Roof, "Building a V-8 Midget."

72. On the history of the MRA, see Genat and Cox, *Birth of Hot Rodding*, 10–17, and Batchelor, *American Hot Rod*, chap. 7.

73. On the 1930s clubs, see Genat and Cox, *Birth of Hot Rodding*, chap. 3, and Batchelor, *American Hot Rod*, chap. 7.

74. Madigan, *Edelbrock*, 55, and Genat and Cox, *Birth of Hot Rodding*, 17 (quote).

75. "Speed Races for Amateur Drivers," *Popular Mechanics*, September 1938, 339–41 and 125A–126A, and Carroll, *Muroc*, passim.

76. Carroll, *Muroc*, 15–16.

77. Batchelor, *American Hot Rod*, 117.

78. On the vital importance of the lakes, see Genat and Cox, *Birth of Hot Rodding*, chap. 2; Batchelor, *American Hot Rod*, chap. 3; and Montgomery, *Hot Rods in the Forties*, chaps. 1 and 4.

79. Ben A. Shackleford, "Masculinity, the Auto Racing Fraternity, and the Technological Sublime: The Pit Stop as a Celebration of Social Roles," in *Boys and Their Toys? Masculinity, Class, and Technology in America*, ed. Roger Horowitz (New York: Routledge, 2001), 229–50.

80. On Veda Orr, see Batchelor, *American Hot Rod*, 22; Drake, *Hot Rodder*, 41–48; and Genat and Cox, *Birth of Hot Rodding*, 22. On Moorhouse's "hot rodding fraternity," see his *Driving Ambitions*, "Racing for a Sign," and "The 'Work' Ethic and 'Leisure' Activity."

81. "Welding and milling" was a common way to increase the compression ratio of flathead engines. Essentially, it involved removing the heads, filling in the combustion chambers with large welds, and machining smaller chambers in their place.

82. Several firms produced overhead-valve conversions for trucks in the mid-1930s, and V8 enthusiasts lucky enough to find a used set eagerly tried them out (Batchelor, *American Hot Rod*, 82–83).

83. As of 1936, only 2 percent of lakes participants used V8s. From 1937 to 1940, however, V8s accounted for a third of lakes entries, on average; and in 1941, they became the majority, with 62 percent. See Montgomery, *Hot Rods as They Were*, 25.

84. Quoted in Drake, *Hot Rodder*, 44 (Kenny's last name was actually "Harman").

85. Quoted in Medley, *Tex Smith's Hot Rod History, Volume One*, 76 (an "A-V8" is a Model A retrofitted with a V8).

86. On Thickstun, see Montgomery, *Hot Rod Memories*, 93 and 121, and Batchelor, *American Hot Rod*, 77.

87. Drake, *Hot Rodder*, 44 and 170; Almquist, "Vic Edelbrock: Excellence in Overdrive," in *Hot Rod Pioneers*, 32–33; Batchelor, *American Hot Rod*, 77; and Madigan, *Edelbrock*, chapter 1.

88. Almquist, "Phil and Joan Weiand: The Ultimate Team," in *Hot Rod Pioneers*, 46–47. Weiand, wheelchair-bound following an accident on the lakes in the mid-1930s, had close friends test his manifolds on their roadsters.

89. On Burns, see Batchelor, *American Hot Rod*, 77, and Montgomery, *Hot Rod Memories*, 119. On Davies, see Montgomery, *Authentic Hot Rods*, 20, and Drake, *Hot Rodder*, 171. On Jack Henry, see Batchelor, *American Hot Rod*, 77; Montgomery, *Authentic Hot Rods*, 92; and Drake, *Hot Rodder*, 60–62 and 171. On Morrison, see Batchelor, *American Hot Rod*, 77, and Montgomery, *Hot Rod Memories*, 117. On Eddie Miller, see Veda Orr, *Hot Rod Pictorial—Featuring Dry Lakes Time Trials of 1946, 1947, 1948* (Los Angeles: Floyd Clymer, 1949), 7, and Batchelor, *American Hot Rod*, 70. On Mal Ord, see Montgomery, *Hot Rod Memories*, 116, and Batchelor, *American Hot Rod*, 16 and 77. On Ted Cannon, see Batchelor, *American Hot Rod*, 77. On Weber, see "Meet the Advertiser: Weber Tool Company," *Hop Up*, July 1952, 26–27. On the Hunt brothers, see Batchelor, *American Hot Rod*, 54. On Belond, see Montgomery, *Hot Rod Memories*, 90; Montgomery, *Authentic Hot Rods*, 16; Medley, *Tex Smith's Hot Rod History, Volume One*, 22; and Almquist, *Hot Rod Pioneers*, 124.

90. Chapel also operated one of the area's—and, by extension, the country's—first speed shops in the 1930s (Batchelor, *American Hot Rod*, 170, and Medley, *Tex Smith's Hot Rod History, Volume One*, 109).

91. See Batchelor, *American Hot Rod*, 96, and Almquist, "Eddie Edmunds: The Manifold Man," in *Hot Rod Pioneers*, 131. Edmunds moved to Los Angeles after World War II.

92. "Permite Aluminum Alloy Heads Ready," *FDaSF*, November 1937, 61.

93. McCulloch was a well-equipped company formed by an independently wealthy airplane and automotive enthusiast, R. G. McCulloch. See "McCulloch Engineering Announces New Supercharger for Ford V-8," *FDaSF*, November 1936, 50; McCulloch Engineering Company advertisement, *FDaSF*, November 1936, 4–5; "Supercharger Kit for Light Cars Fits V-Eight Engines," *Popular Mechanics*, May 1937, 664; Batchelor, *American Hot Rod*, 139; Montgomery, *Hot Rod Memories*, 110, and *Authentic Hot Rods*, 25; and Drake, *Hot Rodder*, 27, 31, and 57. For a detailed look at the McCulloch operation circa 1942, see Joseph Geschelin, "McCulloch Superchargers in the Making," *Automotive Industries*, 1 January 1942, 18–24 and 74.

94. Thickstun, Edelbrock, and Weiand dealt with the same foundry, for example (Batchelor, *American Hot Rod*, 77).

95. It bears repeating, however, that the term "hot rod" was not actually coined until 1944 or 1945.

96. Batchelor, *American Hot Rod*, 181.

CHAPTER 3: FROM HOT RODS TO HOT RODDING, 1945–1955

1. Andrew Hamilton, "Racing the Hot Rods," *Popular Mechanics*, January 1947, 138, and Genat and Cox, *Birth of Hot Rodding*, 33 and 36.

2. On the resumption of time trials in the fall of 1945, see "Hot Rods," *Life*, 5 November 1945, 86–88. On the April 1946 meet, see Genat and Cox, *Birth of Hot Rodding*, 30.

3. See Batchelor, *American Hot Rod*, 22; Drake, *Hot Rodder*, 41–48; Genat and Cox, *Birth of Hot Rodding*, 22; and Madigan, *Edelbrock*, 76.

4. "Hot Rods," *Life*, 5 November 1945, 86–88.

5. The Mercury was available with a flathead V8 engine that was, apart from a handful of minor details and a few more cubic inches of displacement, identical to—and inter-changeable with—the Ford V8.

6. On the informal rules that governed the meaning of "hot rod" immediately after the war, see Montgomery, *Hot Rod Memories*, 11; Montgomery, *Hot Rods as They Were*, 23; Montgomery, *Hot Rods in the Forties*, 7; Pat Ganahl, "The Hot Rod Culture," in Bright, *Customized*, 13; Philip E. Linhares, "Hot Rods and Customs: From the Garage to the Museum," in *Hot Rods and Customs: The Men and Machines of California's Car Culture*, ed. Michael Dobrin, Philip E. Linhares, and Pat Ganahl (Oakland: Oakland Museum of California, 1996), 14; "Hot Rods," *Life*, 5 November 1945, 86–88; and Hamilton, "Racing the Hot Rods," 138–41, 240, 244, and 248.

7. On the overwhelming importance of performance and speed to the postwar rodder, see Pat Ganahl, *The American Custom Car* (St. Paul: MBI Publishing Co., 2001), 12–13; Pat Ganahl, "The California Hot Rod," in Dobrin, Linhares, and Ganahl, *Hot Rods and Customs*, 24; and Hamilton, "Racing the Hot Rods," 244 and 248.

8. On Ak Miller, see Tom Medley, *Tex Smith's Hot Rod History, Volume Two*, 58–62. On California Bill, see Fred W. Fisher, *Chevrolet Speed Manual* (Tucson, Ariz.: Fisher Books, 1995 [1951]), 2.

9. Batchelor, *American Hot Rod*, 119.

10. See Almquist, *Hot Rod Pioneers*, 142–43; Ganahl, "The Hot Rod Culture," 13; Genat and Cox, *Birth of Hot Rodding*, 30; "Hot Rods," *Life*, 5 November 1945, 86–88; and Hamilton, "Racing the Hot Rods," 138–41, 240, 244, and 248.

11. David A. Fetherston, *Moon Equipped: Sixty Years of Hot Rod Photo Memories* (Sebastopol, Calif.: Fetherston Publishing, 1995), 12–17.

12. Genat and Cox, *Birth of Hot Rodding*, 30; Ganahl, *American Custom Car*, 33; and Fetherston, *Moon Equipped*, 12–17.

13. On rodding as a leisure-time hobby rather than an all-consuming passion among older 1940s and 50s enthusiasts, see Moorhouse, "The 'Work' Ethic and 'Leisure' Activity." Moorhouse's take on postwar rodding as an after-work and weekend hobby meshes well with much of the literature on free-time pursuits; see, for example, "Introduction: Context and Theory," in Steven M. Gelber, ed., *Hobbies: Leisure and the Culture of Work in America* (New York: Columbia University Press, 1999), esp. 15–17. On the increasing centrality of the home and the family in the 1950s, see Elaine Tyler May, *Homeward Bound: American Families in the Cold War Era*, rev. ed. (New York: Basic Books, 1999), esp. chaps. 3 and 4.

14. Don Garlits, "Forward," in Almquist, *Hot Rod Pioneers*, xi.

15. Arnie Shuman and Bernie Shuman, *Cool Cars, Square Roll Bars* (Sharon, Mass.: Hammershop Press, 1998), 6.

16. *Convertibles*, while similar in profile to *roadsters*, had padded tops, fixed windshields, and roll-up windows, which made them far more pleasant to drive on a daily basis in climates less predictable than Southern California's.

17. On hot rod and speed equipment theft, particularly in Southern California, see "Hot Rod Hijackers: Steps Taken to Halt Speed Equipment Thefts on the West Coast," *HRM*, November 1948, 24. On illegal street racing, see Batchelor, *American Hot Rod*, 159–60. And on the unscrupulous means with which some rodders obtained their yearly tag renewals, see Shuman and Shuman, *Cool Cars, Square Roll Bars*, esp. chap. 2.

18. Wally Parks, for example, called for harsher penalties for habitual street racers toward the end of the 1940s ("Editor's Column: Street Racing," *HRM*, November 1949, 6), as did Dean Batchelor in the early 1950s ("By the Editor . . .," *Hop Up*, September 1952, 5). Several manufacturers pushed their employees never to race on the streets as well (Madigan, *Edelbrock*, 90). "Shot rods"—a term used by "legitimate" hot rodders to describe the poorly built roadsters of those who gave their hobby a bad name—seems to have first appeared in print in an editorial written by Parks in 1950 ("Editor's Column: Hot Rods, not Shot Rods," *HRM*, May 1950, 7–8).

19. See for example Wally Parks, "Editor's Column: What Price Publicity?" *HRM*, December 1949, 6.

20. On Minnesota, see Wally Parks, "Editor's Column," *HRM*, August 1950, 6; on New Jersey, see Thomas Mannuzza, "Correspondence: Law Says No Duals," *Hop Up*, November 1953, 8; and on Massachusetts, see Shuman and Shuman, *Cool Cars, Square Roll Bars*, 7.

21. In 1953, for example, *Cars* ran a lengthy exposé highlighting the plight of Long Island rodders by contrasting their secret meetings, disguised cars, and frequent arrests with the open and generally cheerful world of their California counterparts. See Jerry Protola and Martin Abramson, "Reign of Terror Against Hot Rods," *Cars*, May 1953, 28–30 and 76–77.

22. On California's "fender law," which went into effect in September 1951, as well as the many ways in which California rodders complied with it, see "Fenders on Hop Ups?" *Hop Up*, October 1951, 3; Don Francisco, "What to Do about Fenders," *HRM*, February 1952, 12–15; and "Album of Best Hot Rods," in *Best Hot Rods—1953*, ed. Eugene Jaderquist and Griffith Borgeson (Greenwich, Conn.: Fawcett Publications, 1953), 41–42.

23. On the clubs' efforts to generate positive publicity, see Almquist, *Hot Rod Pioneers*, 142–43; Mickey Thompson and Griffith Borgeson, *Challenger: Mickey Thompson's Own Story of His Life With Speed* (New York: Signet Key, 1964), 10; Wally Parks, "Editor's Column," *HRM*, February 1950, 6; "Editor's Column: Hot Rods, not Shot Rods," *HRM*, May 1950, 7–8; Dick Van Osten, "Forming a Hot Rod Club," *HRM*, December 1950, 28–29 and 31; and Lee O. Ryan, "The Hot Rod Story," *HRM*, March 1952, 30–31 and 62–63.

24. George Fabry, "The 1949 PRC Reliability Run," *HRM*, January 1949, 8–9, and Ken Pratt, "Roadster Run: Hot-Rodders Find Low-Pressure Competition Can Be Fun—and the Cops Are on Their Side," *Auto Sport Review*, August 1953, 4–5 and 48.

25. "SCTA Again on Display," *HRM*, January 1949, 12.

26. Post, *High Performance*, esp. chaps. 1 and 2; Batchelor, *American Hot Rod*, chaps.

4 and 11; Montgomery, *Authentic Hot Rods*, chap. 5; and Moorhouse, *Driving Ambitions*, chap. 3.

27. See Louis Kimsey, "Saugus Drags," *Hop Up*, October 1951, 12–15; "Santa Ana Drags," *Hop Up*, February 1952, 30–33; and "By the Editor . . .," *Hop Up*, September 1952, 5.

28. See Bob Cameron Jr. [Lee O. Ryan], "Why Not a National Hot Rod Association?" *HRM*, March 1951, 20, and Batchelor, *American Hot Rod*, 121.

29. Post, *High Performance*, passim, and Moorhouse, *Driving Ambitions*, esp. chap. 5.

30. Jaderquist and Borgeson, "NHRA—A Progress Report," in *Best Hot Rods—1953*, 118–21.

31. See, e.g., Ezra M. Ehrhardt, "Deodorizing Drags," *Motorsport*, April 1952, 16–17 and 22 (Ehrhardt was a member of the California Highway Patrol).

32. Ray Garrett, "Above Criticism? Hamilton Shafters Gear Sport Activities with Official Approval for Maximum Civic Torque," *HRM*, September 1953, 52–53; Shuman and Shuman, *Cool Cars, Square Roll Bars*, chap. 2; and Fred Horsley, "Draggin' in the East," *HRM*, July 1953, 52–53 and 73–75.

33. The literature on Wally Parks's career is extensive; for a quick overview, see Batchelor, *American Hot Rod*, 121, and Almquist, *Hot Rod Pioneers*, 66–67.

34. For more on the postwar publications, see Post, *High Performance*, 387–97, and Batchelor, *American Hot Rod*, chap. 13.

35. *Throttle* failed to reemerge after the war (see above, chapter 2).

36. Moorhouse, "The 'Work' Ethic and 'Leisure' Activity," especially 256–57.

37. Drake, *Hot Rodder*, 32, and Shuman and Shuman, *Cool Cars, Square Roll Bars*, 6.

38. Montgomery, *Hot Rods in the Forties*, 136, and Almquist, *Hot Rod Pioneers*, 98–99.

39. For some examples, see Wayne Horning, "Overhead Valves," *HRM*, February 1948, 10 and 23; Harry Weber, "Choosing a Cam," *HRM*, March 1948, 10; and Don Blair, "Engine Stroking," *HRM*, July 1948, 9. Horning, Weber, and Blair were all involved in the fledgling high-performance industry: Horning manufactured cylinder heads, Weber ground performance camshafts, and Blair ran a speed shop and built racing engines to order.

40. The first of the winter–spring series on the construction of a V8 hot rod ran in December of 1948 (Walter A. Woron, "Building a Hot Rod: Classification and Selection," *HRM*, December 1948, 12–13), and the last in July of 1949 (Woron, "Building a Hot Rod: Roadster Completion," *HRM*, July 1949, 12–13, 19, and 21). *Hot Rod*'s first engine conversion article ran in October of 1949 (C. E. Camp, "Cadillac Conversion," *HRM*, October 1949, 16–17).

41. Barney Navarro, "More 'Horses' Thru Chemistry," *Hop Up*, December 1951, 28–29; "No Miracles!" *Hop Up*, February 1952, 4–5; and "The Flame! Combustion Chamber Design and Problems," *Hop Up*, April 1952, 4–5 and 44.

42. See, e.g., Dean Moon, "Build Your Own Hot Ignition," *Car Craft*, May 1954, 36–41; Les Nehamkin, "Here's How: Big Lungs, Small Engine—Opening Up the Ford Four," *Car Craft*, June 1954, 22–25; Edward Munroe, "Build Your Own Chevy Dual Manifold," *Car Craft*, August 1954, 16–19; and Chuck Eddy, "Homemade Tubular Push Rods: Extra Strong and Lightweight for '55 Ford," *Car Craft*, June 1955, 14–15.

43. Typical is Kenneth Kincaid, "If Detroit Won't Do It, Why Don't You?" *Motor Trend*, June 1952, 32–34 and 46–47.

44. Edgar Almquist, *Specialized Automobile Tuning and Customizing Methods* (Brooklyn: Almquist Engineering, 1946); Roger Huntington, *Souping the Stock Engine* (Los Angeles: Floyd Clymer, 1950); and Daniel Roger Post, *Blue Book of Custom Restyling* (Arcadia, Calif.: Post Publications, 1951). See also Fisher, *Chevrolet Speed Manual*; Peter Bowman, ed., *Hot Rod Handbook* (Greenwich, Conn.: Fawcett Books, 1951); Bill Czygan, ed., *How to Build Hot Rods* (Greenwich, Conn.: Fawcett Books, 1952); and Fred W. Fisher, *Ford Speed Manual* (Tucson, Ariz.: Fisher Books, 1995 [1952]).

45. See Montgomery, *Hot Rod Memories*, 89, and Montgomery, *Hot Rods in the Forties*, 136.

46. "The New Engines," in Jaderquist and Borgeson, *Best Hot Rods—1953*, 122–27.

47. "Album of Best Hot Rods," in Jaderquist and Borgeson, *Best Hot Rods—1953*, esp. 60–61, and Don Francisco, "Engine Conversions," in *Hot Rod 1954 Annual* (Los Angeles: Petersen Publishing Company, 1954), 106–108.

48. This we can surmise from the sorts of engine-conversion articles and reader-submitted technical questions that appeared in the popular magazines during these years. See, e.g., C. E. Camp, "Cadillac Conversion," *HRM*, October 1949, 16–17; C. E. Camp, "Engine Conversion: Chevrolet," *HRM*, November 1949, 10–11; C. E. Camp, "Engine Conversions: Studebaker," *HRM*, February 1950, 12–13 and 22–23; Don Francisco, "Engine Conversion: Buick," *HRM*, August 1950, 16–17 and 20–21; "Technical Tips," *Hop Up*, September 1951, 42–43; "Technical Tips," *Hop Up*, October 1951, 42–43; Don Francisco, "Don't Throw Rocks at Your Rocket!" *HRM*, May 1952, 16–19 and 56; California Bill, "Hopping Up the Olds and Cad," *Hop Up*, September 1952, 18–21 and 43; Bill Fisher, "Easy Does It: $100 Chevy Conversion," *Honk*, November 1953, 56–59; and Edward Monroe, "Build Your Own Chevy Dual Manifold," *Car Craft*, August 1954, 16–19.

49. Don Francisco, "Engine Conversions . . . Hudson 8," *HRM*, March 1950, 14–15; Eddie Miller Jr., "Engine Conversions: Pontiac 6," *HRM*, July 1950, 15 and 23–24; "Technical Tips," *Hop Up*, September 1952, 44; Chuck Eddy, "Industrial Dynamite: Horses for the Ford Four," *Car Craft*, June 1954, 16–21, 61, and 65–66; and Bill Schroeder, "Rodding a Rambler," *Speed Mechanics*, April–May 1955, 20–21, 32, and 34.

50. On the enduring popularity of the flathead V8, see "Album of Best Hot Rods," in Jaderquist and Borgeson, *Best Hot Rods—1953*, 40; Roberson, *Middletown Pacemakers*, 41–42; and Almquist, *Hot Rod Pioneers*, 234.

51. "What's New," *HRM*, August 1950, 25–27 and 31–32.

52. A word on detonation, octane ratings, and compression ratios: low-octane fuels burn quickly and tend to ignite spontaneously (i.e., without a controlled engine spark) when exposed to high combustion-chamber pressures and/or heat. Higher-octane fuels burn more slowly and are less prone to ignite on their own. Hence, low-octane fuels require low compression ratios, while higher-octane fuels allow (but do not require) higher compression ratios.

53. A word on superchargers ("blowers"): engine power is directly proportional to the amount of fuel and air an engine burns. Larger engines therefore tend to produce more power than smaller ones because they consume more air and fuel. Superchargers *force* more air into engines than they would otherwise consume (this additional air is pressur-

ized mechanically, and the pressure is generally referred to as "boost"). When coupled with larger carburetor jets or a fuel-pressure increase, superchargers therefore *force* an engine to consume more air and fuel and therefore to produce more power.

54. Drake, *Hot Rodder*, 27 and 57, and Montgomery, *Hot Rod Memories*, 110.

55. "What's New," *HRM*, April 1950, 25–27, and "Trade Topics," *Motor Trend*, February 1951, 34.

56. H. Weiand Bowman, "Super Charging for Speed," *Motorsport*, March 1951, 26–27; Don Francisco, "Packing the Punch," *HRM*, April 1952, 26–29 and 62–63; and California Bill, "Blowing the Chevrolet! An 8 Hour Procedure . . . Ha!" *Hop Up*, January 1953, 8, 10, 43, and 50.

57. McCulloch moved to L.A. toward the end of World War II; see "The New McCulloch Supercharger," *Road & Track*, December 1953, 12–13; Barney Navarro, "Supercharging: The Fast Way to Horsepower," *Motor Life*, January 1954, 22–25 and 29; and Bob Pendergast, "Supercharging the Mercury," *Rod and Custom*, November 1954, 38–41 and 60–61. On the Judson supercharger, see "Supercharging Revival," *Car Life*, February 1955, 44–47.

58. Batchelor, *American Hot Rod*, 76 and 82–83; "Overhead Valve Designs," *HRM*, April 1949, 12–14 and 26; and Don Francisco, "Overhead Valves for Ford 6," *HRM*, March 1951, 16–17, 28–29, and 31. Madis's product was rendered obsolete when Ford introduced its overhead-valve six-cylinder engine in 1952 (W. G. Brown, "132 Easy Horses for Ford Six," *HRM*, July 1953, 26–27 and 70).

59. Contrary to popular belief, the Ardun was not originally developed for use on trucks. From the beginning, Arkus-Duntov wanted to be a hot rod parts manufacturer, and only as it became evident that the market for his conversion among enthusiasts was weak did he turn to the less exciting (and ultimately equally unprofitable) truck market. See Burton, *Zora*, esp. 87–100.

60. See, e.g., "Technical Tips," *Hop Up*, November 1951, 4–5; "Technical Tips," *Hop Up*, July 1952, 2; "Tech Tips," *Rod and Custom*, October 1953, 50–51; and "Tech Tips," *Rod and Custom*, March 1955, 62–63.

61. "Fordebaker," *Hop Up*, May 1953, 34–35; Bob Fendell, "How to Make a Studillac," *Road & Track*, December 1953, 30–31; John Kampp and Harless King, "Midwest 'Studillac,'" *HRM*, March 1954, 24–25 and 50–53; Joe Moore, "Bu-Merc Conversion: Presenting Another First in Engine-Chassis Conversions," *Motor Life*, January 1954, 40–41 and 54; "Fire Powering the Deuce—or, How to Turn 109 in the Quarter Mile," *Rod and Custom*, October 1954, 46–49; "One T-Bone, RARE . . . but well done!" *Rod and Custom*, November 1954, 22–25; and "Reader's Car of the Month," *Rod and Custom*, March 1955, 24–25. A "Studillac" was a Studebaker with an overhead-valve Cadillac mill, a "Fordebaker" was a Ford with an overhead-valve Studebaker V8, and a "Bu-Merc" was a Mercury with an overhead-valve Buick V8.

62. On the finer points of 1950s engine hunting, see Jack Phelps, "Bargains in Horsepower," *Motor Life*, November 1955, 26–27 and 66.

63. Many of the magazines' "engine-conversion" articles, for example, were penned specifically to assist enthusiasts in modifying their engines for use in an engine-swapping project. As a case in point, see Racer Brown, "Soup that Chev!" *HRM*, May 1955, 14–19 and 50–53.

64. On the mounting popularity of drag racing in the early 1950s, see "Draggin' Demons," *Rods and Customs*, May 1953, 14–17, and "Drags," in Jaderquist and Borgeson, *Best Hot Rods—1953*, 21. On the increasingly specialized nature of drag racing—and dragsters—in the 1950s, see Post, *High Performance*, esp. chap. 3; Montgomery, *Hot Rod Memories*, 14; Medley, *Tex Smith's Hot Rod History, Volume Two*, 102; and "Album of Best Hot Rods," in Jaderquist and Borgeson, *Best Hot Rods—1953*, 53, 55, and 67.

65. On the advent of the "lakester," see "The Lakes," in Jaderquist and Borgeson, *Best Hot Rods—1953*, 6–17; John Christy, "The Hot Rod and You: Build for a Purpose," in *Hot Rod 1954 Annual*, especially 4–5; and Montgomery, *Hot Rod Memories*, 14.

66. See Louis Hochman, "Hot Rods and Ends: 200-MPH Hot Rod Problems," *Cars*, May 1953, 75; Montgomery, *Authentic Hot Rods*, chap. 5; Batchelor, *American Hot Rod*, 163; and Genat and Cox, *Birth of Hot Rodding*, 36.

67. On the emergence of the Russetta Timing Association, the SCTA's chief postwar rival (which, not coincidentally, allowed coupes and sedans to participate in its events), see Genat and Cox, *Birth of Hot Rodding*, 30–36, and Batchelor, *American Hot Rod*, 120. On the Bell Timing Association, see "Editor's Column," *HRM*, April 1949, 7, and Louis Kimsey, "Bell Timing Meet," *Hop Up*, October 1951, 24–27.

68. Jack Landrum, "Lakes Meet," *Motorsport*, August 1951, 7 and 29–30, and "Editor's Column," *HRM*, December 1953, 5.

69. On the emergence of the annual Bonneville time trials, see "Bonneville," *Hop Up*, November 1951; "Just Before the Storm: Critical Seconds Before Bonneville," *HRM*, October 1953, 24–27; Genat and Cox, *Birth of Hot Rodding*, 36–39; Montgomery, *Authentic Hot Rods*, chap. 5; Batchelor, *American Hot Rod*, chap. 10; and Embry, "The Last Amateur Sport."

70. On the increasing popularity of oval-track roadster racing among hot rodders, see "Racing Roadsters," in Jaderquist and Borgeson, *Best Hot Rods—1953*, 18–20, and Medley, *Tex Smith's Hot Rod History, Volume Two*, 102. On midget racing, see "The Real Wheels Behind Hot Rodding," *Cars*, December 1953, 16–20, and "Offy Midget Engine," *Hop Up*, October 1951, 35.

71. Dean Batchelor, "The New Hot Rods," *True's Automobile Yearbook*, 1952, 30–31 and 98–99; Eugene Jaderquist, "Hot Rods Are Doomed," *Cars*, November 1953, 30–33 and 62; Christy, "The Hot Rod and You," in *Hot Rod 1954 Annual*, 4–5; Almquist, *Hot Rod Pioneers*, 25; and Medley, *Tex Smith's Hot Rod History, Volume Two*, 102.

72. One famous example of this trend was a fire-breathing yet altogether ordinary-looking 1934 Ford coupe that Dean Moon built in the late 1940s (Fetherston, *Moon Equipped*, 23–29, and Montgomery, *Hot Rod Memories*, 36), although in Moon's case, the car's run-of-the-mill appearance failed to prevent him from being nabbed—and given seven days in prison—for street racing.

73. On customs, see DeWitt, *Cool Cars, High Art*; Dobrin, Linhares, and Ganahl, *Hot Rods and Customs*; Ganahl, *American Custom Car*; and Joe Kress, *Lead Sleds* (St. Paul: MBI Publishing Co., 2002).

74. For the former, see "Street Machines," in *Hot Rod 1954 Annual*, 10–22; Almquist, *Hot Rod Pioneers*, 25; Montgomery, *Hot Rod Memories*, 13; and Montgomery, *Hot Rods as They Were*, 11. Batchelor, on the other hand, forcefully argues for a more inclusive definition of "hot rod" (*American Hot Rod*, 8).

75. For Montgomery's take, see *Hot Rods in the Forties*, 7; *Hot Rods as They Were*, 7; and *Hot Rod Memories*, 13–14.

76. Bill Schroeder, "Rodding a Rambler," *Speed Mechanics*, April–May 1955, 20–21, 32, and 34; *Hot Rodding the Compacts* (Los Angeles: Petersen Publishing Co., 1962); Bob Waar, Steve Smith, and Bill Fisher, *How to Hotrod and Race Your Datsun* (Tucson, Ariz.: HP Books, 1984); and Jeff Hartman, "Illusions of Grandeur," *European Car*, July 1997, 72.

77. Moorhouse, "Racing for a Sign."

78. See "It's in the Bag," *HRM*, October 1949, 6, and "It's in the Bag," *HRM*, December 1949, 32–33.

79. Wally Parks, "The Editor Says," *HRM*, June 1954, 5.

80. Ibid.

81. DeWitt, *Cool Cars, High Art*.

CHAPTER 4: THE CALIFORNIA HOT ROD INDUSTRY, 1945–1955

1. Including manufacturers, distribution centers, mail-order houses, speed shops, engine builders, engine machine shops, and custom-accessory retailers, approximately 290 high-performance companies operated in or near Los Angeles at one point or another during the period in question. This figure is derived from a thorough tabulation of every company that appeared in advertisements, feature articles, and racing coverage over the course of the decade in *Hot Rod, Hop Up, Motor Life, Speed Age, Cars, Car Life, Auto Age, Motor Trend, Rod and Custom*, and several other enthusiast magazines. Company names and dates of operation then were verified against the relevant data from other period sources (how-to manuals, in particular) and secondary accounts to arrive at these totals. However, period sources and enthusiast publications are far less reliable regarding speed shops and other local retail companies outside of the greater Los Angeles area than they are regarding those within. This is an artifact of proximity: most of the periodicals and how-to manuals were written and published in Los Angeles by hot rodders far more familiar with the Southern California scene than with that of any other part of the country. As a result, although I am aware of references as early as 1951 to figures in excess of 2,000 hot-rodding-related companies in the United States as a whole ("Touring the Hot Rod Shops: Production is Theme at Southern California Muffler Company," *HRM*, May 1951, 32–33, for example), I have only managed to verify approximately 300 outside of the L.A. area. Thus, a conservative estimate would place the total number of companies outside Southern California by 1955 at about 300, against 290 within L.A., which translates to a split of roughly 50/50. On the other hand, a more liberal estimate might place the number of firms in the United States as a whole at 2,000 by 1955, at the very most a 10/90 split.

2. Manufacturers—those within Los Angeles itself and those located elsewhere in the United States—are far easier to track than are locally oriented speed shops, engine builders, engine machine shops, and the like. For whereas a tiny neighborhood speed shop located, say, in Peoria, Illinois, might have had no reason whatsoever to cough up the money for an advertisement in *Hot Rod* or *Hop Up*, even the tiniest of *manufacturers* advertised regularly in these nationally circulated periodicals. The figures presented here are derived from a tediously comprehensive survey of period books, racing programs, and enthusiast periodicals as well as the relevant secondary literature. And in my carefully

considered opinion, if a given "manufacturer" fails to show up even once in any of these sources—that is, if none of his products ever show up in technical, racing, or feature-car coverage in the magazines; if his name fails to appear in any of the published memoirs or secondary sources; if his products fail to appear in period mail-order or speed shop catalogs; or, if he failed to advertise even once in any of the enthusiast periodicals published during the course of the period in question—then that "manufacturer" simply was not a part of the high-performance aftermarket, period. Thus, although there may in fact be one or two that I have overlooked, the total cannot possibly be more than can be counted on a single hand, for phantom companies that go out of their way to avoid recognition (and sales) are exceedingly rare.

3. The relatively miniscule rates of growth that these figures from 1948, 1951, and 1954 suggest are somewhat misleading, for they paper over the fact that there was a considerable amount of turnover during the period in question as numerous manufacturers came and went and combined and split. In fact, 182 different equipment manufacturers operated in Southern California at one time or another during the late 1940s and early 1950s, while approximately 75 came and went elsewhere. At any given moment, however, there were far fewer than 257 total firms in operation; the figures cited here reflect this fact.

4. Apart from its own advertising claims (which are often suspect, at best), one reasonably reliable way to gauge the relative importance of a given manufacturer at a given time is to closely monitor (1) the extent to which its products show up in the feature, technical, and racing coverage of the popular magazines and how-to guides; and/or (2) the extent to which their products were *consistently* featured in the advertising spreads of national mail-order chains.

5. Almquist, "The Spalding Brothers: Famous for Ignitions and Cams," in *Hot Rod Pioneers*, 40.

6. Almquist, "Tommy Thickstun: Early Dealer Option to Race Collectible," in *Hot Rod Pioneers*, 29.

7. During the Korean War, Weber did a bit of contract work for the U.S. Army as well ("Meet the Advertiser: Weber Tool Company," *Hop Up*, July 1952, 26–27).

8. Madigan, *Edelbrock*, 70–72, and "The Real Wheels Behind Hot Rodding," *Cars*, December 1953, 19 (quote).

9. "What's New," *HRM*, September 1949, 7, and Witteman Company advertisement, *HRM*, September 1949, 6.

10. Jahns advertisements, *HRM*, May 1950, 28, and *HRM*, December 1951, 66.

11. "Aggressive" may in fact understate Grant's enthusiasm for the high-performance market: every month from 1949 through 1955 (and sporadically for several years beyond), Grant advertisements graced *Hot Rod*'s back cover.

12. On Mitchell, see "Duals for a Chevy," *Rod and Custom*, November 1954, 26–31, 56, and 66, which includes a brief but detailed synopsis of the company's history. On Huth, see "Meet the Advertiser: Huth Muffler Company," *Hop Up*, February 1952, 34–35.

13. "Meet the Manufacturer: Schiefer," *Drag News*, 19 April 1969, 8, and Ray Brock, "Paul Schiefer Eulogy," *Drag News*, 5 September 1970, n.p.

14. See "Meet the Advertiser: Iskenderian," *Hop Up*, April 1952, 26–27; Almquist, "Ed 'Isky' Iskenderian: The Legendary Camfather," in *Hot Rod Pioneers*, 108; and Montgomery, *Hot Rod Memories*, 95.

15. On Navarro, see "California's Big Wheels," in Eugene Jaderquist and Griffith Borgeson, *Best Hot Rods* (Greenwich, Conn.: Fawcett Publications, 1952), 32 (the first in a yearly series of books by the same title), and Almquist, "Barney Navarro: 'Oldfield' of the Hot Rod Industry," in *Hot Rod Pioneers*, 51–52.

16. On Engle, see Dave Wallace Jr., "Jack Engle Walks Softly, Carries Big Bumpstick," *Drag News*, 22 November 1975, 16–17. On Xydias, who actually returned from the war to found a speed shop, not a manufacturing company, see Dick Wells, "Speed Shop History," *Street Rodder*, December 2000, 68–70 and 72; Batchelor, *American Hot Rod*, 170–71; and Christensen, *So-Cal Speed Shop*. And on Richter, who purchased the Bell Auto / Cragar operation from George Wight's widow in 1945, see Bagnall, *Roy Richter*, chap. 1, and "Bell Auto . . . World's First Speed Shop," *Drag Digest*, 11 November 1966, 27.

17. See Almquist, "Speed Equipment," in *Hot Rod Pioneers*, 134, and "The 'Heart' of Rodding," in *Hot Rod Pioneers*, 135–37; Roberson, *Middletown Pacemakers*, 41–42; and Montgomery, *Hot Rod Memories*, 11. However, nowhere is the flathead's domination of the postwar speed equipment industry displayed quite as clearly as in the mail-order catalogs from the period: Bell Auto's 1952 catalog, for example, though quite diverse in its offerings, was overwhelmingly weighted in favor of parts for Ford and Mercury V8s (Root Collection 99A104, Box 6, Watkins Glen International Motor Racing Research Center Archive, Watkins Glen, New York [hereafter, WGR-RC]), as was Sumar Equipment's 1954 catalog (WGR-RC).

18. On Navarro, see Almquist, "Barney Navarro," 51–52, and any of Navarro's advertisements from the period (e.g., *Rod and Custom*, December 1953, 68). On Meyer, see Batchelor, *American Hot Rod*, 65. On Evans, see Montgomery, *Hot Rod Memories*, 92.

19. "Manufacturers News," *HRM*, June 1948.

20. On McGurk, see "What's New," *HRM*, February 1950, 24 and 26. On Horning, see Don Francisco, "The Horning GMC: Converted Truck Engine with Special Cylinder Head Forms Nucleus of Capable Competition Powerplant," *HRM*, April 1951, 20–23, and *HRM*, May 1951, 20–23 and 42–43 (the article ran over the course of two months). On Morgan, see "Technical Tips," *Hop Up*, February 1952, 2. On Brajevich (also known as "Braje"), see "The O-Bones," in Jaderquist and Borgeson, *Best Hot Rods—1953*, 98–103. And on Edelbrock and the V8-60 midgets, see Madigan, *Edelbrock*, 124–25.

21. An excellent reference for the names (and times) of those who raced on the lakes in the 1940s is Veda Orr, *Hot Rod Pictorial*.

22. *First Annual Bonneville National Speed Trials Souvenir Program* (Los Angeles: SCTA, 1949), 4–5, and "Just Before the Storm: Critical Seconds Before Bonneville," *HRM*, October 1953, 24–27.

23. Eugene Jaderquist and Griffith Borgeson, *Auto Racing Yearbook* (Greenwich, Conn.: Fawcett Publications, Inc., 1954), 19 and 24–25, and "The Real Wheels Behind Hot Rodding," *Cars*, December 1953, 19.

24. Iskenderian's V8-powered Model T roadster appeared on the cover of *Hot Rod* in June 1948; Mitchell's 1929 pickup did so a few years later. Moon's 1934 "sleeper" appeared in *Hot Rod* many times during the 1950s, typically in conjunction with an engine-swapping piece (Moon swapped engines in his coupe many times over the years); see Fetherston, *Moon Equipped*, 23–29.

25. Among the many firms with dynamometers were Potvin (advertisement, *Rod and Custom*, June 1954, 63), Braje ("Dyno for Lightweights," *HRM*, October 1953, 32–33 and

74–75), McGurk (Ray Brock, "Dual Exhaust Systems: What Can They Do for '53 Cars?" *HRM*, June 1953, 20–23 and 68–72), Nicson (Ivan T. Galanoy, "Headers: America's Most Misunderstood Speed Equipment," *Motor Life*, February 1954, 35–37 and 63), and Edelbrock (Batchelor, *American Hot Rod*, 85–87, and Madigan, *Edelbrock*, 109). Speedomotive and Evans were exceptional in that they had their own foundries ("This is How It's Done . . . Special Racing Pistons Are Made in an Up to Date Plant," *Rod and Custom*, June 1953, 22–27, and Batchelor, *American Hot Rod*, 173). Instead, most hired outside foundries to handle the rough-casting and/or rough-forging phases of their operations (see, e.g., "Meet the Manufacturer: Venolia," *Drag News*, 22 August 1970, 18–19, and "Meet the Manufacturer: Schiefer," *Drag News*, 19 April 1969, 8); this would be standard practice within the industry for many years to come (Author Interview with Vic Edelbrock Jr., Torrance, California, November 19, 2003).

26. Francisco, "The Horning GMC."

27. On Speedomotive, see "This is How It's Done." On the tooling used by Iskenderian, Harman & Collins, Clay Smith, and other camshaft manufacturers, see Almquist, "Ed 'Isky' Iskenderian," 108–10; "Meet the Advertiser: Iskenderian," *Hop Up*, April 1952, 26–27; Almquist, "Cliff Collins and Kenny Harman," 130; Clay Smith advertisement, *Rod and Custom*, February 1954, 63; Medley, *Tex Smith's Hot Rod History, Volume Two*, 59–60; Almquist, "Howard Johansen: The Quintessential Rodder," in *Hot Rod Pioneers*, 150; and Almquist, "Chet Herbert: Triumph over Adversity," in *Hot Rod Pioneers*, 151.

28. For example, Iskenderian's first camshaft grinder was actually a second-hand cylindrical grinder for which he fabricated a camshaft-grinding attachment (Almquist, "Ed 'Isky' Iskenderian," 108); similarly, Howard Johansen's first cam machine was a homemade monster cobbled together with parts from a post grinder and a lathe (Almquist, "Howard Johansen," 150).

29. "Meet the Advertisers: Southern California Muffler Company," *Hop Up*, November 1951, 24–25, and also "Meet the Advertiser: Iskenderian," *Hop Up*, April 1952, 26–27.

30. In *Endless Novelty*, Philip Scranton emphasizes that specialty manufacturers sought, not to control the market, but rather to manage the uncertainties inherent in it. The same was true of postwar high-performance companies.

31. Egge Machine Company advertisement, *Rod and Custom*, November 1954, 57; Speedomotive advertisement, *Rod and Custom*, March 1955, 61; "Meet the Manufacturer: Venolia," *Drag News*, 22 August 1970, 18–19; "California's Big Wheels," in Jaderquist and Borgeson, *Best Hot Rods*, 31 (on Johansen and Navarro); and Bob Dearborn, "The Clay Smith Story," *Hop Up*, July 1953, 34–37.

32. On sponsorships as write-offs, see "The Competition Season," in Jaderquist and Borgeson, *Best Hot Rods—1953*, 4–5.

33. See Montgomery, *Hot Rod Memories*, 61.

34. Almquist, "Road Warriors: More than Parts Peddlers," in *Hot Rod Pioneers*, 245.

35. Wells, "Speed Shop History," 69, and "Meet the Advertisers: Newhouse Automotive Industries," *Hop Up*, September 1951, 26–27.

36. "What's New," *HRM*, April 1950, 25–27; "Meet the Advertisers: Southern California Muffler Company," *Hop Up*, November 1951, 24–25; and Fetherston, *Moon Equipped*, 19.

37. Robert Lee Behne, "Speed Equipment," in Jaderquist and Borgeson, *Best Hot Rods—1953*, 128.

38. See below, chapter 6.

39. G. G. Gordon, "Hot Rods are Big Business," *Auto Sport Review*, September 1952, 36–39.

40. Period advertisements for equipment manufacturers almost always included their addresses; consequently, I have been able to track the precise location of nearly every California-based company during the entire period in question. A comprehensive review of this data reveals that Iskenderian moved twice, first in the summer of 1949 and then in the summer of 1951. Evans, on the other hand, moved three times: first in the winter of 1948–49, then in the winter of 1950–51, and again in the summer of 1952. Moreover, if we cross-reference this data with the relevant period literature ("Meet the Manufacturer," "Manufacturers News," and "What's New" columns from the enthusiast magazines), we find not only that these and other similar companies moved repeatedly during the late 1940s and the early 1950s but also that they almost always moved into larger facilities.

41. "Touring the Hot Rod Shops: *HRM* Takes a Look at One of the West's Leading Muffler Manufacturers—Douglass Muffler Shop," *HRM*, July 1951, 28–29.

42. Author Interview with Vic Edelbrock Jr., Torrance, California, November 19, 2003, and Wallace, "Jack Engle Walks Softly."

43. Almquist, "Cliff Collins and Kenny Harman," in *Hot Rod Pioneers*, 130.

44. See above, chapter 2.

45. See William J. Tandy, "Speed: Six Dollars a Pound," *Popular Mechanics*, August 1936, 194–96 and 126A, and Almquist, "Robert DeBisschop: A Key Player in the Offy's Struggle for Survival," in *Hot Rod Pioneers*, 273.

46. Almquist, "Fred C. Offenhauser: A Credit to his Namesake," in *Hot Rod Pioneers*, 96–97, and Don Francisco, "Story of the 270 Offenhauser," *HRM*, June 1952, 22–29, 51–53, and 60–63.

47. Almquist, "Fred C. Offenhauser," 96; Francisco, "Story of the 270 Offenhauser"; and Almquist, "Fran Hernandez: Funny-Car Baptizer," in *Hot Rod Pioneers*, 73. Hernandez was also a key player in the first drag race at Goleta in 1949; see Post, *High Performance*, 1–3.

48. Almquist, "Fred C. Offenhauser," 96.

49. See below, chapters 7, 8, and 9.

50. On the design and performance of Wayne Horning's heads, see Batchelor, *American Hot Rod*, 96 and 115; Almquist, "Chevy 'Stove Bolt' Six," in *Hot Rod Pioneers*, 87; and Francisco, "The Horning GMC." On Marvin Lee, see Drake, *Hot Rodder*, 70.

51. On the split sale of Horning's designs, see Batchelor, *American Hot Rod*, 96 and 115. On the Electronic Balancing Company, see "Manufacturers News," *HRM*, July 1948, 27.

52. Fisher, *Chevrolet Speed Manual*, 2.

53. See, e.g., Fred W. Fisher, "5 Ways to Power!" *Rod and Custom*, June 1953, 57–58, 60–61, and 64; Bill Fisher, "Easy Does It: $100 Chevy Conversion," *Honk*, November 1953, 56–59; and California Bill Fisher, "Got a Driveline Dilemma?" *HRM*, September 1955, 28–31. Fisher's Chevrolet/GMC tuning book, the *Chevrolet Speed Manual*, came out in 1951; he also published a how-to manual for the Ford V8, *Ford Speed Manual*, in 1952.

54. Behne, "Speed Equipment," in Jaderquist and Borgeson, *Best Hot Rods—1953*, 136, and Batchelor, *American Hot Rod*, 96.

55. On Horning's subsequent career in the racing-engine business, see C. E. Camp, "Chevrolet Engine Conversion, Part II," *HRM*, December 1949, 16–17, and Francisco, "The Horning GMC."

56. See below, chapters 5 and 6.

57. Batchelor, *American Hot Rod*, 77.

58. On Meyer's relationship with Winfield, see Almquist, "Ed Winfield: The Reclusive Genius," in *Hot Rod Pioneers*, 5. On the Meyer brothers, see Almquist, "Eddie and Louie Meyer," in *Hot Rod Pioneers*, 19–20; Almquist, "Robert DeBisschop," 273; and Francisco, "Story of the 270 Offenhauser."

59. Montgomery, *Authentic Hot Rods*, 20, and "Flying Water Bugs," *Popular Mechanics*, April 1946, 96–100, 244, and 246.

60. Quoted in Almquist, "Eddie and Louie Meyer," 20.

61. Francisco, "Story of the 270 Offenhauser."

62. On Brown's time at Eddie Meyer Engineering, see Batchelor, *American Hot Rod*, 65. For some examples of Brown's technical articles, see Racer Brown, "Are Exhaust Headers Only a Gag?" *Motor Trend*, February 1953, 28–29; Racer Brown, "Building for Nitro: Sparking the Competition Engine," *Car Craft*, November 1954, 44–47, 61, and 65–66; and Racer Brown, "Do a Better Ring Job," *HRM*, March 1955, 14–19, 58–61, and 66.

63. On Lou Senter's work at Eddie Meyer's shop, see Almquist, "Lou Senter: Racers' Head Guru," in *Hot Rod Pioneers*, 53–54. On Ansen, see "Ansen Automotive," *Drag News*, 2 May 1964, 20–21; Karen Scott, "Meet the Manufacturer: Ansen," *Drag News*, 12 July 1969, 15; and Dick Wells, "Interview with Louie Senter, Jim Deist, and Ed Iskenderian, 7 November 2002," *SEMA "Old Timers" Interviews*, Video Cassette 2, SEMA-RC.

64. Almquist, "Edward Pink: Master Engine Builder," in *Hot Rod Pioneers*, 288–89; Al Caldwell, "Meet the Manufacturer: Ed Pink Racing Engines," *Drag News*, 18 August 1967, 26–27; and Jerry Brandt, "Ed Pink," *Drag Racing*, May 1975, 60–62.

65. On Iskenderian's early training as well as his time at Winfield's shop, see Almquist, "Ed 'Isky' Iskenderian," 108–10, and Dick Wells, "Interview with Louie Senter, Jim Deist, and Ed Iskenderian, November 7, 2002," *SEMA "Old Timers" Interviews*, Video Cassette 2, SEMA-RC.

66. Almquist, "Ed 'Isky' Iskenderian," 108–10, and "Meet the Advertiser: Iskenderian," *Hop Up*, April 1952, 26–27.

67. On Edelbrock's assistance, see Almquist, "Ed 'Isky' Iskenderian," 108–10, and Madigan, *Edelbrock*, 127. On Iskenderian's growth, see "Meet the Advertiser: Iskenderian," *Hop Up*, April 1952, 26–27.

68. On Navarro, see Almquist, "Barney Navarro," 51–52. On Clay Smith, see Dearborn, "The Clay Smith Story." On Karl Orr, see Drake, *Hot Rodder*, 41–43. On Spalding, see Medley, *Tex Smith's Hot Rod History, Volume One*, 77. On Barnes, see Montgomery, *Hot Rod Memories*, 91 and 105. On Hedman, see Almquist, "An 'Exhausting' Business," in *Hot Rod Pioneers*, 124. Finally, on Carrillo and Sparks, see Almquist, "Fred Carrillo: His Rods Make a Better Connection," in *Hot Rod Pioneers*, 250–51.

69. Scranton, *Endless Novelty*, 21. See also Sabel and Zeitlin, "Historical Alternatives to Mass Production," esp. 142–56.

70. Scranton, *Endless Novelty*, 18–21.

71. Here my analysis resonates with Cyrus Mody's recent study of the electron-microscope industry of the 1980s and 90s, for which productive contacts with those who purchased and used these sophisticated scientific instruments were crucial (Mody, "Corporations, Universities, and Instrumental Communities"). On the importance of feedback and user-producer identity, see Misa, *A Nation of Steel*.

72. Sabel and Zeitlin, "Historical Alternatives to Mass Production," 142.

73. Bruno Latour, *Science in Action: How to Follow Scientists and Engineers Through Society* (Cambridge, Mass.: Harvard University Press, 1987), 180.

74. Navarro, "Hopping Up the Chevy V8," *Rod and Custom*, March 1955, 19–23 (quote on 19).

CHAPTER 5: FACTORY MUSCLE, 1955–1970

1. Roberson, *Middletown Pacemakers*, 118.

2. Almquist, "East Versus West: Drag Racing's Surprising Comeuppance," in *Hot Rod Pioneers*, 196–97; Ganahl, "The California Hot Rod," in Dobrin, Linhares, and Ganahl, *Hot Rods and Customs*, 24; Gary and Marilyn Meadors, *Goodguys Hot Rod Chronicles* (Sebastopol, Calif.: Thaxton Press, 1996), 52–53; Cohen, *Consumer's Republic*, 309 (on Wolfe); and Moorhouse, *Driving Ambitions*, 123.

3. This particular suggestion is buried deep within Roger Huntington's otherwise excellent history of OEM performance cars: *American Supercar: Development of the Detroit High-Performance Car* (Osceola, Wis.: Motorbooks International, 1990 [1983]), 36 and 73. In the 1950s and 60s, Huntington was a technical and feature contributor for several periodicals, including *Speed Age*, *Speed Mechanics*, and *Popular Hot Rodding*.

4. Back in 1952, *Auto Sport Review* reported that the twelve largest speed equipment manufacturers alone racked up more than $50 million in sales that year (see above, chapter 4). At a glance, estimates of $36 million for the entire industry as of 1961 thus appear quite low—actually, they seem to indicate at least a 35 percent *decline* in high-performance sales over the course of the 1950s. Closer inspection of the numbers, however, reveals that this was not the case. To wit: the figures for 1961 are estimates based on the recollections of those who participated in *Hot Rod Industry News*'s 1967 survey of manufacturers. In that 1967 survey, respondents were asked to disclose their 1961 and 1966 sales totals, resulting in numbers that the editors then crunched to determine a very conservative estimate of the industry's 1966 aggregate value and its growth over the previous five years. The numbers they came up with were then published in September of 1967 (Cec Draney, "Hot Rod Industry News Survey," *Hot Rod Industry News* [hereafter, *HRIN*], September 1967, 36–42). However, according both to the survey itself and to the publisher's comments that preceded it (Ray Brock, "Publisher's Report," *HRIN*, September 1967, 6), the numbers were low by an estimated 66 percent—retailers, for example, whose numbers also appeared under a separate heading in the write-up, thus were likely to have done not $111 million in gross sales in 1966, as reported, but possibly as much as $333 million. Similarly, the number given for the manufacturing end of the industry for 1961, $36 million, may well have been closer to $108 million; and the figure for 1966 of $148 million, closer to $444 million. At worst, therefore, the industry stood at $36 million in 1961 and $148 million in 1966; at best, $108 million and $444 million, respectively. The aforementioned figure of $1.168 billion for 1970, on the other hand, is spot-on. Though it comes from a survey published in 1971 similar to that published in 1967, an independent survey conducted in the early months of 1969 indicated that the manufacturing end of the industry was already nearing the $1 billion mark, a milestone knowledgeable insiders believed was passed by the end of fiscal 1969; therefore, $1.168 billion for 1970 is reasonably accurate ("Hot Rod Industry News 1971 Industry Sur-

vey," *HRIN*, August 1971, 26–32; Don Prieto, "Detroit Bulletin," *HRIN*, July 1969, 8 and 18; and Dennis Pierce, "A Look into the 70s," *HRIN*, January 1970, 70). Incidentally, if we reject the conservative estimate of $36 million for 1961 and opt instead for the more liberal figure of $108 million, that would mean that the industry grew by more than 1,000 percent between 1961 and 1970. If, on the other hand, we accept the conservative estimate of $36 million, growth approaches an unlikely 3,000 percent for the decade. Therefore, the actual figure for 1961 probably stood much closer to the more liberal estimate of $104 million. Further complicating matters is inflation, although its impact on the magnitude of growth suggested by these figures from the 1960s is minor: adjusted for inflation, the conservative estimate of $36 million for 1961 works out to approximately $46.72 million in 1970 dollars, and the more liberal estimate of $108 million for 1961 to $140.15 million (calculations performed with the Bureau of Labor Statistics CPI Inflation Calculator, available online at http://stats.bls.gov). Compared with the figure of $1.168 billion for 1970, therefore, our inflation-adjusted numbers for 1961 suggest a no less staggering rate of growth of between 850 and 2,500 percent during the 1960s. On American Motors, see "American Flits Ahead," *Time*, 29 November 1971, archived online at http://www.time.com/time/magazine/article/0,9171,877474,00.html. On Chrysler, see Chrysler Corporation, *Annual Report 1970* (Detroit: Chrysler Corporation, 1970), 1. On Ford, see "Fortune 500, 1955–2006: 1971," available online at http://money.cnn.com/magazines/fortune/fortune500_archive/full/1971/ (1971 ratings reflected 1970 total revenue); see also Ford Motor Company, *Annual Report 1971* (Dearborn, Mich.: Ford Motor Company, 1971), 1. On General Motors, see "Fortune 500, 1955–2006, 1971"; General Motors's revenues of $18.75 billion for 1970 were much lower than its revenues of $24.2 billion for 1969 ("Fortune 500, 1955–2006, 1970," available online at http://money.cnn.com/magazines/fortune/fortune500_archive/full/1970/.

5. On government regulations and the high-performance aftermarket, see below, chapters 7, 8, and 9.

6. In 1935 Ford teamed with Harry Miller to produce a set of Indianapolis 500 racers based on the company's then-new flathead V8. See above, chapter 2.

7. Some of these new engines aroused the interest of postwar hot rodders because of the enormous *potential* of their overhead-valve designs. As *delivered*, however, they were relatively tame.

8. Roberson, *Middletown Pacemakers*, 41–42.

9. Almquist, "Hemi Hopping," in *Hot Rod Pioneers*, 183, and "'Full House' Firepower: 310 Horsepower without Supercharging—New Chrysler Achievement," *HRM*, March 1952, 52.

10. Jim McFarland, one of the editors of *Hot Rod* in the mid-1960s and a talented engineer who went on to enjoy a long career within the speed equipment industry, saw Zora Arkus-Duntov's arrival at GM as the turning point for OEM performance (McFarland, "Publisher's Memo," *HRM*, December 1968, 6); see also Jerry Burton, *Zora*, chap. 8. On the "insolent chariots" claim, see Flink, *Automobile Age*, chap. 15.

11. Harry Cushing, "Detroit Looks at the Hot Rods," *HRM*, December 1951, 30.

12. "'Full House' Firepower," 52.

13. The first quote is from Don Francisco, "Stude, Packard, Nash, or Hudson Big 8?" *HRM*, February 1957, 24, and the second is from Ed Almquist, "Soup for the Family Car," *Car Life*, August 1956, 18. Similar claims appeared elsewhere; see, e.g., Louis Hochman, *Hot Rod Handbook* (Greenwich, Conn.: Fawcett Publications, 1958), 4.

14. The National Association of Stock Car Racing, or NASCAR, was (and is) an influential association whose rules generally set the tone for the ways in which the majority of American stock-car races were (and are) conducted.

15. For more on NASCAR's mid-1950s rules and the ways in which OEMs met them, see Huntington, *American Supercar*, 19–20, and Peter Golenbock, *American Zoom: Stock Car Racing—From the Dirt Tracks to Daytona* (New York: Macmillan USA, 1993). See also Ben Shackleford, "Going National while Staying Southern: Stock Car Racing in America, 1949–1979" (Ph.D. diss., Georgia Institute of Technology, 2004).

16. Jerry Titus, "Guide to the '58 Engines," *Speed Age*, February 1958, 16–19 and 69.

17. Huntington, *American Supercar*, 29.

18. "Facts and Forecasts," *Motor Life*, July 1957, 6.

19. Huntington, *American Supercar*, 29.

20. Ibid., 33.

21. As late as 1955, regular gasoline octane ratings in the United States stood at about 80, and static compression ratios in most OEM passenger cars at about 6:1 (W. G. Brown, "Compression Ratios on the Rise," in *Hot Rod 1955 Annual* [Los Angeles: Petersen Publishing Co., 1955], 110–11). By 1960, however, compression ratios had risen considerably—often to 9:1 or even 10:1—as the octane ratings of widely available regular and premium leaded fuels broke into the 90s. In fact, period evidence suggests that the *minimum* octane fuel requirement of *any* domestic car stood at 91 in 1960, while some OEM vehicles actually required fuels rated at a heady 99 octane (Robert Lichty, "What Octane Does Your Car Need?" *Cars*, October 1960, 78–82). And this was only the beginning: by the late 1960s, premium fuels well in excess of 100 octane were widely available throughout the United States, enabling many OEM cars to run with compression ratios in the neighborhood of 11:1 to 12:1. However, during the 1970s the method used in the United States to calculate octane ratings changed, resulting in advertised numbers 3 to 4 points lower. By today's standards, therefore, the fuels of the early 1950s would have rated in the high 70s, those of 1960 at about 87 to 95, and the top-notch fuels of the late 1960s at about 100 octane.

22. Almquist, "Rampaging Ramchargers: Dynasty of Fuelers, Funnies, and Stockers," in *Hot Rod Pioneers*, 236–38.

23. On Ford's new V8, see Dean Brown, "The Best Engines for Hot Rods," *Popular Hot Rodding* (hereafter, *PHR*), August 1962, 10–17 and 70. On Chevrolet's big block, see Huntington, *American Supercar*, 29.

24. "Facts and Forecasts," *Motor Life*, February 1960, 6.

25. On the small-block Chevrolet 327 of 1962, see "Hop-Up Report on the 327 Chevy," *PHR*, November 1962, 26–31. On the new big blocks from Chevrolet, Buick, Oldsmobile, and Ford, see Dean Brown, "The Best Engines for Hot Rods," *PHR*, August 1962, 10–15 and 70. And on Chrysler's 426 Hemi, released for racing applications in 1964, see Ray Brock, "Modern Hemi from Chrysler," *HRM*, April 1964, 44–47.

26. Huntington, *American Supercar*, 35.

27. On the NHRA's stock classes and their rules, which interpreted the term "stock" in a manner far more literal than had even been the case with NASCAR in the 1950s, see Griffith Borgeson and Wayne Thoms, "Super Stocks for '61," in *Hot Rod Ideas* (New York: Arco Publishing Company, 1961), 5–13; *Stock Cars for the Drags* (Los Angeles: Petersen Publishing Company, 1963); and Montgomery, *Supercharged Gas Coupes*, introduction.

28. Huntington, *American Supercar*, 35–36 (quote on 35).

29. Ibid., 73.

30. Fully-machined combustion chambers appeared on aftermarket cylinder heads as early as the 1910s, for example, and high-compression pistons, counterweighted crankshafts, reinforced connecting rods, and even engine-block parts like multiple-bearing crankshaft adapters had also been available from speed equipment firms since the 1910s. See above, chapter 1.

31. Huntington, *American Supercar*, 16.

32. Ray Brock, "Dodge's Hot Options," *HRM*, October 1961, 26–31.

33. FoMoCo advertisement, *HRM*, May 1964, 76.

34. "Hop-Up Report on the 327 Chevy," *PHR*, November 1962, 26–31.

35. Eric Rickman, "Inchin' Up On the 327," *HRM*, January 1967, 62–63.

36. Huntington, *American Supercar*, 49–53, and Roberson, *Middletown Pacemakers*, 118.

37. On the 427 Fords, see Bob Leif, "Detroit Bulletin," *HRIN*, July 1967, 10. On the 455 GM engines, see Steve Kelly, "Mister Muscle of 1970," *HRM*, November 1969, 34–36. On the 426 Street Hemi, see Plymouth advertisement, "The Hot Ones from Plymouth," *HRM*, March 1966, 63–69. On the growth of the muscle-car segment during the mid- to late 1960s, see Don Prieto, "Detroit Bulletin," *HRIN*, October 1969, 16.

38. Huntington's coverage is by far the best, and certainly the most comprehensive, although a number of others do an excellent job as well. See, e.g., *Petersen's Muscle Car Classics* (Los Angeles: Petersen Publishing Co., 1985), and Steve Statham, *Muscle Cars: American Thunder* (Lincolnwood, Ill.: Publications International, 1997).

39. Dodge advertisement, *HRM*, July 1969, 157.

40. By 1970, for example, Ford's vice president and general manager, John B. Naughton, projected that the OEMs would sell more than 750,000 muscle cars that year (Alex Xydias, "Publisher's Report," *HRIN*, January 1970, 20).

41. For an early report on the hop-up potential of Ford's new engine, see Racer Brown, " '54 Ford V8," *HRM*, September 1954, 26–29 and 44–45.

42. For an excellent summary of Edelbrock's (and *HRM's*) early work with the new Chevrolet engine, see Ken Gross, "The Day the Flatheads Died," *Hemmings Muscle Machines*, December 2003, 45.

43. Navarro went into medical-systems research and development, Earl Evans retired, and Jim Kurten went into the aircraft business. On Navarro, see Montgomery, *Hot Rod Memories*, 91–92, and Almquist, "Barney Navarro," 51–52. On Kurten, see Montgomery, *Hot Rod Memories*, 92. And on Evans, see Almquist, "Earl Evans: Terror on the Tracks," in *Hot Rod Pioneers*, 56.

44. On the enduring, albeit diminished popularity of the flathead V8, see Larry Hurd, "Don't Forget the Flathead," *PHR*, July 1963, 32–35 and 81–82; Bob Leif, "Is the Flathead Dead?" *HRIN*, April 1967, 38–39; and Roberson, *Middletown Pacemakers*, 41–42.

45. Hochman, *Hot Rod Handbook*, 5.

46. Jerry McGuire, "Hop Up Secrets for Chevy Engines: The Hot Rodder's Guide to the Chevrolet V-8 Powerplant," *PHR*, June 1963, 28.

47. Dennis Pierce, "An Influential Industry," *HRIN*, April 1967, 40.

48. Ray Brock, "Publisher's Report," *HRIN*, April 1967, 6.

49. On the baby boom and its implications for the high-performance industry of the 1960s and 70s, see Pierce, "A Look Into the 70s," 70.

50. Examples of this "rallying cry" abound; see, e.g., Alex Xydias, "Publisher's Report," *HRIN*, July 1969, 6; Dennis Pierce, "New Car Dealers . . . New Business?" *HRIN*, October 1969, 90; and Don Evans, "Editorially Speaking," *HRM*, March 1970, 8.

51. As late as model-year 1970, for example, mainstream industry insiders estimated that high-performance models (including muscle cars) accounted for no more than 20 percent of overall new-car sales, which meant that more than 80 percent of what they sold were ordinary, low-performance models (Don Prieto, "Detroit Bulletin," *HRIN*, October 1969, 16). John B. Naughton of Ford projected that the OEMs would sell 750,000 high-performance muscle cars in 1970, along with 750,000 low-performance "look-alikes" (Alex Xydias, "Publisher's Report," *HRIN*, January 1970, 20).

52. On Camaro Z-28 cloning, see Bob Leif, "Detroit Bulletin," *HRIN*, May 1967, 8.

53. Ray Brock, "Publisher's Memo," *HRM*, September 1970, 6.

54. Rus Kavich, "Inside the Problem of Elusive Oil," *Drag Racing Magazine*, December 1968, 52–53.

55. John Thawley, "Triple-Duty Dandy," *HRM*, November 1969, 62–63.

56. So good were B&M's high-performance automatics, in fact, that they became a mainstay in a number of popular drag-racing classes. For more on B&M, see below, chapter 6.

57. Aftermarket transistor ignitions first began to appear in the early 1960s (see, e.g., "Transistor Ignition System," *Drag News*, 24 March 1962, 7). Summers Brothers Equipment in Ontario, California, was one of several that devised camshaft gear-drive setups for late-model V8s ("Rodding's '69½ Speed Secrets," *PHR*, April 1969, 27). Among many others, Gotha in Harvey, Illinois, devised a set of lightweight rockers for the Chevrolet V8 as early as 1961 ("What's New," *HRM*, October 1961, 18). Hedman, Cragar, and many others developed performance-enhancing exhaust headers during the 1960s; see, e.g., George Elliot, "Hedman's New Design for Header Efficiency," *PHR*, June 1966, 70–71, and "Manufacturers News," *Drag News*, 11 January 1969, 9. Finally, on the aftermarket fuel-injection systems for late-model V8s, see Bob Vordell, "New Ideas in Speed Equipment: Quick Developments That Keep the Accessory Manufacturers One Step Ahead of Detroit," *PHR*, May 1963, esp. 57.

58. Chevrolet's decision in 1955 to send an example of its yet-to-be-released overhead-valve V8 to Edelbrock for advance testing was one of the few exceptions. See Dain Gingerelli, "Edelbrock: Hot Rodding's First Family," *Street Rodder Magazine*, December 2000, 214, and Madigan, *Edelbrock*, 135.

59. "Hop Up Report on the 327 Chevy," *PHR*, November 1962, 26–31, and Iskenderian advertisement, *HRM*, May 1966, 3.

60. By the time *Hot Rod Magazine* published its review of the all-new Chevy Vega in 1970, for example, Edelbrock, Iskenderian, Offenhauser, Hooker, TRW, and a number of other aftermarket companies already had high-performance items available for it. See Steve Kelley, "Vega: Small-Car Star," *HRM*, November 1970, 36–39.

61. Karen Scott, "Meet the Manufacturer: Fuel Injection Engineering," *Drag News*, 10 May 1969, 8.

62. Schiefer's evolution over time can be traced through the many "meet the manufacturer" articles about the company published in the 1950s and 60s, e.g., Scotty Fenn,

"Meet the Manufacturer: Schiefer," *Drag News*, 26 September 1959, 7; "Meet the Manufacturer: Schiefer Mfg. Co.," *Drag News*, 18 May 1963, 20–21; "Schiefer Manufacturing," *Drag Sport*, 23 April 1966, 10; and "Meet the Manufacturer: Schiefer," *Drag News*, 19 April 1969, 8. On the Crankshaft Company, see "Meet the Manufacturer: Crankshaft Company," *Drag News*, 26 November 1960, 6–7.

63. "A Look at the Old Master's Shop," *Drag Sport*, 6 September 1965, 3–4 and 9; Al Caldwell, "Meet the Manufacturer: Ed Pink Racing Engines," *Drag News*, 18 August 1969, 26–27; Karen Scott, "Meet the Manufacturer: Donovan Precision Products," *Drag News*, 28 June 1969, 23; Arthur H. Irwin III, "'To Breathe or Not to Breathe?' Tubular Automotive," *Drag News*, 18 February 1966, 20; "Meet the Manufacturer: Lakewood Industries," *Drag News*, 20 November 1971, 18–19; and "Meet the Manufacturer: Simpson Safety Equipment," *Drag News*, 5 January 1968, 16–17.

64. See below, chapter 6.

65. Baronian was Iskenderian's custom-grind specialist for years before he founded his own company in 1968 (Dave Wallace Jr., "Meet the Manufacturer: Norris Performance Products," *Drag News*, 17 May 1975, 17).

66. Wallace, "Jack Engle Walks Softly, Carries Big Bumpstick."

67. "Meet the Manufacturer: Edelbrock," *Drag News*, 6 April 1968, 16.

68. Fenton and Eelco were among the equipment companies that mass produced; see Almquist, "Aaron J. Fenton: A Born Salesman," in *Hot Rod Pioneers*, 190; Scott Fenn, "Meet the Manufacturer: Eelco Mfg. & Supply Co.," *Drag News*, 1 August 1959, 7; and "Manufacturers News," *Drag News*, 8 February 1969, 2.

69. On the high-performance market for domestic compacts, see Alex Wallordy, "Soup-Ups for the Compacts," *Cars*, June 1960, 57–59 and 68, and *Hot Rodding the Compacts*.

70. On the turbocharged Corvair Spyder, see Roger Huntington, "Spyder—150 hp!" *Motor Trend*, July 1962, 50–55.

71. See Griffith Borgeson and Wayne Thoms, "More Suds for the Corvair," in *Hot Rod Ideas*, 20–25, and *Hot Rodding the Compacts*, chap. 1.

72. See Don Francisco, "30 Giant Horses for Corvair," *HRM*, October 1961, 34–37, and Ray Brock, "Stroker Kits for Corvair," *HRM*, August 1960, 34–37 and 92. Aftermarket turbos for the Corvair were available from Bell Auto, for example (Ray Brock, "Turbochargers for Non-Spyders," *HRM*, April 1964, 40–43), while superchargers were available from companies like Paxton (Chuck Nerpel, "Mor Air for the Corvair," *Motor Trend*, May 1960, 85).

73. In the 1950s, Volkswagen engines for the American market displaced 1132 to 1192 cubic centimeters (cc) and developed between 25 and 40 horsepower, depending on their model year. By 1966, the mill had grown to 1300cc and 50 horses, and by 1967, 1500cc and 53 horsepower. Displacement rose again in 1970, to 1600cc, and further refinements through the early 1970s brought the engine's power to a peak of 60. See Shuler, *Origin and Evolution*, 165–74.

74. Webber H. Glidden, President, Volkswagen Club of America, "Who Says Volkswagens Are Lousy?" *Speed Age*, May 1957, 24–27, 56, 58, and 60.

75. On Volkswagen's policy regarding the modification of its cars, see *Volkswagen Handbook* (Los Angeles: Petersen Publishing Co., 1963), 2–3.

76. Lee Kelley, "Performance Ideas for VW Engines," *PHR*, January 1969, 36–41.

77. These included, among others, Iskenderian, Weber, Crower, Engle, Cragar, and Jahns; see Karl E. Ludvigsen, *Your Sports Car Engine: Its Maintenance, Tuning, and Modification from Spark Plugs to Supercharging* (New York: Sports Car Press, 1958), 74–76; Iskenderian advertisements, *HRM*, June 1966, 3, and September 1968, 3; Len Griffing, "Small Size . . . Big Surprise!" *Motor Trend*, March 1960, 32–33; Crower advertisement, *HRIN*, August 1967, 5; "Manufacturers News," *Drag News*, 17 May 1969, 2; Engle advertisement, *HRM*, September 1968, 8; "Manufacturers News," *Drag News*, 22 September 1969, 2; and Bud Lang, "VW Dyno Tests: Bolt-On Power," *HRM*, June 1968, 50–52.

78. John Thawley, "Putten der Growl in der Beetle," *HRM*, May 1967, 98–99; Eric Dahlquist, "Small Wonder," *HRM*, July 1967, 32–35; and Patrick J. Bedard, "EMPI GTV: Southern California's Muscle Beetle," *Car and Driver*, July 1968, 64–66 and 93.

79. On Shoemaker, see "Manufacturers News," *Drag News*, 1 March 1969, 2, and "Manufacturers News," *Drag News*, 12 April 1969, 2. On Revmaster, see Ed Orr, "More Power for VWs!" *PHR*, April 1969, 64–67. And on Deano Dyno-Soars, see John Thawley, "43-HP VW Bolt-On: Give That 1500 an Impressive Performance Boost Without Splitting the Cases or Even Yanking the Engine," *HRM*, November 1969, 100–101.

80. On Scat, see Thawley, "Putten der Growl in der Beetle"; "Bolt-On Horsepower for the VW," *PHR*, April 1968, 56–57; Lang, "VW Dyno Tests: Bolt-On Power"; and Ed O'Brian, "Stroker Kits for the VW," *PHR*, April 1970, 34–36 and 103. "Scat" doesn't stand for anything in particular, according to the front office staff of the company's Redondo Beach, California, headquarters. The word *scat* has several (wildly disparate) meanings, all of which are slang: a type of jazz dancing, a type of jazz singing, animal excrement, heroin, or "to move or go off hastily" (*Webster's College Dictionary*, s.v. "scat"). Period evidence suggests that the last of these meanings had at least some currency among performance enthusiasts of the 1960s, and since the company in question was founded in the mid-1960s, the decision in favor of the name "Scat" probably had something to do with this meaning. For an example of the use of the term *scat* to imply "hasty movement" during the 1960s, see Dodge advertisement ("Dodge announces Scat City: The '70 Dodge Scat Pack Is Road Ready"), *HRM*, October 1969, 71–79. The aftermarket company Scat had nothing to do with Dodge.

81. Okrasa and BRM both chose EMPI as their American distributor. See "Special Reports," *Motor Life*, March 1957, 7; Griffing, "Small Size . . . Big Surprise"; and "What's New," *HRIN*, September 1966, 50 and 61.

82. An extensive modification and testing session performed with the assistance of a number of manufacturers at Iskenderian's dynamometer facility in 1968, for example, yielded an average gain of 25 percent for most combinations of simple bolt-ons (Lang, "VW Dyno Tests: Bolt-On Power"); another similar session at Deano Dyno-Soars the following year resulted in gains in excess of 100 percent (Thawley, "43-HP VW Bolt-On").

83. Almquist, "Clay Smith: Mr. Horsepower," in *Hot Rod Pioneers*, 88–89; Almquist, "The Spalding Brothers: Famous for Ignitions and Cams," in *Hot Rod Pioneers*, 40–41; and Huntington, *American Supercar*, 19.

84. Huntington, *American Supercar*, 73.

85. On Miller's work with Ford, see Bob Leif, "Detroit Bulletin," *HRIN*, July 1968, 14; John Thawley, "The Whys, Wise, and Y's of Headers," *HRM*, October 1969, 60–63; and Steve Kelley, "Hot Stuff for Mavericks," *HRM*, January 1970, 90–92. On Hernandez and Ed Pink's 1960s role(s) at Ford, see Jerry Brandt, "Ed Pink," *Drag Racing USA*, May

1975, 61, and Almquist, "Fran Hernandez: Funny-Car Builder," in *Hot Rod Pioneers*, 73. On Keith Black's relationship with Chrysler, see Robert C. Post, "Interview With Keith Black," 22, *Drag Racing Oral History Archive*, National Museum of American History, and "News and Notes," *HRIN*, February 1967, 39. On Navarro's work with AMC, see Almquist, "Barney Navarro," 51–52.

86. Huntington, *American Supercar*, 63 and 99, and "Meet the Manufacturer: Edelbrock," *Drag News*, 6 April 1968, 16.

87. "Rodding Roundup," *PHR*, August 1967, 5.

88. American Motors advertisement, *HRM*, April 1969, 166; "Car Life Road Test: Pontiac Super Stock," *Car Life*, October 1961, 18–21; "Detroit Bulletin," *HRIN*, December 1967, 10; Bob Leif, "Detroit Bulletin," *HRIN*, June 1968, 16; and Hurst-Campbell advertisement, *HRM*, February 1963, 6.

89. Shelby built his first Cobra sports car in the back of Moon's Santa Fe Springs, California, facility in the early 1960s (Fetherston, *Moon Equipped*, 93) before going on to produce high-performance accessories, kits, and packages for Ford (LeRoi Smith, "Bolt On 80 Horsepower," *HRM*, October 1963, 78–81; FoMoCo advertisement, *HRM*, February 1964, 6–7; and "New Cobra Supercharger by Shelby American," *Drag News*, 30 December 1966, 5).

90. Author Interview with Bob Spar, Newbury Park, California, November 13, 2003; Karen Scott, "Meet the Manufacturer: Hooker Headers," *Drag News*, 14 June 1969, 12; "Manufacturers News," *Drag News & Equipment Industry Report*, 6 December 1969, 2; and "Meet the Manufacturer: Mondello Olds Performance," *Drag News*, 2 March 1974, 8.

CHAPTER 6: BOLT-ON POWER, 1955–1970

1. *Hot Rod*'s comprehensive 1971 directory, published in December of 1970, lists 757 individual firms, of which 7 were Canadian and 750 American ("1971 Automotive Specialty Equipment Manufacturers Directory," *HRM*, December 1970, 90–95).

2. Ray Brock, "Publisher's Report," *HRM*, December 1969, 6.

3. Cec Draney, "Hot Rod Industry News Survey," *HRIN*, September 1967, 39.

4. On the industry's geographical distribution in 1948, see above, chapter 4. By 1970, 345 of the country's 750 manufacturers, or 46 percent, were based in Southern California ("1971 Automotive Specialty Equipment Manufacturers Directory").

5. Ibid. Illinois accounted for 7 percent by 1970, Ohio and Michigan for 6 each, New York for 5, New Jersey for 4, and Pennsylvania for 3. No other state accounted for more than 2 percent.

6. Ibid. The Los Angeles area was home to 284 manufacturers in 1970.

7. Ibid.

8. On Weber's move, see Ray Brock, "Stroker Kits for Corvair," *HRM*, August 1960, 35. Weber's original Whiteside Avenue location measured 1,600 square feet (Almquist, "Harry Weber: The Likeable Pioneer Cam Man," in *Hot Rod Pioneers*, 112–13), while the company's new 1960 headquarters in Santa Ana featured 7,500. (In 1965 Weber added 8,500 square feet to its 1960 building, bringing its total square footage to 16,000; thus, Weber's 1960 building originally had approximately 7,500 ["Weber in New Building," *Speed & Custom Equipment News*, July 1965, 1]).

9. Karen Scott, "Meet the Manufacturer: Fuel Injection Engineering," *Drag News*, 10 May 1969, 8.

10. See Rob Kling, Spencer Olin, and Mark Poster, eds., *Postsuburban California: The Transformation of Orange County Since WWII* (Berkeley: University of California Press, 1991).

11. Karen Scott, "Meet the Manufacturer: Hooker Headers," *Drag News*, 14 June 1969, 12.

12. "News and Notes," *HRIN*, August 1967, 20, and "News and Notes," *HRIN*, December 1967, 54.

13. Karen Scott, "Meet the Manufacturer: Hooker Headers," *Drag News*, 14 June 1969, 12, and "Manufacturers News," *Drag News*, 5 July 1969, 2.

14. "Meet the Manufacturer: Lakewood Industries," *Drag News*, 20 November 1971, 18–19.

15. Construction on its new 30,000-square-foot facility began in 1969, and the company moved in 1970 ("Manufacturers News," *Drag News*, 17 May 1969, 2, and "Meet the Manufacturer: Lakewood Industries," *Drag News*, 20 November 1971, 18–19). The company's growth continued well into the 1970s ("Meet the Manufacturer: Lakewood Industries," *Drag News*, 1 June 1974, 22).

16. "Crower: How the Shop Came to Be," *Drag News*, 25 May 1963, 20–21; "Crower Cams in Expansion Move," *Drag News*, 15 April 1966, 21; Crower advertisement, *HRM*, April 1966, 97; and Dave Wallace Jr., "And Now, from the Folks Who Brought You the U-Fab Intake Manifold," *Drag News*, 21 August 1975, 16–17.

17. The number of employees at these companies remained relatively small, averaging 12 to 24 per company during the late 1960s. According to a 1967 survey, aftermarket manufacturers employed an average of 8,343 workers over the course of calendar year 1966, and even if we assume that this figure had doubled by 1970, when *Hot Rod* published its manufacturers' directory, we arrive at an average of no more than 24 per firm (see Cec Draney, "Hot Rod Industry News Survey," *HRIN*, September 1967, 41). Skewing that average, however, were firms like Mickey Thompson, which grew from 4 employees in 1959 to more than 200 in 1969 (Karen Scott, "Meet the Manufacturer: Mickey Thompson," *Drag News*, 3 May 1969, 8); Crane Cams, which grew from a handful of employees in 1953 to more than 80 by 1968 ("Meet the Manufacturer: Crane Cams," *Drag News*, 29 December 1973, 16, and Crane advertisement, *HRM*, July 1968, 117); and Crower, which grew from a staff of 6 in 1960 to more than 75 fifteen years later (Wallace, "And Now, from the Folks").

18. Karen Scott, "Meet the Manufacturer: Fenton Company," *Drag News*, 2 August 1969, 14, and Karen Scott, "Meet the Manufacturer: Cyclone Automotive," *Drag News*, 19 July 1969, 10.

19. For example, Ansen used multiple drill presses and a double boring machine to speed up the hand-machining processes at its Gardena, California, plant ("Ansen Automotive," *Drag News*, 2 May 1964, 20–21, and Karen Scott, "Meet the Manufacturer: Ansen," *Drag News*, 12 July 1969, 15).

20. "Iskenderian: America's Fastest Racing Cams," *Drag News*, 7 March 1964, 24–25.

21. On Schiefer, see Scotty Fenn, "Meet the Manufacturer: Schiefer," *Drag News*, 26 September 1959, 7, and "Manufacturers News," *Drag News*, 15 February 1969, 2. On Crower, see Karen Scott, "Meet the Manufacturer: Crower," *Drag News*, 24 May 1969, 23. On Lakewood, see "Meet the Manufacturer: Lakewood Industries," *Drag News*, 20 November 1971, and "Meet the Manufacturer: Lakewood Industries," *Drag News*, 1 June 1974, 22.

22. Unlike gasoline engines, which use spark plugs to ignite the air-fuel mixture, diesel engines ignite the mixture through high compression; hence "compression-ignition engine," a common technical substitute for the more colloquial "diesel."

23. Barney Navarro, "Blown Corvair," *Motor Life*, September 1960, 34.

24. Roger Huntington, "Spyder—150 hp!" *Motor Trend*, July 1962, 50–55, and "The Steering Column," *PHR*, August 1962, 5.

25. Bell Auto brought out a bolt-on turbo kit for ordinary Corvairs in 1964 (Ray Brock, "Turbocharger for Non-Spyders," *HRM*, April 1964, 40–43), and Rajay did the same for the air-cooled Volkswagen in 1970 (Hugh MacInnes, "Turbocharge!!!" *HRM*, October 1970, 58–60). AiResearch developed turbochargers for racing applications beginning in the early 1960s (Almquist, "Robert DeBisschop," and Drake, *Hot Rodder*, 95).

26. On Winfield, see Medley, *Tex Smith's Hot Rod History, Volume One*, 60; on Hilborn, see Batchelor, *American Hot Rod*, 70, and Almquist, "Stuart Hilborn: Racing's Fuel Injection Pioneer," in *Hot Rod Pioneers*, 68–69.

27. Roger Huntington, "Seven Systems for Detroit: What Types of Fuel Injection Are Auto Makers Considering?" *Motor Life*, January 1957, 24–25. On Chevrolet's introduction of the Rochester system in 1957 (and the skepticism with which it was sometimes met), see Bob Fendell, "Fuel Injection: 1957's Greatest Myth?" *Car Life*, March 1957, 36–37 and 73, and "In 1960, Turbines Will Replace Fuel Injection," *Motor Guide*, March 1957, 36–38.

28. Howard's Cams, Norden, Algon, and Scott all had systems in the works by 1959, as did Enderle by mid-1960 (Bob Pendergast, "New Fields in Fuel Injection," *HRM*, February 1959, 28–35, and "Meet the Manufacturer: Enderle Fuel Injection," *Drag News*, 30 April 1960, 6).

29. See Barney Navarro, "Why Fuel Injection?" *Motor Life*, January 1957, 20–21; "Injecting Fuel into the VW," *Business Week*, 23 September 1967, 44; and "The 1968 SEMA Show," *HRIN*, February 1968, 28.

30. Bob Leif, "Detroit Bulletin," *HRIN*, June 1968, 16, and Bob Leif, "Detroit Bulletin," *HRIN*, July 1968, 14.

31. Bob Pendergast, "New Fields in Fuel Injection," *HRM*, February 1959, 28.

32. Almquist, "Stuart Hilborn," 69.

33. Medley was a member of the *Hot Rod* staff in the 1950s (he was a cartoonist: his "Stroker McGurk" strip was a popular feature in the magazine for much of the 1950s). Here, his source on Winfield's fuel injection system is Kong Jackson, an early equipment manufacturer who knew Winfield well (Medley, *Tex Smith's Hot Rod History, Volume One*, 60).

34. On the Hilborn system, see Almquist, "Stuart Hilborn," 69.

35. Chevrolet advertisement, *HRM*, July 1957, 15.

36. Huntington, "Seven Systems for Detroit," and Almquist, "Stuart Hilborn," 69.

37. See Karen Scott, "Meet the Manufacturer: Mickey Thompson," *Drag News*, 3 May 1969, 8; Dennis Pierce, "Manufacturers Reps," *HRIN*, April 1967, 23–25, 47, and 48; Almquist, "Road Warriors: More than Parts Peddlers," in *Hot Rod Pioneers*, 245; and Dennis Pierce, "A Look into the 70s," *HRIN*, January 1970, 70 and 72.

38. For an overview of the 1960s squabbles among speed shops, chains, and WDs, see Moorhouse, *Driving Ambitions*, 127–31. For some specific examples of the concerted effort by the editors of *Hot Rod Industry News* to convince speed shop owners of the advantages

their status as *specialists* gave them over the larger chains, see Alex Xydias, "Editor's Report," *HRIN*, July 1968, 6; "Is the Speed Shop Dead?" *HRIN*, September 1969, 22–24, 26, 50–52, and 55; and Alex Xydias, "The High Performance Market: An Address to the Automotive Research Council in Philadelphia in October of 1969," *HRIN*, January 1970, 54, 56, 58, 60, and 132.

39. Bob Leif, "Manufacturers Speak: Questions and Answers on Topics of Industry Interest," *HRIN*, May 1968, 20–23 and 50–51; "Bobbins on WD's," *HRIN*, July 1968, 24–26; and Dennis Pierce, "The Role of WD's," *HRIN*, December 1968, 26–29.

40. California Equipment Company advertisement, *Drag News*, 30 April 1960, 11.

41. Author Interview with Delores Berg, Orange, California, November 11, 2003.

42. Mickey Thompson Enterprises advertisement ("An Open Letter to Venolia Pistons"), *Drag News*, 1 June 1966, 5, and Mickey Thompson Enterprises advertisement ("An Open Letter to the Readers of *Drag News*"), *Drag News*, 15 July 1966, 3.

43. For some examples, see Iskenderian advertisement ("We All Know Who Is Number One!"), *Drag News*, 12 August 1966, 7; Crane advertisement ("Mud?"), *Drag News*, 26 August 1966, 25; Iskenderian advertisement ("More Isky Winners at Indy than All Other Cam Mfrs. Combined!"), *Drag News*, 30 September 1966, 11; and Crane advertisement ("Did Mr. Ed's Computer Break Down Again??????"), *Drag News*, 7 October 1966, 17.

44. On the 1950s and 60s "ad wars," see Montgomery, *Supercharged Gas Coupes*, 13–14, 52–53, 67, 74, and 92–94; for more on Iskenderian's hyperbolical advertising, see Post, *High Performance*, 44–46.

45. In the 1970s, for example, Els Lohn sold his company to a conglomerate (Almquist, "Els Lohn: Eelco Left a Footprint," in *Hot Rod Pioneers*, 286–87), as did Harry Weber and several others (Almquist, "Harry Weber," 112–13).

46. On Potvin's sale to Moon, see Al Caldwell, "Go With Moon!" *Drag News*, 28 September 1963, 8–9; "News and Notes," *HRIN*, January 1967, 20; and Fetherston, *Moon Equipped*, 19. On Schiefer and Harman & Collins, see "Meet the Manufacturer: Schiefer Mfg. Co.," *Drag News*, 18 May 1963, 20–21. On McGurk's piecemeal sale, see "News and Notes," *HRIN*, March 1968, 48; "Manufacturers News," *Drag News*, 5 April 1969, 2; and Karen Scott, "Meet the Manufacturer: Iskenderian," *Drag News*, 12 April 1969, 8.

47. Vic Edelbrock Sr., for example, feared that if his company ever got too big for him to manage, he might lose control of it (Author Interview with Nancy and Camee Edelbrock, Las Vegas, Nevada, November 7, 2003; Nancy is Vic Jr.'s wife, and Camee is one of their daughters).

48. Alfred D. Chandler Jr., *Strategy and Structure: Chapters in the History of the American Industrial Enterprise* (Cambridge: MIT Press, 1962).

49. "Manufacturers News," *Drag News*, 5 April 1969, 2.

50. "Mickey Thompson Pistons," *Drag News*, 23 April 1963, 8–9; "Mickey Thompson's Mag Foundry," *Drag News*, 11 May 1963, 20–21; and "M/T Balancing," *Drag News*, 14 September 1963, 20.

51. For example, Mr. Gasket went public in the fall of 1969 with 220,000 shares of common stock ("News and Notes," *HRIN*, 138 and 141).

52. "Meet the Manufacturer: Crane Cams," *Drag News*, 29 December 1973, 16, and Almquist, "Harvey Crane, Jr.: 'Professor' of Camshaft Technology," in *Hot Rod Pioneers*, 290–91.

53. Ibid.

54. "Meet the Manufacturer: Crane Cams," *Drag News*, 29 December 1973, 16, and Post, *High Performance*, 83.

55. Almquist, "Harvey Crane, Jr.," 290; Crane advertisement, *Drag News*, 26 August 1966, 25; Crane advertisement, *HRM*, July 1968, 117; and "Meet the Manufacturer: Crane Cams," *Drag News*, 29 December 1973, 16.

56. See, e.g., Crane advertisement ("The Great Crane Robbery"), *HRM*, July 1968, 117.

57. For a couple of choice examples of Crane's promotion of the East-West rivalry, see Crane advertisement, *Drag News*, 7 October 1966, 17, and Crane advertisement, *HRM*, December 1968, 23.

58. On Garlits, see Post, *High Performance*, esp. 83–103. Garlits, a spirited gentleman I had the privilege of meeting in the fall of 2003, remains an outspoken critic of the lingering West Coast bias still evident within the sport today—see, for example, his introductory comments in Ed Almquist's *Hot Rod Pioneers*, xi.

59. "News and Notes," *HRIN*, February 1967, 18.

60. "News and Notes," *HRIN*, December 1967, 58.

61. See, e.g., Crane advertisement ("The Stoneage Cam Co."), *PHR*, December 1969, back cover.

62. Almquist, "Harvey Crane, Jr.," 290–91.

63. Author Interview with Bob Spar, Newbury Park, California, November 13, 2003. See also Robert C. Post, "Henry Kaiser, Troy Ruttman, and Madman Muntz: Three Originals," *Technology and Culture* 46 (2005): 773–77.

64. Author Interview with Bob Spar, Newbury Park, California, November 13, 2003.

65. Ibid.

66. Ibid. On the origins of B&M, see also Karen Scott, "Meet the Manufacturer: B&M," *Drag News*, 5 April 1969, 6.

67. Author Interview with Bob Spar, Newbury Park, California, November 13, 2003.

68. Ibid., and Greg Curtis, "Stick vs. Automatic Transmission," *PHR*, April 1963, 40–43 and 89–90. On "gassers," see Montgomery, *Supercharged Gas Coupes*; Montgomery, *Those Wild Fuel Altereds*; and Post, *High Performance*, chap. 7.

69. Author Interview with Bob Spar, Newbury Park, California, November 13, 2003.

70. Ibid., and Karen Scott, "Meet the Manufacturer: B&M," *Drag News*, 5 April 1969, 6.

71. John Thawley, "The Automatic Dyno: Perhaps the Greatest Advance Since the Racing Automatic Itself," *HRM*, December 1969, 32–35, and Author Interview with Bob Spar, Newbury Park, California, November 13, 2003.

72. Author Interview with Bob Spar, Newbury Park, California, November 13, 2003.

73. T. L. Christian and Antwoine Alferos, "Gene Berg: A Force to be Reckoned With," *VW Trends*, April 1993, 34–37 (an interview that Christian and Alferos did with Gene Berg in 1993; quote p. 34).

74. Ibid., and Author Interview with Delores Berg, Orange, California, November 11, 2003.

75. Author Interview with Delores Berg, Orange, California, November 11, 2003, and Christian and Alferos, "Gene Berg."

76. Author Interview with Delores Berg, Orange, California, November 11, 2003.

77. Ibid., and Christian and Alferos, "Gene Berg."

78. See for example Thawley, "43-HP VW Bolt-On."

79. Author Interview with Delores Berg, Orange, California, November 11, 2003.

80. So well-respected was Gene among enthusiasts that upon his untimely death in 1996, several groups joined forces to promote an annual Gene Berg Memorial Cruise (Don Bulitta, "The Gene Berg Memorial National Cruise," *VW Trends*, December 1996, 36–39, and Don Bulitta, "Cruisin', Cruisin', Cruisin'—The 1997 Gene Berg Memorial National Cruise," *VW Trends*, December 1997, 34–38).

81. Pierce, "A Look into the 70's," 70.

CHAPTER 7: THE SPEED EQUIPMENT MANUFACTURERS ASSOCIATION

1. Lee Kelley, "Can Performance Cars be Saved?" *PHR*, October 1971, 32–33 (quote on 32).

2. For a review of the 1973 351 HO Mustang, see Tom Madigan, "Last of the Street Super Cars?" *PHR*, October 1972, 74–75 and 107 (the compact, low-performance Mustang II was announced in the fall of 1973). There were other holdouts—the Dodge Challenger and the Chevrolet Camaro, for example—but their performance capabilities, like that of the Mustang, fell off considerably in the early to mid-1970s.

3. See, for example, Dave Hetzler, "Here Come the '70s!" *1001 Custom and Rod Ideas*, Winter 1969, 51–59, and Steve Kelly, "Hot 70s: First of the New Muscle Cars," *HRM*, September 1969, 34–37, both of whom waxed enthusiastic about Detroit's new performance offerings for 1970 without any hint of the rapid decline to come.

4. See Roger Huntington, "Don't Let Legislation Wipe Out Hot Rodding: Men Who Know Nothing About the Thrill of Acceleration, Cornering or Braking Are Out to Set Government Regulations on What the Automobile Can Do for You!" *PHR*, July 1968, especially 43; Don Evans, "Editorially Speaking, *HRM*, December 1969, 8; Ray Brock, "Publisher's Report," *HRIN*, October 1966, 6; and Ray Brock, "Publisher's Memo," *HRM*, September 1970, 6.

5. Kelley, "Can Performance Cars be Saved?"; Huntington, *American Supercar*, chap. 10; and Almquist, *Hot Rod Pioneers*, 324–25.

6. Kelley, "Can Performance Cars be Saved?" (quote on 32). "13s" and "14s" refer to quarter-mile elapsed times (E.T.s); hence, a car that could barely cover a quarter-mile dragstrip in under 15 seconds (in the 14s) was considerably slower than one that could easily do it in under 14 (in the 13s).

7. See Almquist, *Hot Rod Pioneers*. Almquist proceeds chronologically; tellingly, the title for his section on the 1960s is "The Sexy Sixties," and the title for his section on the 1970s is "The Sad Seventies."

8. This quote appeared in *Hot Rod* later in the decade, toward the middle of an editorial with a less-than-subtle subtitle: "I'm mad as hell and I'm not going to take it anymore" (Lee Kelley, "Editorially Speaking," *HRM*, July 1978, 5).

9. Moorhouse, *Driving Ambitions*, 141.

10. Following the publication of its second performance industry survey in 1971 ("Hot Rod Industry News 1971 Industry Survey, *HRIN*, August 1971), industry-wide sales data disappears from *Hot Rod Industry News* until 1975, when its third survey was published. An annual "Outlook" feature that appeared in 1972, 1973, and 1974, however, does offer a glimpse of the industry's fortunes during those years. Although they included no hard

292 NOTES TO PAGES 146-149

figures, these features consisted of state-of-the-trade statements penned by industry leaders like Vic Edelbrock Jr., Bob Spar, Roy Richter, and others; and their basic point was clear: 1971 was much better than 1970, 1972 much better than 1971, and 1973 much better than 1972 ("Outlook for '72," *HRIN*, January 1972, 28–32, 36, and 134–43; "Outlook '73," *HRIN*, October 1972, 30–32; and "Outlook '74," *HRIN*, November 1973, 60, 68, and 76). The 1975 performance industry survey (based on 1974 figures) reported industry-wide sales gains of 29 percent for 1974, on average, over 1973 ("Hot Rod Industry News 1975 Performance Industry Survey," *HRIN*, Summer 1975, 37–44); the 1976 survey reported average gains of 24 percent ("Hot Rod Industry News 1976 Performance Industry Survey," *HRIN*, Summer 1976, 31–38); and the 1977 survey also reported gains of 24 percent ("Hot Rod Industry News 1977 Performance Industry Survey," *HRIN*, April 1977, 31–38).

11. Almquist, *Hot Rod Pioneers*, 325.

12. This is the subtitle of his book.

13. Period evidence teems with military metaphors; e.g., see "Why All the Flag Waving? Because There's a War On," *HRIN*, August 1967, 22–23, and Jack Duffy, "War on High Performance," *HRIN*, May 1972, 18, 20, and 38–40.

14. Not until the mid-1960s would wholesale distributors begin to assume the vital role of middleman within the industry's distribution system. See above, chapter 6.

15. Period and secondary evidence is ambiguous regarding the name of this group; most refer to it as the "credit managers group" or "credit managers association," but two declare that it was in fact known as the "Speed Equipment Manufacturers Credit Association" (Madigan, *Edelbrock*, 190, and Dick Wells, "SEMA History," undated and unpublished manuscript, 7, in Dick Wells's private files, Santa Ana, California [hereafter, DWF]). However, because the bulk of the evidence clearly indicates that this was an informal group, it is unlikely that it ever had such a formal name.

16. Author Interview with Bob Spar, Newbury Park, California, November 13, 2003. See also Wells, "SEMA History," 7.

17. Wells, "SEMA History," 8–9; Author Interview with Carl Olson, Pomona, California, April 4, 2003; and Author Interview with Bob Spar, Newbury Park, California, November 13, 2003.

18. Bob Spar claims that this occurred at the NHRA Nationals (held every year in September) in 1961 or 1962 (Author Interview with Bob Spar, Newbury Park, California, November 13, 2003); Dick Wells, on the other hand, claims that it was actually in 1960 ("SEMA History," 8), as does Kevin C. Osborn ("The Specialty Equipment Market Association: A Synopsized History," unpublished manuscript [September 1988], History File, SEMA-RC [by 1988, the "Speed Equipment Manufacturers Association" had been renamed twice; hence, "Specialty" and "Market" in Osborn's title—see below, chapters 8 and 10]). In light of the balance of the evidence and the events that followed on the heels of the meet in question, Wells and Osborn are almost certainly right: the episode must have actually taken place at the 1960 Nationals in September of 1960.

19. On scattershields, a relatively new device in 1960 that became increasingly common as the 1960s wore on, see Eric Rickman, "Why Be Half Safe? Contain the Results of an Overly Enthusiastic Engine with an All Steel Bell Housing / Scattershield," *HRM*, September 1961, 44–45; Jack Hart, "Armorplate," *HRM*, October 1961, 88–89 and 112; and "Scattershields for Safety," *PHR*, February 1965, 63.

20. Author Interview with Bob Spar, Newbury Park, California, November 13, 2003, and Wells, "SEMA History," 8.

21. Author Interview with Bob Spar, Newbury Park, California, November 13, 2003; Wells, "SEMA History," 8–9; and Osborn, "The Specialty Equipment Market Association," 4.

22. Dick Wells, "Special to *SEMA News*—SEMA: Reasons for Being," undated and unpublished manuscript, 2, DWF, and Osborn, "The Specialty Equipment Market Association: A Synopsized History," 4–5.

23. "Industry History," anonymous and unpublished manuscript (2003), 2–3, History File, SEMA-RC; Osborn, "The Specialty Equipment Market Association," 5; and Author Interview with Bob Spar, Newbury Park, California, November 13, 2003.

24. "Industry History" (2003), 2–3.

25. Author Interview with Bob Spar, Newbury Park, California, November 13, 2003; Author Interview with Vic Edelbrock Jr., Torrance, California, November 19, 2003; and Fetherston, *Moon Equipped*, 20.

26. Each of SEMA's charter members paid yearly dues of only $25 in 1963; therefore, finances were extremely tight. See Osborn, "Specialty Equipment Market Association," 5, and "Original SEMA Members," undated and unpublished document, History File, SEMA-RC.

27. "Industry History" (2003), 3.

28. Author Interview with Bob Spar, Newbury Park, California, November 13, 2003. Osborn, "Specialty Equipment Market Association," 10–11; Wells, "Special to *SEMA News*—SEMA: Reasons for Being," 2; and "SEMA Technical Committee: Their Job's for Safety's Sake & Nothing Else!" *HRIN*, January 1967, 36 and 38.

29. "SEMA Technical Committee," and Wells, "Special to *SEMA News*—SEMA: Reasons for Being," 2.

30. Author Interview with Bob Spar, Newbury Park, California, November 13, 2003.

31. "SEMA Technical Committee."

32. "Original SEMA Members," and Author Interview with Bob Spar, Newbury Park, California, November 13, 2003.

33. Author Interview with Carl Olson, Pomona, California, April 4, 2003.

34. Ibid.

35. "Roddin' at Random," *HRM*, October 1965, 100.

36. Dick Wells, "Special to *SEMA News*—SEMA: Reasons for Being," and Bob Leif, "Meets SEMA Specs," *HRIN*, January 1968, 24–27 and 36–37. Organizationally, the new "SEMA Specs" program was a subcommittee of the SEMA Technical Committee, which continued to oversee its progress for several more years (see below, chapter 9). In 1967 SEMA also changed its name, replacing "speed" with "specialty" (see below, chapter 8).

37. Author Interview with Carl Olson, Pomona, California, April 4, 2003.

38. See, e.g., "News and Notes, *HRIN*, October 1967, 50–53; "SEMA's Wheel Committee," *HRIN*, January 1969, 66–67, 68, 76, 86, and 88; and "SEMA News," *HRIN*, August 1973, 36–37. See also below, chapter 8.

39. Osborn, "The Specialty Equipment Market Association," 10–11; "SEMA Scene," *HRM*, August 1978, 17; and Author Interview with Carl Olson, Pomona, California, April 4, 2003.

40. On the relationship many SEMA members saw between the successes of the "Specs" program and the association's growing membership, see Osborn, "The Specialty Equipment Market Association," 11. By 1972, one year shy of its ten-year anniversary, SEMA had 450 dues-paying members nationwide, more than ten times as many as it had at its inception (*SEMA: Serving the Automotive High Performance and Custom Industry*, informational pamphlet [1972], History File, SEMA-RC).

41. See below, chapter 8.

42. "SEMA: Coming of Age," *HRIN*, September 1966, 36–37.

43. On SEMA's health insurance program, see "News and Notes," *HRIN*, March 1967, 14, and "News and Notes," *HRIN*, August 1967, 20. On SEMA's "seminars," see Willie Garner, "Comments from the President—Why a Seminar?" *HRIN*, March 1968, 38–39. And on SEMA's efforts to better serve its East Coast constituents, see "News and Notes," *HRIN*, April 1967, 16–17, and "SEMA Indy Seminar," *HRIN*, October 1967, 46.

44. This was the official name of the show for its first few years; today, the week-long affair is officially known as "Automotive Aftermarket Industry Week."

45. See "Motoring Magic on Display," *HRM*, January 1951, 8–11 and 21–22 (on Motorama), and Author Interview with Vic Edelbrock Jr., Torrance, California, November 19, 2003 (on AAMA). Similar replacement-parts shows had long attracted a handful of speed equipment companies—Rajo, for one, sent representatives to man a booth at the annual Chicago Show in the 1920s ("Equipment Makers' Demand for Show Space Exceeds Any Previous Exhibit," *Motor Age*, 5 January 1922, 10–11).

46. See "Post Entry," *HRIN*, September 1966, 8 and 10; "Post Entry," *HRIN*, October 1966, 48–51; Ray Brock, "Publisher's Report," *HRIN*, September 1967, 6; and Alex Xydias, "Publisher's Report," *HRIN*, October 1968, 8.

47. Robert Petersen, "Welcome to the 1st Annual High Performance & Custom Trade Show," *HRIN*, January 1967, 28.

48. *Hot Rod Industry News* did have some competition: a magazine geared almost exclusively toward the *retail* (as opposed to the *manufacturing*) end of the industry known as *Speed & Custom Equipment News* had been launched by a rival company back in 1964. Petersen therefore needed a way to give *his* new publication a boost. See Ray Brock, "Publisher's Report," *HRIN*, May 1967, 6.

49. Author Interview with Dick Wells, Santa Ana, California, April 2, 2003.

50. Ibid.

51. Ibid., and "1967 High Performance and Custom Equipment Trade Show," *HRIN*, February 1967, 22.

52. Ray Brock, "Publisher's Report," *HRIN*, May 1967, 6, and "Announcing the SEMA Show," *HRIN*, June 1967, 22–23.

53. Ray Brock, "Publisher's Report," *HRIN*, February 1968, 8.

CHAPTER 8: "INK-HAPPY DO-GOODERS," 1960–1978

1. For an example of an early investigation of this haze, see N. Robert Heyer, "Smog Sleuths," *Popular Mechanics*, April 1949, 178–81, 252, and 254. For a more complete discussion of the smog research of the 1950s, see Krier and Ursin, *Pollution and Policy*, 6–8; Gins-

burg and Abernathy, *Government, Technology, and the Future of the Automobile*, 406–407; Volti, "Reducing Automobile Emissions in Southern California"; and below, chapter 9.

2. Rae, *American Automobile Industry*, 136–39.

3. Flink, *Car Culture*, 191.

4. Famous though this dictum was by 1960, Wilson never actually claimed that "what is good for General Motors is good for America." An executive at General Motors during the 1940s who was later nominated by Eisenhower to serve as Secretary of Defense, Wilson actually proclaimed, at his Senate confirmation hearings in 1953, that he believed that "what's good for the country is good for General Motors, and vice versa." Pundits immediately pounced on "vice versa," however, and soon, an extrapolated misquote became *the* quote. See Rae, *American Automobile Industry*, 107.

5. "Spotlight on Detroit," *Motor Trend*, October 1965, 6–9.

6. See, for example, Daniel P. Moynihan's "Epidemic on the Highways," which was originally published in 1959 in *The Reporter* before appearing verbatim in the general-interest automotive publication *Motor Life* in February 1960 (pp. 62–67); see also Flink, *Car Culture*, chap. 7.

7. On the AMA's racing ban, see above, chapter 5. On the VESC's origins, see Flink, *Automobile Age*, 290.

8. Krier and Ursin, *Pollution and Policy*, 6–8, and Ginsburg and Abernathy, *Government, Technology, and the Future of the Automobile*, 406–407.

9. See above, chapter 5.

10. Flink, *Car Culture*, 215.

11. See Krier and Ursin, *Pollution and Policy*, 6–9, 156–60, and 174–75; Robert W. Crandall et al., *Regulating the Automobile* (Washington, D.C.: The Brookings Institute, 1986), 85–86; and below, chapter 9.

12. Ibid.

13. Rae, *American Automobile Industry*, 136.

14. Ibid., 138–39 (on these pages, Rae includes a comprehensive table of highway fatalities and death rates, compiled from government data, that covers 1913–80).

15. An example of the former, at least in Daniel P. Moynihan's eyes, was the National Safety Council, a non-governmental group that meant well but whose morbid statistics and "emphasis on individual responsibility for accidents" tended to mislead the general public into believing that a few bad (and possibly drunken) apples were responsible for the escalating death rate (Moynihan, "Epidemic on the Highways," 63). An example of the latter, according to *Hot Rod*'s editor, Bob Greene, was the Commonwealth of Pennsylvania's Commissioner of Traffic Safety, O. D. Shipley, who published a lengthy editorial in the *Saturday Evening Post* in the spring of 1962 in which he claimed that motorsports—NHRA-style drag racing, in particular—fostered "competitive driving habits" and therefore contributed directly to the highway death toll (Greene, "The Editor Says," *HRM*, July 1962, 5, and Greene, "The Editor Says," *HRM*, August 1962, 5).

16. Nader, *Unsafe at Any Speed: The Designed-In Dangers of the American Automobile* (New York: Grossman, 1965).

17. See Flink, *The Car Culture*, 216, and Rae, *American Automobile Industry*, 136–137. On the shortcomings of Nader's *Unsafe at Any Speed* as well as his later book, *The Volks-*

wagen: An Assessment of Distinctive Hazards (Washington, D.C.: Center for Auto Safety, 1971), see John Tomerlin, "Ralph Nader vs. Volkswagen: A *Road & Track* Report," *Road & Track*, April 1972, 25–33. Unlike Nader, of course, *Road & Track*'s bias leaned toward the automobile industry.

18. Rae, *American Automobile Industry*, 137; Flink, *Automobile Age*, 384; and Crandall et al., *Regulating the Automobile*, 47–49. Flink dates the GSA's original standards to model-year 1967, which means that they must have been in place by early 1966, while Rae implies that they were already established and well known *before* Ralph Nader's book first hit the shelves in 1965. Either way, the GSA's standards definitely preceded—and directly informed—NHTSA's 1966 requirements.

19. Robert Herzberg, "Washington Report," *HRIN*, July 1968, 8. Herzberg was the first Washington, D.C. correspondent for *Hot Rod Industry News*.

20. Robert Herzberg, "Washington Report," *HRIN*, April 1967, 12; Robert Herzberg, "Washington Report," *HRIN*, September 1968, 10; and "Inside Detroit: Used Car Safety," *Motor Trend*, November 1968, 11. To this day, however, a handful of states still have no annual inspections.

21. See for example Rae, *American Automobile Industry*, chap. 11; Flink, *Car Culture*, chap. 7; and Flink, *Automobile Age*, chap. 20.

22. Crandall, *Regulating the Automobile*, 47–49, and Earl W. Kintner, "SEMA Bulletin on Safety Legislation," *HRIN*, July 1967, 53–54 and 56.

23. On NHTSA's proposal to mechanically limit automobile speeds, see Robert Herzberg, "Washington Report," *HRIN*, May 1967, 10; Don Evans, "Editorially Speaking," *HRM*, February 1971, 6; Robert Herzberg, "Washington Report," *HRIN*, June 1971, 10; and "Legislation," *HRIN*, May 1977, 14–15. On muscle car premiums in the late 1960s and early 70s, see Don Evans, "Editorially Speaking," *HRM*, December 1969, 8, and Alex Xydias, "Publisher's Report," *HRIN*, April 1970, 8 and 63. On the re-emergence of rodding's rogue image, see Robert Herzberg, "Washington Report," *HRIN*, September 1966, 18 and 20, and "Post Entry," *HRM*, October 1969, 12, 14, 16, and 18. On the ways the rodding fraternity combated this image in the past, see above, chapter 3.

24. A more complete discussion of this issue follows later in this chapter.

25. See above, chapter 3.

26. Bob Greene, "The Editor Says," *HRM*, May 1966, 5.

27. John P. Keelan, M.E., P.E., "Post Entry: It Can't Happen Here—Can It?" *HRM*, September 1966, 6 and 8.

28. See, e.g., Huntington, "Don't Let Legislation Wipe Out Hot Rodding," 42–44, and Jim McCraw, "Editorially Speaking," *HRM*, April 1975, 10.

29. "Rodding Readers: Do-Gooder!" *PHR*, November 1972, 10 and 16; see also "Rodding Readers: Don't Worry?" *PHR*, October 1972, 6 (a milder critique).

30. See, e.g., Jim McCraw, "Editorially Speaking," *HRM*, September 1975, 7; John G. Rako, "Hot Rodders' Bill of Rights," *HRM*, March 1976, 12; Alex Xydias, "Publisher's Report," *HRIN*, June 1971, 4; and "Rodding Readers: More Sport Saving," *PHR*, March 1972, 6 and 15.

31. See especially Almquist, *Hot Rod Pioneers*, 265, and Moorhouse, *Driving Ambitions*, chap. 6.

32. One could easily count on one hand the number of times SEMA appeared in the enthusiast publications before 1966.

33. On the NHRA, the UDRA, and the AHRA in the 1950s and 60s, see Post, *High Performance*, chap. 7.

34. "*HRIN* Interview: SEMA President Garner," *HRIN*, August 1967, 30 (Garner was elected SEMA president in 1967).

35. "News and Notes," *HRIN*, February 1968, 38, and Osborn, "The Specialty Equipment Market Association," 13.

36. Wells, "Special to *SEMA News*—SEMA: Reasons for Being," 2.

37. Ibid.; "News and Notes," *HRIN*, June 1967, 48; and "SEMA Indy Seminar," *HRIN*, October 1967, 46.

38. Earl W. Kintner, "SEMA Bulletin on Safety Legislation," *HRIN*, July 1967, 53–54 and 56.

39. Throughout the 1960s and well into the 1970s, street racing remained one of the dirty little secrets of the high-performance community (and it remains one to this day, particularly among the tuner crowd). See Rob Ross, "The Subterranean World of Los Angeles Street Racing," *UCLA Daily Bruin*, 15 December 1965, and "Big Willie . . . King of the Street," *Drag Racing Magazine*, December 1968, 42–47.

40. "*HRIN* Interview: SEMA President Garner," *HRIN*, August 1967, 31.

41. See, e.g., Bob Leif, "Manufacturers Speak: Questions and Answers on Topics of Industry Interest," *HRIN*, May 1968, esp. 51.

42. Don Francisco, "Hot Rodding, SEMA, and You," *Drag News*, 13 October 1967, 19. Francisco was well-known among rodders, particularly for the detailed articles he wrote during a lengthy tenure as *Hot Rod*'s technical editor.

43. Robert Herzberg, "Washington Report," *HRIN*, December 1967, 14 and 48.

44. "News and Notes," *HRIN*, October 1967, 50–53; Bob Leif, "Meets SEMA Specs," *HRIN*, January 1968, 24–27 and 36–37; and Author Interview with Bob Spar, Newbury Park, California, November 13, 2003.

45. "*HRIN* Interview: Dale Herbrandson on the SEMA Specs Program," *HRIN*, September 1970, 22, 42–43 and 48–49, and Author Interview with Bob Spar, Newbury Park, California, November 13, 2003.

46. SEMA's Technical Committee continued to administer the "Specs" program for another couple of years, however ("SEMA Section," *HRIN*, May 1970, 60, and "*HRIN* Interview: Dale Herbrandson on the SEMA Specs Program"). Some maintain that SEMA's board of directors formalized the program in 1967 (Wells, "Special to *SEMA News*—SEMA: Reasons for Being," 2), and indeed, Bob Leif's use of the phrase "Meets SEMA Specs" in his January 1968 *HRIN* article on the program suggests that Wells is right. On the other hand, evidence from *Hot Rod Industry News* in 1970 suggests that the program was in fact formalized earlier *that* year ("SEMA Section," *HRIN*, May 1970, 60, and "*HRIN* Interview: Dale Herbrandson on the SEMA Specs Program"). In any event, what is clear is that by the beginning of 1970, at the very latest, "Meets SEMA Specs" was a formal, stand-alone project under the larger organizational umbrella of SEMA.

47. "Industry History" (2003), 4, and "*HRIN* Interview: SEMA President Garner," *HRIN*, August 1967, 30.

48. "Why All the Flag Waving? Because There's a War On," *HRIN*, August 1967, 22–23.

49. For examples of SEMA's late-1960s recruitment efforts, see Willie Garner, "Comments from the President," *HRIN*, July 1968, 40–41, and "SEMA Section," *HRIN*, June 1970, 46. On SEMA's decision to loosen its membership requirements, see Alex Xydias, "Publisher's Report," *HRIN*, August 1969, 6.

50. *SEMA: Serving the Automotive High Performance and Custom Industry*, informational pamphlet (1971), History File, SEMA-RC.

51. "Why All the Flag Waving," 23.

52. Eric Grant, "The High Performance Market Today and in the '70s," *HRIN*, November 1970, 24; "Outlook for 71," *HRIN*, January 1971, especially 65–66, 72, and 94; A. B. Shuman, "Editorially Speaking," *HRM*, October 1971, 6; and "Who's Doing What for HP?" *HRIN*, July 1972, 26–28.

53. See "SEMA Policy Statement," *HRIN*, November 1972, 26, and below, chapter 9.

54. Robert Herzberg, "Washington Report," *HRIN*, May 1967, 10.

55. Robert Herzberg, "Washington Report," *HRIN*, February 1971, 10 and 35. Toms, a Nixon appointee, replaced Haddon in 1969.

56. For some examples of the OEMs' response to Toms's idea, see Robert Herzberg, "Washington Report," *HRIN*, March 1971, 12 and 32, and Robert Herzberg, "Washington Report," *HRIN*, June 1971, 10.

57. "Rodding Readers: 1973—a 95 mph Limit?" *PHR*, March 1971, 12.

58. "SEMA Section," *HRIN*, April 1971, 34–35, and Robert Herzberg, "Washington Report," *HRIN*, July 1971, 14 and 35.

59. Ultimately, all that came out of this idea was a decision made in the late 1970s to limit speedometer readings on new cars to eighty-five mph. See "Legislation," *HRIN*, May 1977, 14–15.

60. "SEMA Section: DOT's Toms and Ford's Jensen 'Tell It Like It Is' at SEMA Luncheon," *HRIN*, March 1971, 38.

61. James M. Beggs, "The Administration Speaks: Remarks by U.S. Under Secretary of Transportation James M. Beggs before the Specialty Equipment Manufacturers Association Annual Meeting," *HRIN*, February 1972, 10 and 34.

62. "Hot Rod Reports on the SEMA Scene: Doug Toms Named SEMA's Man of the Year," *HRM*, March 1972, 22.

63. See below, chapter 9.

64. "Roddin' at Random," *HRM*, June 1968, 112; "News and Notes," *HRIN*, January 1969, 118 and 124; Osborn, "The Specialty Equipment Market Association," 13; Author Interview with Carl Olson, Pomona, California, April 4, 2003; and Author Interview with Bob Spar, Newbury Park, California, November 13, 2003. Roulston's résumé was impressive: he had been the editor of the popular periodical *Car Craft*, he had served as the publicity director for the NHRA, and he had also spent some time in the promotions department at a colossal Pennsylvania-based equipment manufacturer, Hurst Performance Products.

65. See below, chapter 9.

66. Author Interview with Vic Edelbrock Jr., Torrance, California, November 19, 2003.

67. Author Interview with Carl Olson, Pomona, California, April 4, 2003, and Osborn, "The Specialty Equipment Market Association," 13.

68. "News and Notes," *HRIN*, October 1967, 52, and "SEMA's Wheel Committee," *HRIN*, January 1969, 66–68, 76, 86, and 88.

69. See "Hot Rod Reports on the SEMA Scene," *HRM*, November 1972, 38. Earlier iterations of the 5-1 Spec had simply required participating firms to submit their own ongoing testing data for the purposes of annual renewal.

70. "SEMA Bulletin," *HRIN*, April 1973, 67–68.

71. "The SEMA Scene: New Quality Control Program Assures Safer Custom Wheels," *HRM*, July 1973, 38.

72. See, e.g., "SEMA News," *HRIN*, August 1973, 36, and "SEMA Test Lab: Another Step Forward," *HRM*, October 1973, 38.

73. On Toms's departure from NHTSA (and SEMA's response), see Terry Cook, "Editorially Speaking," *HRM*, March 1973, 10–12, and "The SEMA Scene: If Only Somebody in Washington Understood Us . . .," *HRIN*, April 1973, 40.

74. "The SEMA Scene: New Quality Control Program Assures Safer Custom Wheels."

75. Russ Deane, "Legislation," *HRIN*, November/December 1977, 32, 34, 36, and 43.

76. Author Interview with Dick Wells, Santa Ana, California, April 2, 2003, and Author Interview with Carl Olson, Pomona, California, April 4, 2003.

77. For example, SEMA dealt with a seatbelt-retrofit proposal and a drive-by noise proposal in California in the fall of 1967 (Ray Brock, "Publisher's Report," *HRIN*, December 1967, 6), and it also began to deal with the Golden State's Motor Vehicle Pollution Control Board earlier that same year (see below, chapter 9).

78. Willie Garner, "Comments from the President," *HRIN*, July 1968, 41.

79. "Post Entry: SEMA Speaks," *HRM*, June 1968, 10.

80. Quoted in A. B. Shuman, "Editorially Speaking," *HRM*, October 1971, 6.

81. See, e.g., "Post Entry," *HRM*, April 1969, 12 and 14; "Post Entry," *HRM*, October 1969, 12, 14, 16, and 18; A. B. Shuman, "Editorially Speaking," *HRM*, October 1971, 6; "The SEMA Scene," *HRM*, October 1971, 124; "Post Entry," *HRM*, April 1972, 18 and 22; and Fred Gregory, "PPC Editorial," *HRM*, May 1973, 16–17.

82. See Roger Huntington, "Don't Let Legislation Wipe Out Hot Rodding," esp. 44. Perhaps this reluctance on the part of the editors of *Popular Hot Rodding* to endorse the SEMA program had something to do with SEMA's close ties their major rival, Robert Petersen of the Petersen Publishing Company.

83. "SEMA News," *HRIN*, May 1973, 58–59 and 62, and "Outlook '74," *HRIN*, November 1973, 60, 68, and 76.

84. "The SEMA Scene," *HRM*, January 1972, 110.

85. "The SEMA Scene: Hot Rods May Get the Axe on Wednesday," *HRM*, February 1973, 32; "SEMA News," *HRIN*, May 1973, 58–59 and 62; and "Outlook '74," *HRIN*, November 1973, 60, 68, and 76.

86. On the origin and evolution of these Pennsylvania problems (and the failure of SEMA officials and organized enthusiasts, through the winter of 1972, to open a meaningful dialogue with Penn-DOT on them), see "Hot Rod Reports on the SEMA Scene," *HRM*, February 1972, 124; Steve Green, "The Pennsylvania Crisis!" *HRM*, May 1972, 26–27; and "Rodding Readers: The Right Way," *PHR*, January 1973, 11. A "rod run" is an organized, often long-distance cruise event in which dozens of enthusiasts drive their rods to a designated location (and back).

87. On the revamped VESC's duties, see for example "SEMA News," *HRIN*, September 1973, 28, and "The SEMA Scene," *HRM*, November 1974, 20.

88. "SEMA News," *HRIN*, September 1973, 28, and "The SEMA Scene," *HRM*, November 1973, 22.

89. Terry Cook, "Editorially Speaking," *HRM*, May 1974, 10 and 127.

90. "The SEMA Scene," *HRM*, May 1974, 26.

91. "The SEMA Scene," *HRM*, November 1974, 20.

92. On the resolution of the Pennsylvania crisis (and Brainard's role in it), see "The SEMA Scene," *HRM*, November 1973, 22, and "The SEMA Scene—Update: The Pennsylvania Crisis . . . a Brighter Picture," *HRM*, December 1973, 30.

93. See *SEMA: Serving the Automotive High Performance and Custom Industry*, informational pamphlet (1972), History File, SEMA-RC.

94. On the NSRA, see Author Interview with Dick Wells, Santa Ana, California, April 2, 2003, and below, chapter 10.

95. "SEMA News," *HRIN*, September 1973, 28; Terry Cook, "Editorially Speaking," *HRM*, May 1974, 10 and 127; and Author Interview with Dick Wells, Santa Ana, California, April 2, 2003.

96. The same was true of noise-related difficulties in Illinois in 1973, 1974, and 1975 (see "The SEMA Scene: SEMA's Good Guys—Who They Are," *HRM*, September 1973, 32; Fred M. H. Gregory, "Big Noise from Illinois," *HRM*, January 1974, 54–55; "HRIN Opinion: Panic Turns to Progress," *HRIN*, May 1974, 24; "The SEMA Scene: Positive Progress Announced on Illinois Noise Problem," *HRM*, June 1974, 32; and "SEMA News," *HRIN*, Winter 1975, 66, 68, and 74).

97. "Legislative Front," *HRIN*, November 1974, 50.

98. See Lou Baney's comments in "Performance Industry Leaders Project Business Trends for the Coming Year," *HRIN*, Winter 1975, 12, and also "SEMA News," *HRIN*, March 1975, 14–15.

99. *SEMA '78*, informational pamphlet (1978), History File, SEMA-RC; "SEMA News," *HRIN*, May 1977, 10–13; and Author Interview with Dick Wells, Santa Ana, California, April 2, 2003.

CHAPTER 9: "THIS DREADFUL CONSPIRACY," 1966–1984

1. The phrasing of Section 27156 appeared frequently both in the minutes of the MVPCB (later restructured as CARB, the California Air Resources Board) and in the pages of *Hot Rod*, *Hot Rod Industry News*, *Speed and Custom Equipment Dealer*, and *Popular Hot Rodding*. However, the clearest, most accessible summary of its original wording (as well as that of its later, much-amended iterations) appears in the minutes of a CARB meeting that took place in August 1977 (State of California, "Meeting Summary: Air Resources Board," August 25, 1977, www.arb.ca.gov/board/mi/mi.htm [accessed March 2000]).

2. For additional details on this meeting (and the reactions of those who attended it), see Jim McFarland, "SEMA Scene," *HRM*, November 1979, 17 and 98; Jim McFarland, "Clean Air Costs Money! How Fast . . . ?" *HRIN*, July 1971, 19–20 and 22–23; and Vic Edelbrock Jr., "Political Activity," *HRIN*, June 1971, 24–27 and 38.

3. Krier and Ursin, *Pollution and Policy*, 6–9, 156–60, and 174–75; Crandall, *Regulat-*

ing the Automobile, 85–86; Rae, *American Automobile Industry*, 133–35; and Flink, *Automobile Age*, 386–87.

4. Ibid. See also Volti, "Reducing Automobile Emissions in Southern California," 281.

5. Jim McFarland, "SEMA Scene," *HRM*, November 1979, 17 and 98; McFarland, "Clean Air Costs Money"; and Vic Edelbrock Jr., "Political Activity," *HRIN*, June 1971, 24–27 and 38.

6. Robert Herzberg, "Washington Report," *HRIN*, October 1966, 10 and 14 (quote).

7. See above, chapter 8.

8. Ray Brock, "Publisher's Report," *HRIN*, February 1967, 6, and Ray Brock, "Publisher's Report," *HRIN*, March 1967, 6.

9. Joseph M. Callahan, "What Smog Control Will Cost You," *Motor Trend*, September 1967, 60–61 and 70–73; Ray Brock, "Publisher's Report," *HRIN*, March 1967, 6; *Motor Trend* staff, "Who's Kidding Who? Part 1," *HRIN*, July 1970, 24 and 68–69; and Tex Smith, "Performance-Tuning the 'Smoggers,'" *PHR*, April 1971, 46. Incidentally, the OEMs agreed among themselves to install PCVs on all cars sold in the United States in 1963, five years before the federal government would require them to do so.

10. Eric Dahlquist, "One for the Road," *HRM*, May 1966, 32–35; Callahan, "What Smog Control Will Cost You"; and *Motor Trend* staff, "Who's Kidding Who? Part 1."

11. Even then, EGR was phased in gradually: not until 1972 would all cars sold in California be equipped with EGR devices. See Callahan, "What Smog Control Will Cost You"; "Over the Counter 'Clean Air,'" *HRIN*, June 1973, 48–53; and C. J. Baker, "Emissions Systems," *HRM*, July 1974, 112, 114, 116, and 118.

12. Ray Brock, "Publisher's Memo," *HRM*, March 1967, 6.

13. Ray Brock, "Publisher's Report," *HRIN*, February 1967, 6; Ray Brock, "Publisher's Report," *HRIN*, March 1967, 6; and Ray Brock, "Publisher's Memo," *HRM*, March 1967, 6.

14. Ray Brock, "Publisher's Report," *HRIN*, March 1967, 6.

15. Ray Brock, "Publisher's Report," *HRIN*, November 1967, 6.

16. See Ray Brock, "Publisher's Report," *HRIN*, March 1967, 6; Dennis Pierce, "Information File," *HRIN*, April 1967, 45–46; and "*HRIN* Interview: SEMA President Garner," *HRIN*, August 1967, 30–31. For some evidence of the stance of the popular periodicals on these matters, see Ray Brock, "Publisher's Memo," *HRM*, March 1967, 6; "Detroit's Hot Cars for '68," *PHR*, September 1967, 30–33; and Don Francisco, "Hot Rodding, SEMA, and You," *Drag News*, 13 October 1967, 19.

17. See, for example, "Detroit's Hot Cars for '68," *PHR*, September 1967, 30–33, and Ray Brock, "Publisher's Memo," *HRM*, March 1967, 6.

18. "Why All the Flag Waving? Because There's a War On," *HRIN*, August 1967, 22–23.

19. Noel Carpenter, "The Industry Scene," *HRIN*, May 1973, 8. *Hot Rod Industry News* was a closed-circulation magazine that only went out to members of the industry (see above, chapter 7). Consequently, 18,500 could be—and, in Carpenter's case, was— taken as a *rough* estimate of the size of the industry (manufacturers, speed shops, wholesale distributors, and other retailers combined).

20. Quoted in Author Interview with Carl Olson, Pomona, California, April 4, 2003. On Baney's rise within the association, see, e.g., "News Makers," *HRIN*, February 1971, 12;

"Hot Rod Reports on the SEMA Scene," *HRM*, October 1972, 34–35; and "Performance Industry Leaders Project Business Trends for the Coming Year," *HRIN*, Winter 1975, 12.

21. See "Performance Industry Leaders Project Business Trends"; *SEMA '76: Get in the Spirit*, informational pamphlet (1976), History File, SEMA-RC; and *SEMA '78*, informational pamphlet (1978), History File, SEMA-RC.

22. Krier and Ursin, *Pollution and Policy*, 178–79. This resolution took effect in the winter of 1967–68.

23. Ibid. Eventually, the federal government decided to allow other states to *adopt* California's stricter measures if they wanted to, but to this day, only California has the power to *develop* guidelines stricter than those of the EPA.

24. Ray Brock, "Publisher's Memo," *HRM*, December 1969, 6.

25. Krier and Ursin, *Pollution and Policy*, 204.

26. Quoted in ibid., 204.

27. Ibid., 204–207. The engine-modification prohibition was to begin immediately (though an enforcement mechanism had yet to be established), but the warranty provision was to take effect at the discretion of the EPA ("Late News File: Washington, Detroit, the World," *HRIN*, February 1973, 20–21, and "The SEMA Scene," *HRM*, June 1973, 38).

28. See Jerry M. Flint, "Auto Industry, Changing Strategy, Opens Counterattack on Environmental and Consumer Movements," *New York Times*, 18 November 1970, C29; Jonathan Spivak, "Battle Looms between U.S., Auto Firms over New Proposals to Limit Pollution," *Wall Street Journal*, 27 July 1970, A2; and Flink, *Automobile Age*, 376–403.

29. "Board Meeting Minutes + Attachments," February 17, 1971, 1–2, "Agendas, Resolutions, Minutes, 1971" binder, ARB Archive, California EPA Headquarters, Sacramento, California (hereafter, ARB-A).California remained free to set its own pollution-control standards, but they had to be at least as strict as the federal standards, and for all of their work during the course of the 1960s to clean up the air in Southern California, Haagen-Smit recognized that the statutory requirements of the federal Clean Air Act of 1970 were far stricter than the ARB's had ever been. Hence his concern.

30. Spivak, "Battle Looms"; Flint, "Auto Industry, Changing Strategy"; Spencer Rich, "More Time Asked on Auto Pollution," *Washington Post*, 19 November 1971, A3; "Why Your New Car Stalls," *New York Times*, 2 April 1972, 28; Burt Schorr, "Pipe Dream?" *Wall Street Journal*, 14 December 1972, 1 and 31; "Poor Automobile Fuel Economy Linked to Pollution Controls," *New York Times*, 25 March 1973, 54; "Mileage vs. Pollution," *Wall Street Journal*, 15 May 1974, 27; and Flink, *Automobile Age*, 387–88.

31. "Late News File: Washington, Detroit, the World," *HRIN*, June 1973, 9–12 (on Honda); "Review and Outlook: Shut Down Motown!" *Wall Street Journal*, 18 October 1976, 24; Bernard Asbell, "The Outlawing of Next Year's Cars," *New York Times Magazine*, 21 November 1976, 219; and "Legislation," *HRIN*, April 1977, 20 (quote).

32. Eric Grant, "The High Performance Market Today and in the '70s," reprinted in *HRIN*, November 1970, 20, 24, and 26–27.

33. Quoted in "SEMA Section," *HRIN*, March 1971, 38.

34. Vic Edelbrock Jr., "Political Activity," reprinted in *HRIN*, June 1971, 26–27 (emphasis in original).

35. For a complete discussion of the leaded-fuel phase-out and its repercussions for the hot rodding fraternity, see below, chapter 10.

36. McFarland, "Clean Air Costs Money!"

37. "Super Tarantula," *PHR*, September 1971, 46–47; "Industry Notes: Edelbrock Emits," *HRIN*, October 1971, 32; Edelbrock advertisement, *PHR*, September 1972, 2; and Steve Green, "Clean and Mean: Who says you can't put horsepower into the air . . . and still breathe?" *HRM*, November 1971, 152. See also Offenhauser advertisement ("Clean, Mean, and Legal Dual Port 360"), *PHR*, October 1973, 4, and C. J. Baker, "Brave New Manifold," *HRM*, November 1974, 59.

38. Don Prieto, "Clean Air Engine," *HRIN*, December 1971, 23–27 (quote on 25).

39. At the time, the "7-mode test" was the standard test used by the EPA and the ARB on new vehicles. This particular 1968-model Corvette was unregulated in terms of NOx emissions, so Weggeland and Morris only tested its CO and HC levels. See "Corvette Clean-Up," *HRIN*, July 1972, 24 (quotes) and 44, and Steve Green, "Clean-Air Corvette," *HRM*, September 1972, 112–113.

40. "Build a Low-Emission Street Engine," *PHR*, January 1973, 24–25; "Build a Low-Emission Hi-Performance Chevy Engine," *PHR*, February 1973, 36–39 and 107; "Build a Low-Emission Hi-Performance Chevy Engine, Part 3," *PHR*, April 1973, 50–51; and "87 HP Chevy Hop-Up!" *PHR*, July 1973, 24–27. See also Steve Kelly, "Rated G," *HRIN*, June 1973, 46–47.

41. "87 HP Chevy Hop-Up!" These figures reflect the project engine's performance compared with California's strictest requirements for 1966–69 models, those for cars with air-pumps. Compared with the slightly less-restrictive standards set for 1966–69 California cars without air-pumps, the project engine did even better, producing 66 percent fewer HC and 88.6 percent fewer CO emissions.

42. Ibid. Here, the legal maximum was 350 ppm HC and 4.0 percent CO.

43. Ibid. These numbers reflect a comparison of the project engine's HC and CO output with California's strictest, air-pump–equipped new-vehicle standards for 1972–1973.

44. Vic Edelbrock Jr., "Acceptance Speech," reprinted in *HRIN*, September 1971, 34 and 38–41, and Vic Edelbrock Jr., "Hot Rod Reports on the SEMA Scene," *HRM*, June 1972, 28.

45. Excerpts from this statement were published in July 1972 ("Who's Doing What for HP?" *HRIN*, July 1972, 26–28), and the entire declaration was reprinted in "SEMA Policy Statement," *HRIN*, November 1972, 26.

46. "Who's Doing What for HP?"

47. "Late News File," *HRIN*, February 1973, 20–21.

48. California Air Resources Board, "Policy on Replacement Parts," May 16, 1973, "Minutes, 1973" binder, ARB-A. See also "The SEMA Scene: Some Good News for a Change," *HRM*, August 1973, 26.

49. "The SEMA Scene: Some Good News for a Change."

50. "SEMA Bulletin," *HRIN*, April 1973, 67–68. This five-year, 50,000-mile emissions warranty should not be confused with the ordinary new-car warranties offered by most manufacturers, for they are separate contracts.

51. "The SEMA Scene: Aftermarket Parts to be Illegal," *HRM*, June 1973, 38 (this is a summary discussion of the entire episode published several months after the fact).

52. "SEMA Bulletin," *HRIN*, April 1973, 67–68.

53. See Vic Edelbrock Jr., "SEMA News," *HRIN*, July 1973, 36 and 39–40, and "The SEMA Scene: Legislative Outlook for 1974—Good News or Bad?" *HRM*, January 1974, 30.

54. "SEMA News," *HRIN*, October 1974, 32–33, and "SEMA News: SEMA Vehicle Combination Emission Testing Program," *HRIN*, November 1974, 62.

55. Jim McCraw, "The Four Most Important Used Cars in California," *HRM*, September 1975, 88–90 and 92 (emphasis in original).

56. See above, chapter 7, and Author Interview with Carl Olson, Pomona, California, April 4, 2003.

57. McCraw, "The Four Most Important Used Cars in California."

58. The results of phase two were summarized in Donna Imrie, "SEMA Scene," *HRM*, October 1976, 31.

59. Also present at these meetings were representatives of another organization, the Automotive Service Industry Association, a standard-replacement-parts group that also lobbied the EPA on the warranty issue in the mid-1970s. See M. S. Greicus, *Automotive Service Industry Association, The First Twenty-Five Years: Study of an Association, History of the Aftermarket* (Chicago: Automotive Service Industry Association, 1984), 81–82.

60. By the mid-1970s, a number of cities and states required periodic emissions inspections, and SEMA fought to make sure that these programs were performance- rather than design-based.

61. "SEMA News," *HRIN*, July 1977, 13–19 and 58, and Russ Deane, "Legislation," *HRIN*, November/December 1977, 32, 34, 36, and 43. This 1977 act made it illegal for OEMs to void an *emissions* warranty if that customer installed non-OEM parts (whether standard-duty or high-performance), but it did not prevent the OEMs from voiding *new-car* warranties for the same reason. In the 1980s, the Magnusson-Moss Warranty Act made it illegal for the OEMs to void new-car warranties for the use of replacement OEM-style parts as well, although this later act does *not* prevent the OEMs from doing so when high-performance products are involved (Ray and Tom Magliozzi, "Car Talk: High-Performance Replacement Parts Can Void Warranty," *Atlanta Journal-Constitution*, 7 July 2000, S1).

62. Some have claimed that this warranty victory was SEMA's *first* regulatory triumph (see Osborn, "The Specialty Equipment Market Association: A Synopsized History," 13, and Wells, "Special to *SEMA News*—SEMA: Reasons for Being," 2–3). However, this is true only if we require that a victory be *legislative* in nature for it to count, for by 1977, SEMA had already enjoyed a number of *interpretive* triumphs in its dealings with NHTSA, the VESC, the MVPCB, the ARB, and the EPA.

63. Recall that the ARB's 1973 ruling was temporary; as of 1977, the board had yet to finalize its plans for an aftermarket parts-certification program.

64. State of California, "Meeting Summary: Air Resources Board," August 25, 1977, archived online at www.arb.ca.gov/board/mi/mi.htm (accessed March 2000).

65. Russ Deane, "Legislation," *HRIN*, November/December 1977, 32, 34, 36, and 43; "SEMA Scene: Four Problem Areas," *HRM*, January 1978, n.p.; and Author Interview with Dick Wells, Santa Ana, California, April 2, 2003. See also Pat Ganahl, "Can They Outlaw Hot Rods?" *HRM*, October 1984, 22–26, 28, 33–34, 36, and 112–13.

66. See, e.g., "Washington Mail," *HRM*, June 1968, 124; Bill Hahn, "Post Entry: Onward and Upward," *HRM*, April 1969, 12 and 14; "The SEMA Scene," *HRM*, July 1974, 28; and "The SEMA Scene," *HRM*, August 1974, 32.

67. See Miles Brubacher, "Motorists United: Save Fun Cars," *PHR*, March 1972, 82 and 84–85, and "Motorists United: Will Hot Rods Be Nader's Next Target?" *PHR*, July

1972, 100. Brubacher—who along with SEMA's Eric Grant was once with the MVPCB—headed Motorists United.

68. See, e.g., NSRA advertisement, *Street Rodder*, January 1977, 66.

69. By 1977, a variety of new noise and safety regulations had cropped up across the country, many of which were very tough on modified vehicles. SEMA therefore boosted its state-level lobbying in an attempt (at which it was successful in the majority of cases) to secure rodder-friendly modifications to these new laws. In addition, restrictive emissions-inspection programs had emerged in certain areas of the country, and these also required attention. All of this placed an enormous financial strain on SEMA's state-level system for the management of local issues; hence the formation of the Enthusiast Division. See "SEMA News," *HRIN*, May 1977, 10–13.

70. According to an update from the summer of 1977, the Enthusiast Division was attracting new members at a rate of only 150 per month—less than 6 percent of the monthly figure of 2,667 required to achieve Wells's projection ("SEMA News," *HRIN*, July 1977, 13–19 and 58).

71. Beginning in the late 1960s, the California Highway Patrol did perform emissions spot-checks on vehicles stopped for moving violations, but this was by no means a systematic enforcement program. For more on California's lax enforcement of 27156 prior to 1984, see Ganahl, "Can They Outlaw Hot Rods?"

72. "SEMA Scene: Four Problem Areas," *HRM*, January 1978, n. p.

73. See "SEMA Scene," *HRM*, December 1978, 92; Donna Imrie, "Post Entry," *HRM*, October 1979, 13; and Lee Kelley, "Editorially Speaking, *HRM*, January 1981, 4.

74. See above, chapter 7, and below, chapter 10.

75. Lee Kelley, "Editorially Speaking," *HRM*, February 1980, 4.

76. "Board Meeting Minutes + Attachments," February 16, 1977, 1, "Minutes, 1977" binder, ARB-A.

77. Dave Wallace, "Endangered Species—Annual Inspections: Clean Air or Dirty Deal?" *HRM*, November 1980, 48–50; Kevin Boales, "The Politics of Clean Air," *HRM*, March 1984, 98–99; and Ganahl, "Can They Outlaw Hot Rods?"

78. Lee Kelley, "Editorially Speaking," *HRM*, January 1981, 4. In 1979, SEMA changed its name again, substituting "Market" for "Manufacturers." See below, chapter 10.

79. McFarland, "Clean Air Costs Money!" and "SEMA Bulletin," *HRIN*, April 1973, 67–68. Performance-based tests were favored by the speed equipment industry because they measured the level of tailpipe pollution and nothing else; aftermarket parts were perfectly acceptable as long as the vehicle's emissions were within the legal limits. By 1980, performance-based inspection programs had been established in a number of other states, and SEMA hoped that California would follow suit.

80. Boales, "The Politics of Clean Air," and Ganahl, "Can They Outlaw Hot Rods?"

81. The exact phrasing of this warning varied, but its basic thrust remained the same (the wording quoted here is from Edelbrock advertisement, *HRM*, March 1982, 3). For more on these warnings, see Boales, "The Politics of Clean Air," and "Post Entry," *HRM*, July 1984, 13.

82. Again, the exact phrasing varied; this particular example is from Performance Automotive Wholesale, Inc. advertisement, *HRM*, August 1985, 84–85. By the early 1990s, some firms had combined the two, resulting in shorter warnings like "Legal in California only for racing vehicles which may never be used upon the highway" (Car Custom advertise-

ment, *Dune Buggies and Hot VWs*, December 1991, 104–105) or "Legal in California for racing vehicles which may never be used upon the highway" (Fast Freddy's advertisement, *VW Trends*, March 1992, 41).

83. Author Interview with Delores Berg, Orange, California, November 11, 2003.

84. See "Bug Mail," *Dune Buggies and Hot VWs*, November 1983, 7–8, and "Bug Mail," *Dune Buggies and Hot VWs*, November 1984, 10.

85. John Baechtel, "Street Outlaw: 12-Second Suspects That Skirt the Law with Technology," *HRM*, January 1986, 16 and 19–23.

86. Scores of EO exemptions were issued by the end of the 1980s, according to a comprehensive list of EO numbers included in a reference volume known as *The Black Book* that SEMA makes available to its members for $150 (State of California Air Resources Board, "Modifications to Motor Vehicle and Emissions Control Systems exempted Under Vehicle Code Section 27156" [1994], reprinted in SEMA, *The Black Book*, 2nd ed. [Diamond Bar, Calif.: SEMA, 1996], i–xxxiv and 1.1–25.1, SEMA-RC).

87. Leonard S. Reich, "Ski-Dogs, Pol-Cats, and the Mechanization of Winter: The Development of Recreational Snowmobiling in North America," *Technology and Culture* 40 (1999): esp. 514.

88. Spivak, "Battle Looms Between U.S., Auto Firms."

89. Here, I am appropriating the words that Moorhouse uses to describe the hot rodding fraternity's experience with government regulation (*Driving Ambitions*, 141), for they are far more appropriate for the OEMs' experience than for that of the speed equipment industry.

90. Rae, *American Automobile Industry*, 135.

91. "Post Entry: No New Z for Me," *HRM*, March 1982, 6–8.

92. See above, chapter 3.

93. On this "elemental" form of enthusiasm, see Ferguson, "Enthusiasm and Objectivity in Technological Development"; Ferguson, "Elegant Inventions"; Post, *High Performance*, chap. 14; and above, Introduction.

94. Hughes, *American Genesis*, 3.

95. Examples of this sort of justification abound, e.g., see "Detroit Never Satisfies Them," *Popular Mechanics*, June 1948, 114–20, and Kenneth Kincaid, "If Detroit Won't Do It, Why Don't You?" *Motor Trend*, June 1952, 32–34 and 46–47.

96. Hughes, *American Genesis*, 3.

97. Whether the average American's faith in technological progress actually vanished is an open question. Certainly, Congress did not lose its faith. After all, vital to the federal approach to the air pollution crisis was the assumption that, somehow, the OEMs would find a technological "fix" that would enable them to meet the government's standards.

98. See, e.g., "SEMA Scene—Your Engine Compartment: Off-Limits?" *HRM*, February 1978, 9, and "Ruckelshaus: 'The Public Must Start Paying,'" *Business Week*, 24 February 1973, 62–64.

CHAPTER 10: THE BEST OF TIMES, THE WORST OF TIMES, 1970–1990

1. In November 1978, SEMA's board of directors voted 17–2 to replace "Manufacturers" in the SEMA acronym with "Market"; see Donna Imrie, "SEMA-Gram," *Specialty &*

Custom Dealer, January 1979, 36 and 38. On the growth of the speed equipment industry in the 1970s and 80s, individually and collectively, see "Hot Rod Industry News 1971 Industry Survey," *HRIN*, August 1971, 26–32; "Outlook for '72," *HRIN*, January 1972, esp. 30, 31, and 135–36; Edelbrock advertisement, *HRIN*, January 1973, 20; "Performance Industry Leaders Project Business Trends for the Coming Year," *HRIN*, winter 1975, esp. 12–14 and 59; "Roddin' at Random," *HRM*, April 1976, 115; "1977 Performance Industry Survey," *HRIN*, April 1977, 31–38; and (on the 1980s) Specialty Equipment Market Association, "2002 Automotive Specialty Equipment Industry Update," 1, SEMA-RC. On the growth of SEMA during the 1970s, see *SEMA: Serving the Automotive High Performance and Custom Industry*, informational pamphlet (1972), History File, SEMA-RC; "Performance Industry Leaders Project Business Trends for the Coming Year," *HRIN*, winter 1975, esp. 12; Al Dowd, "Publisher's Note," *HRIN*, July 1977, 3; *SEMA '78*, informational pamphlet (1978), History File, SEMA-RC; and "SEMA Ready to Build New Office," *Specialty & Custom Dealer*, January 1979, 42. On the growth of the SEMA Show (which moved from Anaheim, California, to Las Vegas, Nevada, in 1977), see "Outlook for '72," *HRIN*, January 1972, esp. 31; Alex Xydias, "Publisher's Report," *HRIN*, January 1973, 4 and 20; "Performance Industry Leaders Project Business Trends for the Coming Year," *HRIN*, winter 1975, 12; and Al Dowd, "Publisher's Note," *HRIN*, July 1977, 3.

2. On the cautious optimism of the 1970s, see "Outlook for '72," *HRIN*, January 1972, especially the comments by Vic Edelbrock Jr. (p. 30), Roy Richter (p. 31), Joe Cornelske (pp. 32, 36, and 134), Harvey Crane Jr. (pp. 134–35), and Richard J. Deney (pp. 135–36); "Detroit Hotline," *HRIN*, February 1974, 18; "Performance Industry Leaders Project Business Trends for the Coming Year," *HRIN*, winter 1975, especially the comments by Els Lohn (p. 59) and Bob Spar (pp. 122–23); Alex Xydias, "HRIN Opinion," *HRIN*, summer 1976, 4; and Lee Kelley, "Editorially Speaking," *HRM*, December 1978, 5–7.

3. Almquist, *Hot Rod Pioneers*, 324–25. See also Moorhouse, *Driving Ambitions*, 141, and Terry Cook, "Editorially Speaking," *HRM*, May 1973, 8 and 10.

4. "Rodding Readers: Do Something!" *PHR*, July 1972, 12.

5. See, e.g., "Post Entry," *HRM*, April 1972, 18 and 22; Fred M. H. Gregory, "Big Noise from Illinois," *HRM*, January 1974, 54–55; "The SEMA Scene," *HRM*, June 1974, 32; "Scare Tactics," *HRM*, July 1977, 23; and Dave Wallace, "Hot Rodders—An Endangered Species?" *HRM*, July 1979, 52–53.

6. On the horsepower excise-tax proposals, see "Washington Report," *HRIN*, August 1970, 8 and 49; "White House May Seek to Cut Gasoline Use by Cutting Horsepower, Hoisting Gas Tax," *Wall Street Journal*, 21 December 1972, 3; and "Post Entry," *HRM*, April 1972, 18 and 22. On the automobile-limit and internal-combustion phase-out proposals, see "Roddin' at Random," *HRM*, September 1975, 8, and "SEMA Scene," *HRM*, May 1978, 21.

7. Fred M. H. Gregory, "PPC Editorial: Organized Racing—It's Time to Organize," *HRM*, November 1973, 8–9; Steve Alexander, "Fair Treatment for Auto Racing," *HRIN*, February 1974, 24–26; and John Baechtel, "Down the Road," *HRM*, March 1983, 16.

8. For some examples of this sense of disappointment and abandonment, see any of the enthusiast-penned installments of the 1976 "Hot Rodders' Bill of Rights" series published in *Hot Rod*.

9. Ray Brock, "Publisher's Memo," *HRM*, July 1970, 6. See also above, chapter 7.

10. See Miles Brubacher, "Motorists United: Save Fun Cars," *PHR*, March 1972, esp. 82; "Roddin' at Random: End of the V8?" *HRM*, August 1977, 18; "Roddin' at Random," *HRM*, April 1978, 22–23; Dave Wallace, "New Cars 1980: *Hot Rod*'s Guide to Survival in the 'Efficiency Eighties,'" *HRM*, October 1979, 65–69; and "Roddin' at Random," *HRM*, June 1981, 21.

11. "California's New Hard Line on Smog," *Business Week*, 7 April 1975, 25, and Dave Wallace, "Hot Rodders: An Endangered Species?" *HRM*, June 1979, 67–70.

12. See "A Healthy Aspen," *HRM*, February 1977, 66–68; "The Thrill is Back," *HRM*, October 1977, 62–65; Lee Kelley, "Editorially Speaking," *HRM*, June 1980, 4; Gray Baskerville, "Getting Sideways," *HRM*, December 1980, 21–26; "Chevy V6 Bolt-on Power," *HRM*, March 1982, 30–32; and "4 Bangers," *HRM*, August 1986, 20–21. *Popular Hot Rodding* ran similar stories; see Leonard Emanuelson, "Hopping Up Buick's V6," *PHR*, January 1976, 26–30, 35, and 84, and Alex Walordy, "Ford's Hemi Four . . . A Look Inside," *PHR*, January 1981, 24–27 and 77–78.

13. See, e.g., "Post Entry," *HRM*, March 1982, 6–8, which consists of enthusiasts' negative reactions to *Hot Rod*'s claim (C. J. Baker, "The Best One Yet!" *HRM*, December 1981, 20–23) that the 1982 Z-28 Camaro was the best Camaro ever to hit the streets; see also "Reaction Time," *HRM*, August 1986, 8–11, which includes enthusiasts' negative reactions to *Hot Rod*'s claim (Rick Titus, "Shelby: The GLHS Surpasses the Legend," *HRM*, April 1986, 22–25) that Carroll Shelby's new GLHS, based on a compact Dodge Omni, was a better performer than the legendary Shelby GT350 Mustang of the 1960s.

14. California Air Resources Board, "Board Meeting Minutes + Attachments," March 18, 1970, 2–9, "Minutes, 1968 thru 1970" binder, ARB-A; California Air Resources Board, "Board Meeting Minutes + Attachments," January 20, 1971, 9–16, "Agendas, Resolutions, Minutes, 1971" binder, ARB-A; Kent Carlton, "Can Hot Rodding Survive?" *PHR*, January 1971, 18–19 and 76–79; and Ronald Chiswell, "Tech Tips: Lead Blues," *PHR*, January 1972, 88.

15. On the introduction of the catalytic converter, see Ray Brock, "Publisher's Memo," *HRM*, July 1970, 6; Flink, *Automobile Age*, 388; and Volti, "Reducing Automobile Emissions in Southern California," 283–86.

16. "Roddin' at Random: Leaded Fuel Ban," *HRM*, August 1984, 24; "Pit Stop: Leading Us Astray?" *HRM*, January 1985, 75; and "News Briefs," *HRM*, May 1988, 19. Leaded fuel was only banned from highway use; to this day, high-octane racing fuels often contain lead.

17. On the quality of late 1960s gasoline, see "Tech Tips," *PHR*, April 1969, 106 (in this piece, *PHR*'s technical editor advised that the premium leaded fuels of 100 to 103 octane that were widely available at the time were more than adequate for high-compression muscle cars; because the methods used to rate fuel quality in the late 1960s differed somewhat from those that came into widespread use in the 1970s [see above, chapter 5, note 21], I have taken the liberty of adjusting the 1960s "100 to 103" figure that appears in this 1969 article to "97 to 101" in order to render it comparable with the other figures I cite). On the best-available unleaded fuels of the 1970s and 80s, see California Air Resources Board, "Board Meeting Minutes + Attachments," March 18, 1970, esp. 2–9, "Minutes, 1968 thru 1970" binder, ARB-A; Robert Herzberg, "Washington Report," *HRIN*, October 1971, 12; David Vizard, "Sky High," *Dune Buggies and Hot VWs*, February 1980, 70–73; and David

Vizard, "The Octane Dilemma," *PHR*, October 1981, 34–36, 77–78, and 80–81. For an excellent retrospective look at octane requirements and pump-gas ratings, see Jeff Hartman, "Engine Octane Number Requirement (ONR)," *European Car*, October 1997, 45.

18. Flink, *Automobile Age*, 383. See also Fred M. H. Gregory, "PPC Editorial—Here We Go Again: More Car Laws," *HRM*, October 1973, 8–9, and "Legislation," *HRIN*, April 1977, 20.

19. See Ronald Chiswell, "Tech Tips: Lead Blues," *PHR*, January 1972, 88, and "Detroit Hotline," *HRIN*, August 1974, 12. Today, 94 octane is available in the Northeast and the Midwest, but for much of the country, 93 octane unleaded remains the best that money can buy. Stations in much of the American West, however, only offer "91 pump piss," in the words of *European Car*'s Pablo Mazlumian and Les Bidrawn ("E36 M3 Exhaust Test," *European Car*, January 2003, 84).

20. See "Bench Racing: Big Foot Vega," *Super Chevy*, July–August 1980, 5; "Post Entry," *HRM*, December 1980, 11; "C. J. Talks Tech," *HRM*, March 1981, 116; "Tech Talk," *Dune Buggies and Hot VWs*, August 1981, 22 and 78; and "Backfire: No-Lead Crisis," *Car Craft*, May 1989, 8.

21. "1975 Performance Industry Survey," *HRIN*, Summer 1975, 37–44; and "1976 Performance Industry Survey," *HRIN*, Summer 1976, 31–38.

22. For some examples of 1940s and 50s water-injection systems, see "What's New," *HRM*, July 1949, 29, and Vapojet advertisement, *Popular Mechanics*, December 1949, n.p.

23. On Edelbrock's system, see "Vara-Jection," *PHR*, September 1980, 28–30; Edelbrock advertisement, *HRM*, October 1980, 13; and Edelbrock advertisement, *HRM*, June 1981, 92. On Crower's system, see Crower advertisement, *Super Chevy*, July 1981, 75. And on Holley's system, see Holley advertisement, *Super Chevy*, October 1981, 76.

24. See "The SEMA Scene," *HRM*, February 1974, 36; Roger Huntington, "Economy and Performance: It Can Be Done," *PHR*, March 1974, 80–83 and 108; Gray Baskerville, "Selling the Economy-Performance Package—Part One," *HRIN*, August 1974, 34–35; Baskerville, "Selling the Economy-Performance Package—Part Two," *HRIN*, September 1974, 30–31; and Baskerville, "Selling the Economy-Performance Package—Part Three," *HRIN*, October 1974, 30–31.

25. See, e.g., Edelbrock advertisement, *HRM*, February 1974, 6; "A Manifold for All Reasons," *Drag Racing USA*, April 1974, 59–61; C. J. Baker, "The Great Header Emission, Mileage, and Noise Flap," *HRM*, May 1974, 35–37; and Edelbrock advertisement, *HRM*, January 1975, 14.

26. See "Energy Crisis," *HRIN*, January 1974, 16–21; Iskenderian advertisement, *HRM*, January 1975, 26; and Crane advertisement, *HRM*, June 1980, 49.

27. Terry Cook, "Editorially Speaking," *HRM*, May 1973, 8 (emphasis in original).

28. "SEMA Scene—Your Engine Compartment: Off Limits?" *HRM*, February 1978, 9.

29. Jim McFarland, "The SEMA Scene," *HRM*, November 1979, 98 (emphasis in original).

30. See Roger Huntington, "'71 'Super' Car Power Boost," *PHR*, February 1971, 68–71; Edelbrock advertisement, *HRM*, April 1975, 24; and above, chapter 9.

31. "Mini Buyers' Guide," *HRIN*, May 1973, 40–45; Will Hertzberg, "Industry Profile: Cannon Industries," *HRIN*, April/May 1974, 28–29; "New Products: Mini-Cars," *HRIN*,

April 1977, 58; C. J. Baker, "Just One Look," *HRM*, October 1980, 34–35; C. J. Baker, "Chevy V6 Bolt-On Power," *HRM*, March 1982, 30–32; and above, chapter 9.

32. See Ray Brock, "Publisher's Memo," *HRM*, October 1970, 6; C. J. Baker, "Henry's Hemi," *HRM*, July 1980, 59–60; and Marlan Davis, "Pinto Ponies," *HRM*, March 1981, 75–79.

33. Davis, "Pinto Ponies," 75. See also Terry Cook, "Editorially Speaking," *HRM*, May 1972, 6; Lee Kelley, "Editorially Speaking," *HRM*, June 1980, 4; and Leonard Emanuelson, "Editorially Speaking," *HRM*, August 1983, 6. Speed equipment manufacturers often needed to be reminded of this too; see "Late News File: Washington, Detroit, The World," *HRIN*, May 1972, 13.

34. See Bruce Crower, "Hot Rod Forum: Turbocharging—The Only Game in Town for Small Engines (or, Keeping the Go-Fast Syndrome Alive)," *HRM*, December 1980, 116, and "4 Bangers," *HRM*, August 1986, 20–21. Equipment manufacturers were often told the same thing; see John Christy, "Small Cars Are Big Business," *HRIN*, May 1973, 56–57 and 72.

35. Leonard Emanuelson, "Hopping Up Buick's V6," *PHR*, January 1976, 26–30, 35, and 84; Crower, "Hot Rod Forum: Turbocharging," 116; Walordy, "Ford's Hemi Four"; and Titus, "Shelby: The GLHS Surpasses the Legend." See also Bud Lang, "Project Anglia Makes a Comeback," *HRM*, August 1974, 90, and "Roddin' at Random," *HRM*, October 1980, 20–21.

36. See Alex Xydias, "Publisher's Report," *HRIN*, November 1972, 6; Steve Kelly, "Calling It," *HRIN*, May 1973, 10; and Noel Carpenter, "The Industry Scene," *HRIN*, August 1973, 6. See also "1977 Performance Industry Survey," *HRIN*, April 1977: as of 1976, compact or "mini car" parts accounted for only 5 percent of the industry's aggregate wholesale-level sales (p. 36).

37. "Roddin' at Random," *HRM*, April 1977, 20.

38. On the rise of Honda, Toyota, and Datsun, see Flink, *Automobile Age*, 327.

39. Terry Cook, "Editorially Speaking," *HRM*, May 1972, 6; "What's New From . . .," *HRIN*, June 1972, 56–57; and Hertzberg, "Industry Profile: Cannon Industries."

40. See above, chapter 5.

41. Steve Coonan, "Across 49th Street," *Street Rodder*, November 1976, 68–69, and "The 12-Second VW Hits the Street," *Drag Racing USA*, January 1975, 16–19. See also Kevin Clemens, "Forward Progress: Thirty Years of Automotive Technology," *European Car*, December 1999, 44–49, esp. 46.

42. Clemens, "Forward Progress," 46.

43. By the end of the 1970s, most of the Rabbits sold in the United States were produced, not in Germany, but in Westmoreland, Pennsylvania (Flink, *Automobile Age*, 325).

44. John Rettie, "VW Rabbit GTI," *Dune Buggies and Hot VWs*, December 1982, 38–41 and 66.

45. See "Second Generation Parade," *Dune Buggies and Hot VWs*, June 1979, 44–47 and 82–83; "VW Tuner Timeline: 20 Years of Tweaks," *VW Trends*, January 1997, 82–85, esp. 82 and 83; and Clemens, "Forward Progress," 46–47.

46. "Second Generation Parade," 44–47, and "VW Tuner Timeline," 82–83.

47. John Rettie, "16-Valve VW Head," *Dune Buggies and Hot VWs*, October 1980, 33. On the Okrasa kit, see above, chapter 5.

48. See "VW Tuner Timeline," 82–83, and above, chapter 5.

49. "VW Tuner Timeline," 82–83. On Offenhauser, Meyer, and Drake, see above, chapter 4.

50. "Parts Palace," *Dune Buggies and Hot VWs*, February 1980, 62.

51. "VW Tuner Timeline," 82–84.

52. See John T. Faraczelle, "Hot Rodders' Bill of Rights," *HRM*, February 1976, 12; "1977 Van Market Survey," *HRIN*, May 1977, 26–29; and "Exclusive Consumer Report to *HRIN* Readers," *HRIN*, November/December 1977, 94, 96, and 98.

53. On van meets, see for example Phil Carpenter, "West Coast Van Nationals," *PHR*, January 1974, 62–65 and 81; "National Truck-In," *PHR*, November 1975, 24–29 (east-of-the-Mississippi vanners often referred to their vehicles as "trucks"; see Terry Cook, "Remember Butler," *HRM*, December 1974, 50–52); and Robert K. Smith, "Truckin'," *Street Rodder*, November 1976, 66–67. For a couple of particularly choice examples of how-to and project-van pieces, see Phil Carpenter, "Vantastic Van Tricks!" *PHR*, July 1974, 56–59, and Jay Amestoy, "Introducing Project Van," *PHR*, August 1975, 74–77 and 87.

54. Noel Carpenter, "The Industry Scene," *HRIN*, August 1973, 6, and Carl Olson, "SEMA Scene," *HRM*, May 1976, 12. See also "1975 Performance Industry Survey," *HRIN*, summer 1975, 37–44.

55. See, e.g., Dan Harms, "Rodding Readers: Not Performance Oriented?" *PHR*, July 1976, 7 and 9, and John G. Dzwonczyk, "Post Entry: Are Vans Hot Rods?" (letter no. 1), *HRM*, January 1977, 9.

56. John C. Freitsch, "Post Entry," *HRM*, April 1977, 15.

57. Cliff Aston, "Post Entry: Are Vans Hot Rods?" (letter no. 2), *HRM*, January 1977, 9. See also Kenneth Hamm, "Post Entry," *HRM*, March 1977, 10.

58. Leonard Emanuelson, "Editorially Speaking," *HRM*, June 1984, 8 (emphasis in original).

59. Indeed, 1980s compact street-truck enthusiasts were often part of the lowrider crowd that Brenda Jo Bright analyzes in "Style and Identity," "Mexican American Low Riders," and *Customized.*

60. See Robert Genat and Robin Genat, *Hot Rod Nights: Boulevard Cruisin' in the USA* (Osceola, Wis.: Motorbooks International, 1998), 74–75; Doug Price, "Street Nationals News," *PHR*, February 1973, 108–11; and "*Hot Rod Magazine* Street Machine Market Survey," *HRIN*, June 1977, 46–47.

61. Martin Austin, "Hot Rodders' Bill of Rights," *HRM*, July 1976, 7; "*HRIN* Predicts," *HRIN*, June 1977, 3; and "*Hot Rod Magazine* Street Machine Market Survey," 46–47.

62. "1975 Performance Industry Survey," *HRIN*, summer 1975, 37–44; "1976 Performance Industry Survey," *HRIN*, summer 1976, 31–38; and "1977 Performance Industry Survey," *HRIN*, April 1977, 31–38.

63. DeWitt, *Cool Cars, High Art,* esp. 112; Ganahl, *American Custom Car,* 9; and Timothy Remus, *Custom Cars and Lead Sleds: America's Best Customs, '50s–'90s* (Osceola, Wis.: Motorbooks International, 1990), 8.

64. DeWitt, *Cool Cars, High Art,* 113.

65. See, e.g., Dave Wallace, "Graffiti Nights," *HRM*, November 1982, 24–26 and 28–29, and Pat Ganahl, "Tough Sledding," *HRM*, May 1983, 84–86. See also *Hot Rod's* June 1983 cover spread ("'50s Flashback: Classic Chevys and Early Fords").

66. On the decline of the dual-use rod, see above, chapters 3–6. For more on "cool rods," see Dean Batchelor, "The New Hot Rods," *True's Automobile Yearbook*, 1952, 30–31 and 98–99; Eugene Jaderquist, "Hot Rods Are Doomed," *Cars*, November 1953, 30–33 and 62; and "One T-bone: RARE . . . but well done!" *Rod and Custom*, November 1954, 22–25.

67. Medley, *Tex Smith's Hot Rod History—Volume Two*, 102, and Author Interview with Dick Wells, Santa Ana, California, April 2, 2003.

68. See, e.g., *Street Rod Pictorial* advertisement, *HRM*, December 1974, 20; "'50 Phaeton," *HRM*, March 1977, 74–75; Gray Baskerville, "Resemblance of Things Past," *HRM*, December 1983, 74–76; and Author Interview with Dick Wells, Santa Ana, California, April 2, 2003.

69. On the street rod–street machine feud, see Leonard Emanuelson, "Editorially Speaking," *HRM*, October 1983, 3; Leonard Emanuelson, "Editorially Speaking," *HRM*, May 1984, 6; and Joe Mayall, "NSRA News," *Street Rodder*, January 1986, 27.

70. See "Hot Rod Magazine Street Rod Market Survey," *HRIN*, June 1977, 20–21; "Hot Rod Magazine Street Machine Market Survey," *HRIN*, June 1977, 46–47; and "Exclusive Consumer Report to HRIN Readers," *HRIN*, November/December 1977, 94, 96, and 98. The average age of a street-machine enthusiast in 1976 was 28.

71. Because the original customs of the 1950s were in his view a form of modern art, DeWitt interprets their recreation in the early 1980s as a form of postmodern expression (*Cool Cars, High Art*, 118). Likewise, because they represented an attempt to recreate (with the technology of the 1970s and 80s) the cool rods of the 1950s, early 1970s street rods were also postmodern.

72. This sketch is based on data from "Hot Rod Magazine Street Rod Market Survey," *HRIN*, June 1977, 20–21, as well as from "Exclusive Consumer Report to HRIN Readers," *HRIN*, November/December 1977, 94, 96, and 98. See also "Smooth Operator," *PHR*, September 1971, 44–45; Don Emmons, "Four for a Vicky," *1001 Custom and Rod Ideas*, December 1976, 24–25; and "Plans of Action," *HRM*, February 1977, esp. 62.

73. On "repro rods," see Gray Baskerville, "How to Build a Repro Street Rod," *HRM*, October 1977, 47–51; Bruce Crower, "Hot Rod Forum," *HRM*, November 1981, 95; Leonard Emanuelson, "Editorially Speaking," *HRM*, June 1982, 2; Pat Ganahl, "Old Gold," *HRM*, July 1984, 95–99; and Dan Burger and Robert Genat, *Retro Rods* (Osceola, Wis.: MBI Publishing Co., 2001). On "repro customs," see Gray Baskerville, "Newstalgia," *HRM*, July 1977, 39 and 41–42. On cruise nights, see Wallace, "Graffiti Nights," and Ganahl, "Tough Sledding."

74. Medley, *Tex Smith's Hot Rod History—Volume Two*, 105, and Author Interview with Dick Wells, Santa Ana, California, April 2, 2003. On the increasing scarcity of genuine prewar bodies (and the resulting demand for fiberglass reproductions), see Andrew Morland, *Street Rods: Pre-'48 American Rods in Color* (London: Osprey Publishing, 1983), 5. Fiberglass bodies and body panels weren't new in the 1960s, of course. Chevrolet's Corvette had always been fiberglass-bodied, for example, and fiberglass panels were available from a number of aftermarket firms by the early 1950s (see, e.g., Thalco advertisement, *Hop Up*, January 1953, 10).

75. See Gray Baskerville, "New Old Stuff," *HRM*, November 1976, 48–50 and 52; Terry Cook, "New Rails for the Model A," *PHR*, January 1977, 58–61; Tom Senter, "Street Rod

Chassis Buyer's Guide," *PHR*, May 1977, 60–63, 92, 96, and 100; Baskerville, "Bolt-On Bonanza," *HRM*, August 1978, 90–92, 94, and 96–100; Baskerville, "How to Build a Repro Street Rod"; and "Repro Report," *HRM*, March 1982, 42–50.

76. See "Piecemeal," *HRM*, April 1977, 94 and 96–97, and Mike Johnson, "Project Lo-Buck Lo-Boy," *HRM*, February 1985, 18–19 and 21–25. See also Bob Leif, "Kit Cars," *HRIN*, August 1967, 24–27, 31, and 40–41.

77. Lawrence Donald, "9 Reasons Why You Should Be in the Street Rod Business," *HRIN*, June 1977, 22–29. See also Bud Bryan, "Your Target . . . The Rodder," *HRIN*, July 1973, 12; Noel Carpenter, "The Industry Scene," *HRIN*, October 1973, 8 and 39; and John Duke, "Editor's Note," *HRIN*, June 1977, 4.

78. On street rod meets, see, e.g., Ed O'Brien, "'72 Street Rod Nationals," *PHR*, October 1972, 22–25, 82, 84, and 86; Bud Lang, "Northwest Mini-Nats," *HRM*, December 1974, 98–100; "Dixie Nats," *Street Rodder*, August 1976, 14–18; and "*Hot Rod* Supernationals," *HRM*, September 1982, 19–25.

79. On the early origins of the NSRA, see Medley, *Tex Smith's Hot Rod History— Volume Two*, 105–109, and Author Interview with Dick Wells, Santa Ana, California, April 2, 2003.

80. Author Interview with Dick Wells, Santa Ana, California, April 2, 2003.

81. See above, chapter 8.

82. Doug Price, "Street Rod Nationals," *PHR*, February 1973, 108–11.

83. Import enthusiasts eventually took to the Honda during the 1990s, but during the 1970s and 1980s, it was not a viable option for the performance-minded.

84. On the emergence of computerized ignitions, carburetors, and knock sensors, see General Motors, "News For Release: Monday, February 28, 1977" (my thanks to Victor McElheny for providing me with a copy of this press release); Cam Benty, "The Return of High-Compression Trucking," *HRM*, August 1981, 76–77; Martin Wingate, "Post Entry: Turn of the Tide," *HRM*, January 1982, 7; Marlan Davis and Jeff Tann, "SEMA: What's New for '82," *HRM*, February 1982, 30–33; and Marlan Davis, "SEMA: What's to Be in '83," *HRM*, February 1983, 26–29.

85. See "Injecting Fuel into the VW," *Business Week*, 23 September 1967, 44; "The 1968 SEMA Show," *HRIN*, February 1968, 28; Shuler, *Origin and Evolution*, 146 and 173; and *Gasoline Fuel-Injection System L-Jetronic* (Stuttgart: Robert Bosch GmbH, 1997), 1.

86. On GM's TBI system, see C. J. Baker, "Throttle Body Fuel Injection: Improvement or Impairment?" *HRM*, October 1981, 46–48. On the development of multi-port systems at GM, see John Baechtel, "Down the Road," *HRM*, September 1983, 16.

87. See, e.g., Leonard Emanuelson, "Editorially Speaking," *HRM*, August 1981, 6; Emanuelson, "Editorially Speaking," *HRM*, January 1982, 2; and John Baechtel, "Down the Road," *HRM*, February 1983, 14.

88. Roger E. Crawfork, "Tech Talk," *Dune Buggies and Hot VWs*, July 1983, 21.

89. For an accessible period discussion of this matter, see Jim McFarland, "*PHR* Fundamentals: On-Board Computer Performance," *PHR*, June 1987, 12.

90. Edelbrock advertisement, *HRM*, February 1975, 6; John Baechtel, "SEMA Parts Preview: Holley Fuel Injection," *HRM*, November 1985, 33; Rick Voegelin, "In the Chips: A Hot Rodder's Guide to Electronic Fuel Injection," *HRM*, November 1986, 36–40; Doug Marion, "Computer Electronics for '60s Performance," *Super Chevy*, January 1983,

64–65 and 76; Pat Ganahl, "Computer Hot," *HRM*, February 1985, 32–33, 35, and 39; and Ganahl, "Computer Check," *HRM*, March 1985, 74.

91. Baker, "Throttle Body Fuel Injection"; Ganahl, "Computer Hot"; Hypertech advertisement, *Car Craft*, August 1986, 32; Coast Performance Engineering advertisement, *HRM*, October 1986, 113; Voegelin, "In the Chips"; and Dave Emanuel, "Champion Chips," *HRM*, July 1987, 62, 64, 68–69, 71, and 128. See also James Sly, "Plug-in Legal Performance," *European Car*, November 1991, 76–77.

92. Emanuel, "Champion Chips."

93. Voegelin, "In the Chips," 36.

94. On this inherent limitation of P-ROMs, see Baker, "Throttle Body Fuel Injection"; Ganahl, "Computer Hot"; Voegelin, "In the Chips"; McFarland, "*PHR* Fundamentals"; and Emanuel, "Champion Chips." On the emergence of more versatile chips in the late 1980s, see, e.g., Jeff Smith, "Starting Line," *HRM*, May 1988, 7.

95. "Tips and Tricks," *HRM*, July 1988, 32, and "Chip Shot," *HRM*, March 1989, 21. See also John Baechtel, "Street Outlaws: 12-Second Suspects That Skirt the Law with Technology," *HRM*, January 1986, 16 and 19–23.

CONCLUSION

1. "2002 Automotive Specialty Equipment Industry Update," 1, SEMA-RC, and "Market Watch," *SEMA News*, August 1992, 22–23.

2. Sales of speed equipment for street machines accounted for only 45 percent of the industry's aggregate total for 1976, down from 60 percent the previous year and 61 percent in 1974. See "1975 Performance Industry Survey," *HRIN*, summer 1975, 37–44; "1976 Performance Industry Survey," *HRIN*, summer 1976, 31–38; and "1977 Performance Industry Survey," *HRIN*, April 1977, 31–38.

3. Ibid. See also John Duke, "Editor's Note," *HRIN*, May 1977, 8; "2002 Automotive Specialty Equipment Industry Update," 1; and above, chapter 10.

4. Lee Kelley, "Editorially Speaking," *HRM*, June 1980, 4.

5. "Reaction Time," *HRM*, August 1986, 10.

6. This is the substance of another GLHS naysayer's response (ibid.).

7. On VW's sixteen-valve engine, see James Sly, "8V or 16V?" *European Car*, March 1992, 30–31 and 110; Autothority advertisement, *European Car*, March 1992, 117; James Sly, "VW Performance Options," *European Car*, April 1992, 32–36, 98, and 102; and "Volkswagen Passat," *Road & Track*, October 1992, 110.

8. On the VR6, see James Sly, "Volkswagen VR-6," *European Car*, November 1991, 88–89 and 91; "1992 Corrado SLC," *European Car*, April 1992, 52–56; and Michael Jordan, "Volkswagen Corrado SLC," *Automobile*, August 1992, 56–61.

9. See, e.g., Les Bidrawn, "Giant Killers," *European Car*, May 1996, 36–39; Les Bidrawn, "Twin-Turbo VR6 New Beetle," *European Car*, May 2003, 70; and Les Bidrawn, "VR6 Rabbit," *European Car*, May 2003, 72.

10. On the VR6 GTI as an overweight and less-than-nimble sled (even in the eyes of those who adored it), see James Sly, "1.8T," *European Car*, June 2000, 44–46 and 48–51.

11. For a brief overview of the VTEC system and its development, see Mike Ancas, *Honda & Acura Performance Handbook* (St. Paul: Motorbooks International, 1999), chap. 2.

12. The names of some of the more popular import-tuner magazines bear this out: *Import Tuner* (1998–present), for example, focuses on Japanese brands, while *European Car* (1991–present, formerly *VW & Porsche*), *Eurotuner* (2002–present, formerly *Max Speed*), and *VW Tuner* (1999–present) focus on European makes.

13. "2002 Automotive Specialty Equipment Industry Update," 6.

14. By the early 2000s, one no longer needed to replace an OEM ECU chip to modify its parameters. This is because most new cars were by then equipped with ECUs that could be reprogrammed via federally-mandated on-board diagnostics (OBD) ports (the government mandated these ports in the early 1990s to enable emissions inspectors to scan in-use ECUs for evidence of emissions-related failures). Thus, by plugging a laptop or hand-held computer into a vehicle's OBD port, rodders can now upload recalibrated ECU maps directly into their cars' black boxes without ever turning a single screw or melting a single drop of solder. For an accessible look at flash tuning, see Hubert Zimara, "TT Tuning," *Eurotuner*, January 2007, 88–90.

15. Jim McFarland, "The SEMA Scene," *HRM*, November 1979, 98. See also above, chapter 10.

16. Franz, *Tinkering*, 163.

17. Even among 1970s street rodders—whose cars were by definition among the most radically transformed, mechanically and aesthetically, in the history of the activity—more than 90 percent built their cars themselves or, at most, with the help of a friend ("Hot Rod Magazine Street Rod Market Survey," *HRIN*, June 1977, 21). For similar, more recent numbers, see "Market Watch," *SEMA News*, August 1992, 22.

GLOSSARY

advance: slang for *ignition advance.*

air cooling: a method of regulating engine temperatures with air, often through the use of a belt-driven fan.

balance: to ensure, through an iterative process of careful weighing and machining, that each of an engine's pistons (and connecting rods and crankshaft throws) are of equal weight, in order to reduce destructive engine vibrations at higher rpms.

bank: slang for *cylinder bank.*

barrel: 1. slang for a carburetor throttle body; 2. slang for an engine cylinder, particularly on air-cooled models in which the cylinders are not cast as integral parts of the engine block.

engine bay: the space within the body of a car in which the engine is installed.

beefed-up: slang for "strengthened."

bellhousing: a large, bell-shaped case that fits between the transmission and the flywheel-end of the engine block and houses the clutch and pressure plate.

bent eight: slang for V8.

big block: describes a given automobile company's larger class of V8 engines, typically those displacing more than 350 cubic inches.

bench racing: that which occurs when two or more enthusiasts get together and talk about races (especially drag races) that they have run, that they have seen run, that they would like to run, or that they would like to see run.

Big Four: once-common slang for the mainstream American automobile industry that includes American Motors, Chrysler, Ford, and General Motors.

Big Three: common slang for the mainstream American automobile industry that includes Chrysler, Ford, and General Motors.

block: see *engine block.*

blower: slang for *supercharger.*

blown: 1. describes an engine equipped with a supercharger; 2. describes an engine that has ceased to function, typically in a catastrophic manner.

boards: slang for the wood-surfaced racetracks that were common in the 1910s to 1940s.

boost: slang for the pressurized intake air that superchargers and turbochargers generate.

bore: 1. the diameter of an engine's cylinders; 2. the internal diameter of a throttle body; 3. to increase the size of an engine's cylinders.

bolt-on: describes speed equipment that can easily be added to a production engine with simple hand tools—i.e., equipment that can be added without performing major and/or irreversible modifications to an engine. For example, a high-performance intake manifold is almost always considered "bolt-on," while a stroked crankshaft, which might require engine-block machining and clearancing, typically is not.

breathe: slang for the process whereby an internal combustion engine consumes an incoming charge of fuel and air and subsequently expels the burned exhaust gases.

buff: slang for "enthusiast."

build-up: slang for "engine project."

bump up: slang used to refer to increases in engine compression or displacement.

bumpstick: slang for "camshaft."

cam: slang for "camshaft."

carbureted: an engine equipped with one or more carburetors (as opposed to fuel injection).

carburetion: refers to the carburetor (or the system of carburetors) used on an engine.

charge: slang for the incoming fuel-air mixture on an internal combustion engine.

clock: 1. slang for the act of measuring the speed of a vehicle, especially during a dry lakes race or a drag race (e.g., "my rod was clocked at 121.57 mph"); 2. the device used to measure the speed of a racing vehicle.

combustion chamber: the cavity in which the fuel-air mixture of a cylinder burns in an internal combustion engine.

compression ratio: a numerical ratio of volumes used to express the extent to which the fuel-air mixture is compressed during an internal combustion engine's compression stroke (for example, a compression ratio of 9:1 indicates that the fuel-air mixture is compressed by a factor of nine).

connecting rods: metal rods with round journals on both ends that connect the pistons to the crankshaft in an internal combustion engine, transferring the reciprocating motion of the pistons to the offset throws of the crankshaft.

counterweighted: describes a crankshaft whose offset throws have been balanced to prevent vibration and flexing at higher engine speeds.

crank: slang for "crankshaft."

crankcase: the portion of an engine block in which the crankshaft is housed.

cross-flowing: an engineering term used to describe a cylinder head in which the intake ports and the exhaust ports are located on opposite sides of the head, resulting, conceptually, in a flow of intake and exhaust that passes across the cylinder head.

cubes: slang for *cubic inches of displacement*.

cubic inches of displacement: a common means of expressing the size of an engine.

cylinder head: a metal casting that bolts to the engine block directly above the cylinders.

cylinder bank: a set of inline cylinders that makes up one half of a V-type engine (i.e., a V8).

detonation: an undesirable explosion in a combustion chamber that results from the premature ignition of the fuel-air mixture, typically due to an excessive compression ratio, the use of low-octane gasoline, or both.

Detroit: slang for the mainstream American automobile industry.

differential: a set of gears that distributes driven power to an automobile's wheels.

displacement: the total swept volume of a given engine's cylinders, expressed in cubic inches, cubic centimeters, or liters and calculated by multiplying the length of the crankshaft stroke by the area of a cross-section of the piston and then multiplying the result by the number of cylinders.

distributor: a mechanical device with a rotating internal assembly that distributes high-voltage charges to each of an engine's spark plugs.

double-barrel: describes a carburetor with two throttle bodies.

double overhead camshaft: an engine configuration in which two camshafts are mounted in the cylinder head and are driven by a chain, a belt, or a set of bevel gears.

double-well: see *double-barrel*.

downdraft: a type of carburetion in which the incoming air passes downward through the carburetor, where it is mixed with fuel.

drag: slang for *drag race*.

drag race: a type of automobile contest in which two cars line up and race in a straight line, typically for one-quarter mile (1,320 feet).

dragster: an automobile specifically constructed (or reconstructed) for drag racing.

dragstrip: a paved track at which off-road drag races occur.

drivetrain: the entire set of components and systems related to the mechanical powering of a vehicle, including the engine, the transmission, the driveshaft(s), and the differential.

duration: the length of time, expressed in degrees of rotation, during which a camshaft holds a valve open.

dynamometer: a testing apparatus that measures the output of an engine (usually in foot-pounds of torque, which is then converted into horsepower).

dyno: slang for *dynamometer.*

eight overhead valve: a four-cylinder head that features one intake valve and one exhaust valve per cylinder, both of which are positioned above the piston.

elapsed time: the length of time it takes a car to complete a drag race.

engine block: the large metal casting in which major engine components reside.

exhaust gas recirculation: an emissions-control device that reduces oxide-of-nitrogen emissions by lowering engine combustion temperatures through the controlled admission of small quantities of exhaust gas into the air-fuel mixture.

exhaust header: a type of exhaust manifold, typically found on high-performance or modified engines, that is precisely calibrated to more efficiently evacuate exhaust gas.

exhaust manifold: a series of tubes that gathers exhaust gas from the exhaust ports and directs it to the exhaust system.

exhaust port: a passageway, integral either with the cylinder head (in overhead-valve engines) or with the engine block (in L-head engines), through which engine exhaust gas passes on its way to the exhaust manifold.

F-head: an engine configuration in which the exhaust valves or the intake valves are situated in an overhead-valve configuration, while the other set of valves are situated in-block.

factory: slang for unmodified cars, engines, or parts.

flathead: 1. any engine in which the intake and exhaust valves are located in the engine block, adjacent to the cylinders. 2. slang for the iconic Ford V8, produced from 1932–1953.

flywheel: a heavy metal wheel bolted to the end of the crankshaft that smoothens the rotation of the engine by masking the effects of the jerking motions of the reciprocating pistons.

four banger: a four-cylinder engine.

four barrel: describes a carburetor with four throttle bodies.

four overhead valve: see *F-head.*

fuel injection: a type of induction system that delivers measured, pressurized fuel to the incoming stream of air through one or more metering nozzles or fuel injectors.

gear: slang for *speed equipment.*

gearbox: slang for "transmission."

gearhead: slang for "enthusiast."

glasspack: a type of high-performance muffler that uses fiberglass packing to muffle the sound of the engine.

gow job: pre-WWII slang used to describe what would eventually be known as a hot rod.

head: slang for *cylinder head.*

header: slang for *exhaust header*.

hemi: 1. slang for *hemispherical combustion chamber;* 2. (usually capitalized) slang for Chrysler's hemispherical-chamber V8.

hemispherical combustion chamber: a combustion-chamber configuration in which each cylinder's valves and spark plug are arranged to give a hemispherical shape to the top of the chamber.

hop up: 1. slang used to describe the act of modifying a production·automobile for improved performance; 2. (rare, usually "hop-up") a slang alternative to *hot rod*.

hopped-up: slang for "modified."

hot rod: 1. any production automobile that has been modified for improved performance; 2. to modify a production automobile or automobile engine for improved performance.

hot rodder: one who modifies a production automobile for improved performance.

hot rodding: the act or activity of modifying production automobiles for improved performance.

ignition advance: the amount of time, measured in degrees of engine rotation, between the firing of a cylinder's spark plug and the point at which its piston reaches the end of its compression stroke. Generally speaking, the faster an engine is rotating, the more ignition advance it will require for optimum combustion efficiency, because the amount of time it takes to burn the fuel-air mixture does not vary, whereas the speed at which a piston travels does.

injection: slang for *fuel injection*.

inline four: a four-cylinder engine in which the cylinders are arranged in a single row.

intake manifold: a part that consists of a series of passages that distribute the incoming fuel-air charge (on carbureted and certain types of fuel-injected engines) to the intake ports.

intake port: a passageway, integral either with the cylinder head (in overhead-valve engines) or with the engine block (in L-head engines), through which the incoming air-fuel mixture passes on its way to the cylinder.

intercooler: a finned core, similar to a radiator, through which the compressed air exiting a turbo or a supercharger is directed in order to cool it prior to its entry into the cylinders.

knock: slang for *pre-ignition* or *detonation*.

L-head: see *flathead*.

lakes: slang for the dry lakes of the Southern California high desert.

lakes pipes: slang for a type of exhaust system that expels the exhaust gases directly into the air, typically underneath the engine or to one or another side of the car, rather than allowing the gases to pass through a muffler first.

lead sled: slang (usually derogatory) for a 1950s custom car whose body has been so extensively reworked that it is slowed down by the weight of its lead body filler.

lift: the amount, usually expressed in fractions of an inch or in millimeters, that an engine's valves open as a result of the action of the camshaft.

lifters: metal cylinders that transfer the action of a camshaft's lobes to the engine's valves, either directly (by acting on the valve stems) or indirectly (by acting on pushrods).

lobe: an eccentric metal protrusion on a camshaft.

magneto: an ignition distributor that uses a series of magnets to generate its own electricity.

mags: slang for "mag wheels," which are lightweight custom wheels. Originally, "mag wheel" referred specifically to a wheel cast from magnesium, but it has since come to be a more widely applicable (and, thus, less specific) label for custom wheels of all sorts.

manifold: see *intake manifold* and *exhaust manifold*.

mill: slang for "engine."

motor around: to casually drive a car.

octane rating: a measure of the detonation or pre-ignition resistance of a grade of gasoline.

OEM: 1. slang for unmodified cars, engines, or parts; 2. general shorthand for "original equipment manufacturer," a mainstream automobile company such as Ford.

overhead camshaft: an engine configuration in which the camshaft (or, in engines with multiple camshafts, at least one of them) is mounted in the cylinder head (and driven by a chain, a belt, or a set of bevel gears) rather than in the engine block.

overhead valve: an engine-and-cylinder-head configuration in which at least some of the intake and/or exhaust valves are mounted above the piston (in the cylinder head itself) rather than in the engine block.

overlap: the extent, usually expressed in degrees of rotation, to which a camshaft holds both the intake and the exhaust valves open at the same time.

positive crankcase ventilation (valve): an emissions control device that lowers hydrocarbon emissions by admitting engine crankcase vapors into the incoming air-fuel mixture rather than allowing them to escape through a draft tube.

performance tuner: a company that specializes in the production of high-performance automotive parts, accessories, and packages.

performance tuning: the act or activity of modifying production automobiles for improved performance.

ping: slang for *pre-ignition* or *detonation*.

pot: slang for "carburetor."

powerband: the operating range of an automobile engine, expressed in revolutions per minute.

powerplant: slang for "engine."

powertrain: see *drivetrain.*

pre-ignition: an undesirable explosion in a combustion chamber that typically results from the use of low-octane gasoline in an engine with substantial ignition advance.

progressive carburetor: a double-barrel carburetor in which the first (or primary) chamber is used to deliver the fuel-air mixture at low engine speeds (to maximize fuel economy), and the secondary chamber kicks in at higher speeds to supplement the primary chamber (for more top-end power). Progressive four-barrel carburetors with two primary chambers and two secondary chambers also exist (and are popular among rodders).

pushrod: a metal rod used to transfer the action of a camshaft's lobes to the rocker arms in overhead-valve engines. Pushrods are often (but not always) hollow, performing the additional function of carrying engine oil from the crankcase to the rocker assembly.

pushrod tube: a hollow metal tube that covers and protects a pushrod, often also performing the function of carrying engine oil from the rocker assembly back to the crankcase.

quad: slang for a four-barrel carburetor.

quarter-mile: slang for *dragstrip.*

raked: slang for a particular type of suspension setup, common on modified 1960s muscle cars, in which the rear of the car sits higher than stock and the front lower than stock.

ramp-up: describes the speed, usually expressed in degrees of rotation, with which a camshaft opens the valves.

ratio rocker: a high-performance rocker arm that multiplies valve lift by translating the upward motion of the pushrod into the downward force on the valves at a ratio greater than that of the OEM rocker arms.

ride: slang for "car."

rocker arm: a metal see-saw lever that mounts on a rocker shaft and converts the upward motion of the pushrods into the downward motion that acts on the valves.

rocker assembly: the entire assembly of the rocker arms and the rocker shaft.

rocker shaft: a metal rod on which the rocker arms are mounted.

rocker: slang for *rocker arm.*

rod: slang for *hot rod.*

rodder: slang for *hot rodder.*

rodding: slang for *hot rodding.*

roller bearing: a type of crankshaft or camshaft bearing that uses ball bearings in place of a more conventional smooth steel or alloy shell.

screamer: slang for an engine capable of extremely high rpms (or for an automobile equipped with such an engine).

single-barrel: describes a carburetor with a single throttle body.

shoe: slang for "driver."

shoehorn: slang that refers to the installation, in a given car, of a larger engine from another make or model.

shot rod: slang for poorly-built, worn-out, or unsafe roadsters that 1950s enthusiasts did not want journalists, policemen, and others to confuse with their well-built hot rods.

sidedraft: a type of carburetion in which the incoming air passes horizontally through the carburetor, where it is mixed with fuel.

sixteen overhead valve: a four-cylinder head that features two intake valves and two exhaust valves per cylinder, all of which are positioned above the piston.

skirt: the cylindrical section of metal that extends below the surface of a piston.

sled: slang for "car."

sleeper: slang for a vehicle whose run-of-the-mill appearance disguises the fact that it has a highly modified engine.

slicks: slang for racing tires which have little or no tread.

slug: slang for "piston."

slushbox: slang for "automatic transmission."

small-block: 1. describes a given automobile company's smaller class of V8 engines, typically those displacing less than 350 cubic inches; 2. slang for "small-block Chevrolet V8."

single overhead camshaft: an engine configuration in which a single camshaft is mounted in the cylinder head (and driven by a chain, a belt, or a set of bevel gears) rather than in the engine block.

soup up: to modify a production automobile for improved performance.

speed equipment: automobile parts and accessories (especially engine parts and accessories) that can be used on a production automobile in place of corresponding original parts in order to improve the automobile's performance.

steelpack: a type of high-performance muffler that uses steel-wool packing to muffle the sound of the engine.

stock: slang for unmodified cars, engines, or parts.

straight eight: an eight-cylinder engine in which the cylinders are arranged in a single row.

straight six: a six-cylinder engine in which the cylinders are arranged in a single row.

strip: slang for *dragstrip*.

stripped: slang used to refer to an automobile that has been lightened (in order to achieve a more favorable power to weight ratio) through the removal of those parts its owner deems unnecessary, often including (but certainly not limited to) fenders, running boards, chrome trim, and interior upholstery.

stroke: 1. the distance that a piston travels each way as it reciprocates; 2. to increase the distance that a piston travels as it reciprocates by offset-grinding the connecting-rod journals of the crankshaft.

stroker: slang for an engine the displacement of which has been increased by offset-grinding the connecting-rod journals of the crankshaft.

supercharger: a mechanical device, powered by the crankshaft (either directly or via a tensioned belt), that forces an engine to consume more air and fuel, producing more power.

throttle body: a metal tube through which the incoming flow of air must pass (on both fuel-injected and carbureted engines) en route to the intake ports.

throttle butterfly: a flap of metal within a throttle body (on both carbureted and fuel-injected gasoline engines) that regulates the incoming flow of air by pivoting on a central shaft connected (electronically or by means of a cable) to the vehicle's accelerator pedal.

throws: the offset beams on a crankshaft to which the connecting rods are bolted.

time: slang, common among early dry-lakes racers, that refers to the top speed — *not* the elapsed time — that a car achieves at a meet. (This was commonly expressed as follows: "my rod's best time is 121.57 mph.")

tool around: to casually drive a car.

traps: 1. slang for the end of a dragstrip, where a car's speed is measured; 2. slang for that point along a dry lakes course where a car's speed is measured, a point typically located *not* at the end of the course, but somewhere in its middle; 3. slang for the mechanical triggering system used in early dry-lakes racing to measure a car's top speed.

tuned: slang for "modified."

tuner: 1. slang for *performance tuner;* 2. 1990s and 2000s slang for *hot rodder;* 3. 1990s and 2000s slang for *hot rod.*

tuning: slang for *performance tuning.*

turbo: a type of supercharger that uses the flow of exhaust gases to drive a small turbine that forces additional air into an engine.

twin-barrel: see *double-barrel.*

underslinging: slang (from the Model T era) that refers to the act of lowering a vehicle through the modification of its suspension.

updraft: a type of carburetion in which the incoming air passes upward through the carburetor, where it is mixed with fuel.

V6: an engine in which two banks of three inline cylinders are joined at a common crankshaft, forming, when viewed from the front, the shape of a V.

V8: an engine in which two banks of four inline cylinders are joined at a common crankshaft, forming, when viewed from the front, the shape of a V.

valve cover: a metal housing that covers the rocker assembly on overhead-valve engines.

valve guide: a metal tube in which a valve stem reciprocates.

valve springs: metal coil springs that keep an engine's valves closed until the camshaft forces them to open.

valve stem: the slender, shaft-like portion of an engine valve that is situated outside of the combustion chamber.

valvetrain: the entire engine assembly related to the valves (including the camshaft, lifters, pushrods, rocker assembly, and the valves themselves).

warm over: to modify a production automobile engine for improved performance.

water-cooling: a method of regulating engine temperatures through the use of a liquid coolant medium which flows through the engine in dedicated passages.

wheels: slang for "car."

winged: slang for an automobile, especially a 1990s or 2000s import, that has been fitted with an oversized rear spoiler.

work over: to modify a production automobile engine for improved performance.

When the inexpensive paper used in magazines approaches thirty to fifty or more years of age, it begins to rot. Magazines produced on newsprint-grade stock tend to deteriorate much faster than those on glossy stock, but eventually, unless they have been sealed in little plastic sleeves since the day they were printed, *all* of them will slowly turn to dust. Their pages will fade from white to yellow and from yellow to brown. Their spines will stiffen and protest when opened. Once-innocent corner thumb-creases will become clean tears and cracks, and individual pages will actually shatter into miniscule fragments if they are turned with too much zeal. By far the most distinctive attribute of decomposing periodical stock is the smell, a rancid and acidic odor that often appears in magazines that are only ten or fifteen years old. But this stench is evident only at close range. Stand a couple of feet away from a stack of monthlies printed in the 1950s and you'll barely notice it, but if you sit down and flip though them page by page, it's virtually impossible to ignore.

This I know because I've spent the better part of the last nine years carefully reading through thousands of magazine and trade-journal issues—many of which were printed fifty, seventy, or even ninety years ago—at very close range. I've shattered my fair share of pages, cracked my fair share of spines, and breathed in more than my fair share of rotting fiber dust. And I've loved every minute of it.

Or at least, I have been deeply grateful, from the moment that I first conceived of this project back in 1999, that so many periodicals have survived the ravages of time more or less intact. For as I have written this history of automotive performance tuning and speed equipment manufacturing in the United States from the days of the Model T to those of the import tuner, enthusiast periodicals and industry trade journals have been absolutely indispensable. Some primary and secondary monographs, collections, how-to manuals, and edited volumes have been vital as well, as have several caches of documents and a handful of one-on-one interviews with key figures. But without the magazines, this project never would have progressed beyond a brainstorm.

MAGAZINES

During the Model T and Model A eras of the 1910s, 20s, and 30s, dedicated enthusi-ast periodicals did not yet exist. However, general-interest automotive magazines regu-larly published photographs and features of modified cars, how-to articles for those who wanted to learn how to improve their cars' performance and handling, and reader forums in which performance-tuning matters were often discussed at great length.

By far the most useful of these general-interest periodicals—for my purposes, at least—was *The Fordowner*, put out by the Hallock Publishing Company of Cleveland, Ohio. During the 1910s, this magazine was owner-oriented and was therefore filled with content that was useful for consumers. But over the course of the 1920s, 30s, and early 40s, it gradually evolved into a trade-oriented publication, with tips for shop owners, dealers, retail-parts outlets, and investors. Its name changed several times during these years too, as its editors worked to better match their masthead with their content: *The Fordowner* became *Ford Owner and Dealer* in 1920, *Ford Dealer and Owner* in 1925, *Ford Dealer and Service Field* in 1926, and *Ford Field* in 1939. Through-out the 1910s, 20s, and 30s, however, there was at least one constant in this magazine's front office: Murray Fahnestock, its technical editor. Fahnestock must have been an enthusiast himself, for he regularly wrote with noticeable passion about organized (and unorganized) motorsports, overhead-valve conversions, underslinging methods, high-speed cams, racing bodies, and, perhaps most importantly, the companies and the men who manufactured aftermarket high-performance parts for Fords and other inexpensive makes.

Motor Age, another general-interest publication from the days of the Model T and the Model A, also gradually shifted its focus from end-users to shop owners, dealers, and parts producers. Although it did not have a technical editor as rabidly devoted to the pursuit of speed as *The Fordowner*'s Murray Fahnestock, *Motor Age* did neverthe-less run a number of how-to stories about Model T and Model A performance tuning, and news about aftermarket companies frequently appeared within its pages alongside (typically much longer) stories about the goings-on at Ford, Chevrolet, Overland, Peerless, and the rest of the OEMs of the period. On the other hand, *Automotive Industries—The Automobile* (shortened in 1925 to *Automotive Industries*) was through and through a trade-oriented publication. Nevertheless, its frequent coverage of af-termarket firms—in particular, its coverage of the bankruptcies and buyouts of the 1930s—served as an invaluable reference for me as I studied the early high-perfor-mance industry. So too did the coverage of aftermarket innovations and automobile racing (including the then-new Southern California dry-lakes events) that appeared in *Popular Mechanics* during the 1930s and 40s.

After World War II, consumer-oriented automobile magazines flooded American newsstands, particularly in the early 1950s, as did publications written specifically for automotive enthusiasts and hot rodders. Among the latter type, *Speed Age* debuted in 1947, as did the more specialized *Hot Rod* in 1948, *Hop Up* in 1951, and both *Rod and Custom* (briefly known as *Rods and Customs*) and *Honk* (quickly re-dubbed *Car Craft*) in 1953. Each of these early enthusiast-oriented monthlies was filled with how-to articles, reader forums, feature stories, and speed equipment advertisements, all of which were absolutely vital to me as I worked to understand the postwar rodding scene.

So too were some of the more specialized drag-racing periodicals that emerged during the later 1950s and the 1960s, including *Drag News*, *Drag Sport*, *Drag Digest*, and *Drag Racing USA*. Particularly useful among these was the tabloid *Drag News*, which first appeared in 1955 and consistently ran "meet the manufacturer"–style articles about individual speed equipment companies throughout the 1960s and well into the 70s. (For a brief while, in fact, the editors of *Drag News* toyed with the notion of allowing their newspaper to evolve into a trade-oriented publication, beefing up its industry-news section and even changing its title to *Drag News & Equipment Industry Report* in 1969 before reverting to the tried-and-true, enthusiast-friendly *Drag News* in 1970.)

In 1962, *Popular Hot Rodding* debuted as an upstart rival to what was by then the most widely circulated and influential of them all, *Hot Rod*. Rich with how-to features, speed equipment advertisements, and industry news (both aftermarket and OEM), *Popular Hot Rodding* served as a vital resource for this project as it moved into the muscle car era of the second half of the 1960s and the regulatory era of the 1970s and early 80s. The same was true of the Petersen Publishing Company's closed-circulation *Hot Rod Industry News* (which first appeared in speed shop, equipment manufacturer, and wholesale distributor mailboxes in 1966) and to a lesser extent its 1960s competitor, *Speed & Custom Equipment News*. For the 1970s, 80s, and 90s, *Specialty & Custom Dealer* and *SEMA News* were occasionally useful to me too.

From the 1960s through the early 2000s, a number of niche-oriented magazines appeared as well, most notably *Street Rodder*, *Street Scene*, and *Rod Action* for street rodders; *Street Machine*, *Super Chevy*, and other brand-specific titles for street machine buffs; *Custom Craft*, *Customs Illustrated*, and *Popular Customs* for custom car fans; *Dune Buggies and Hot VWs*, *VW Trends*, *VW & Porsche* (which became *European Car* in 1991), and *Eurotuner* for European-car enthusiasts; and *Import Tuner* and *Modified Mag* for Japanese-import tuners. Of these, *Street Rodder*, *Super Chevy*, *Dune Buggies and Hot VWs*, *VW Trends*, and *European Car* were particularly vital to this project's final chapter.

In an effort to cast as wide a net as I could manage, I also spent a considerable amount of time reading through full (or very nearly full) runs of 1950s, 60s, 70s, and

80s general-interest automotive magazines. These included *Road & Track*, which appeared in 1947; *Motor Trend*, Petersen Publishing's second magazine, which debuted in 1949; *Hop Up and Motor Life*, which succeeded *Hop Up* in early 1953 as the editors of that magazine moved away from their earlier, exclusive focus on hot rods; *Motor Life*, which succeeded *Hop Up and Motor Life* in 1954, completing its transition; *Car Life*, which debuted in 1953; *Cars*, which also debuted in 1953; *Auto Age*, another 1953 addition; *Car and Driver*, which first appeared in 1961 (succeeding the more specialized *Sports Cars Illustrated*, which debuted in 1955); and *Motor Guide*, which debuted in 1956. Of these, *Car and Driver*, *Motor Life*, *Motor Trend*, and *Road & Track* were particularly useful sources for the horsepower race of the 1950s and the muscle car boom of the 1960s.

The problem with most of these periodicals is that few of them are available as full runs in any single library or archive. The exceptions, in my experience, are as follows. The Free Library of Philadelphia (which I visited in 2002) has an excellent automotive collection that includes a complete run of *The Fordowner* and its many successors. Likewise, the Benson Ford Research Center at The Henry Ford (which I visited in the fall of 2006 and the spring of 2007) has a full run of *Motor Age*; the Periodicals Archive at the Don Garlits Museum of Drag Racing (a closed-access collection I was fortunate enough to be allowed to visit in the fall of 2003) has complete runs of *Drag News* and *Popular Hot Rodding*; the Retrospective Collection of the MIT Library System (which I visited repeatedly in 2002 and 2003) has a complete run (or at least, complete for the 1910s through the 1930s) of *Automobile Industries—The Automobile* and its successor, *Automobile Industries*; and the main library at MIT, in Building 14, has a full run of *Popular Mechanics* that I accessed regularly in 2002 and 2003. Nearly complete runs of two particularly important magazines are housed at the Library of Congress as well: though unlisted in its online portal (as of my visit in 2000), it has a bound run of *Hot Rod Industry News* that is better than 90 percent complete for the 1960s and 70s, and it also has a bound run of *Hot Rod* that is better than 95 percent complete for the 1960s and 70s.

Needless to say, the major challenge that I faced during the research phases of this project was locating partial runs of all of the other magazines mentioned in this essay (and, of course, piecing together a research strategy that would enable me to view full runs of nearly all of them by looking at a few issues here, a handful of others there, and so forth). Fortunately, partial runs of many of them were available at libraries within a reasonably short distance of my research bases. To fill in the gaps, I traveled a bit farther afield on numerous occasions, and I also borrowed hundreds of exceedingly rare issues of *Hot Rod*, *Motor Life*, *Car Life*, *Car and Driver*, and *Motor Trend* from an uncle whose picturesque car barn behind his house near Buffalo, New York, contained dozens of boxes of decomposing magazines. In addition, I racked up many hours on eBay during the early stages of my research. There I was able to purchase individual issues and even full-year runs—on more than one occasion, full-decade runs—of these magazines for but a tiny fraction of what it would have cost to travel to far-flung libraries to view them. Besides, actually owning these periodicals (or, in the case of those that I borrowed from my uncle, having them close at hand) meant that I was able to read them carefully, over and over and over again, on my own schedule and in the comfort of my own office.

BOOKS AND OTHER SOURCES

Since the 1940s, how-to manuals have served as vital sources of rodding know-how for enthusiasts across the United States. This was especially true in the 1950s. For the historian, therefore, these manuals, like the magazines, are excellent sources of primary data that are particularly helpful if you're trying to get into the average rodder's head: What sorts of problems did enthusiasts encounter when modifying their cars, and what sorts of solutions did (often self-appointed) experts within the rodding fraternity suggest to deal with them? What modifications were considered exotic in, say, 1950? What about in 1955? What level of theoretical engine-tuning detail was available to the average enthusiast in the 1940s and 50s? Fortunately, scores of these how-to manuals have been published over the years; those from the 1950s, while no longer abundant, are nevertheless available at a number of libraries, used bookstores, and, of course, on eBay. Some of them are available as reprints as well. Among the most useful of them for this project were Edgar Almquist's *Specialized Automobile Tuning and Customizing Methods* (1946); Roger Huntington's *Souping the Stock Engine* (1950); Peter Bowman's *Hot Rod Handbook* (1951); "California Bill" Fisher's *Chevrolet Speed Manual* (1951, reprinted in 1995) and *Ford Speed Manual* (1952, also reprinted in 1995); and Bill Czygan's *How to Build Hot Rods* (1952).

In a similar vein, many of the companies that published popular magazines put out annual collections during the 1950s, as did a number of other publishing houses. Because these collections essentially reviewed the goings-on among drag racers, lakes racers, customizers, and street-scene hot rodders in a particular year, many of them were useful resources for this project. This was especially true of those put out by the editors of *Hot Rod Magazine*, including *Hot Rods* (1951); *Hot Rod Annual* (published annually in the 1950s); and *Custom Cars* (also published annually in the 1950s). Also useful in this vein were Eugene Jaderquist and Griffith Borgeson's annual collections from the early to mid-1950s, particularly their *Best Hot Rods* and their *Auto Racing Yearbook*. Other published primary sources worth consulting include Veda Orr's *Hot Rod Pictorial* (1949), Hank Elfrink's *Special Racing Cars and Hot Rods of the World* (1950), and Mickey Thompson and Griffith Borgeson's *Challenger* (1964).

The secondary literature on hot rodding and automotive enthusiasm is exceedingly vast, although its academic component is for the most part limited to H. F. Moorhouse's *Driving Ambitions* (1991), Robert C. Post's *High Performance* (1994), John DeWitt's *Cool Cars, High Art* (2002), and Brenda Jo Bright's *Customized* (2000). On the other hand, popular books on the subject abound. Many of these nonacademic titles are little more than nicely bound, coffee-table-style collections of photographs, but there are a number of popular histories available that are serious, well-researched, and fully worth our attention. Among the best (and, for this project, most useful) are Dean Batchelor's *The American Hot Rod* (1995), which is based on Batchelor's many decades of experience covering the world of the hot rodder as a popular-magazine journalist; Ed Almquist's *Hot Rod Pioneers* (2000), a massive collection of biographical sketches of prominent racers, rodders, and speed equipment gurus; Tom Medley and LeRoi Smith's two books, *Tex Smith's Hot Rod History, Volume One: The Beginnings* (1990) and *Tex Smith's Hot Rod History, Volume Two: The Glory Years* (1994), each of which contains a mix of biographical data and useful period sketches; the prolific Don Montgomery's six contributions about 1940s, 50s, and 60s hot rodding and drag racing, *Hot Rods in the Forties* (1987), *Hot Rods As They Were* (1989), *Hot Rod Memories* (1991), *Supercharged Gas Coupes* (1993), *Authentic Hot Rods* (1994), and *Those Wild Fuel Altereds* (1997); Peter Vincent's collection of essays and photographic stills, *Hot Rod: An American Original* (2000); Robert Genat and Don Cox's excellent history of early dry lakes racing, *The Birth of Hot Rodding* (2003), which also includes a number of beautiful color photographs of the lakes scene from the 1940s; Albert Drake's *Hot Rodder!* (1993), a collection of interviews with prominent rodding personalities from the 1940s, 50s, and 60s; Roger Huntington's detailed history of the horsepower race of the 1950s and the muscle cars of the 1960s, *American Supercar* (1983, revised and reissued in 1990); and, last but certainly not least, William Carroll's *Muroc, May 15, 1938* (1991), a detailed look at the first official SCTA meet that contains useful reprints of the programs from this and several subsequent meets.

Excellent popular biographies are available as well. Though hard to find, *Roy Richter: Striving for Excellence* (1990) is a thoughtful and well-executed account of Richter's life written by Art Bagnall, a man who worked for him at Cragar for many years. I also consulted *Zora Arkus-Duntov* (2002), Jerry Burton's biography of an engineer who produced an overhead-valve conversion for the flathead V8 in the 1940s before going to work at Chevrolet in the 1950s (where he worked on the Corvette and on the small-block V8). A handful of biographical treatments of individual speed-equipment firms have appeared over the years as well, most notably Mark L. Dees's *The Miller Dynasty* (1981); David A. Fetherston's *Moon Equipped* (1995); Mark Christensen's *So-Cal Speed Shop* (2005); and Tom Madigan's *Edelbrock* (2005).

In the realm of unpublished and archival sources, I was fortunate to have access to several important collections. These included SEMA's Research Center at its Diamond Bar, California, headquarters; the ARB Archive at the California EPA's offices in Sacramento, California; the Romaine Trade Catalog Collection at the University of California, Santa Barbara; the collection of automotive trade catalogs housed at The Henry Ford's Benson Ford Research Center in Dearborn, Michigan; the Periodicals Archive at the Don Garlits Museum of Drag Racing in Ocala, Florida, which in addition to its thousands of magazines also contains a number of catalogs and other ephemera; and the Root Collection at the Watkins Glen International Motor Racing Research Center Archive in Watkins Glen, New York. In addition, I was able to conduct personal interviews with key aftermarket personalities whose insights were invaluable, including Delores Berg of Gene Berg Enterprises; Kathy Flack of JE Pistons; Vic Edelbrock Jr., Nancy Edelbrock, and Camee Edelbrock, all of the Edelbrock Equipment Company; Bob Spar of B&M Automotive; Carl Olson, SEMA's technical and legislative coordinator in the mid-1970s; and Dick Wells, who organized the first SEMA Show in the 1960s, headed the NSRA in the early 1970s, and served as SEMA's executive director in the late 1970s. (Dick was also kind enough to grant me access to some documents and unpublished manuscripts from his personal files.)

Finally, though it did not contribute directly to the content of this book in any meaningful way, the 2003 SEMA Show in Las Vegas, Nevada, was a vital indirect resource. Dick Dixon of Access RPM, whom I first met through Dick Wells at a parking-lot car show at the NHRA Museum in April of 2003, was kind enough to obtain a pass for me to attend the closed-access SEMA Show that November. For an entire week, I wandered among the booths at this colossal aftermarket exposition, chatting up industry insiders, absorbing the current-day business culture of the speed equipment industry, and slowly coming to grips with the massive scale and scope of the modern automotive aftermarket; the contacts that I made at that show also led to a couple of significant interviews back in Southern California later that year.

Although the 2003 SEMA show was by no means vital as a source of content, it most certainly was crucial as a source of contacts. And the way that I got past the gates at that show illustrates the importance of finding and meeting the right people. In fact, with this in mind, the best advice I can give to anyone who might want to follow up on this work is to pick up the phone and get in touch with someone — *anyone* — associated with the industry. Better yet, hop on a plane that's bound for LAX, and hit the pavement once you're in L.A. If you do, I guarantee that you'll find exactly what I found: down-to-earth people who are friendly, helpful, and eager to share their stories about hot rods, racing, and the speed equipment business.